The Downy Mildews - Genetics, Molecular Biology and Control

A. Lebeda · P.T.N. Spencer-Phillips · B.M. Cooke

The Downy Mildews - Genetics, Molecular Biology and Control

Reprinted from European Journal of Plant Pathology, Volume 122, No. 1, 2008

Springer

A. Lebeda
Palacký University in Olomouc
Czech Republic

P.T.N. Spencer-Phillips
University of the West of
England, Bristol, UK

B.M. Cooke
University College Dublin
Ireland

ISBN 978-90-481-8039-4 e-ISBN 978-1-4020-8973-2

Cover photo captions

1. Abaxial side of *Lactuca serriola* leaf with *Bremia lactucae* sporulation
2. *Bremia lactucae* conidiosporangia
3. Nitric oxide (NO) probe DAF-FM DA distributed within epidermal cell of susceptible *Lactuca sativa* (cv. Cobham Green) penetrated by haustoria of *B. lactucae*
4. Immunolocalization of alpha-tubulin by FITC-conjugated secondary antibody showing rearrangement of microtubules in epidermal cell of *L. sativa* (UCDM2) during plasma membrane invagination due to *B. lactucae* penetration
5. 2D-DIGE gel of the pea leaf proteome showing changes in abundance of proteins following infection by *Peronospora viciae* (red = increased; green = decreased; yellow = no change)

Printed on acid-free paper

9 8 7 6 5 4 3 2 1

springer.com

European Journal of Plant Pathology

Volume 122 · Number 1 · September 2008

Special Issue: The Downy Mildews - Genetics, Molecular Biology and Control
Edited by: A. Lebeda · P.T.N. Spencer-Philipps · B.M. Cooke

Instructions for Authors for *Eur J Plant Pathol* are available at http://www.springer.com/10658.

Eur J Plant Pathol (2008) 122:1
DOI 10.1007/s10658-008-9324-z

Foreword

Aleš Lebeda · Peter T. N. Spencer-Phillips · B. M. Cooke

Received: 29 April 2008 / Accepted: 29 April 2008

The downy mildews—genetics, molecular biology and control

This book contains chapters presented by keynote speakers at The Second International Downy Mildews Symposium, held in July 2007 at Olomouc, Czech Republic. The conference focused on the evolution, taxonomy, biology, genetic variation, host resistance, population diversity, epidemiology, chemical and biological control of downy mildews and some related plant-pathogenic oomycetes. Regular oral presentations were published in a separate volume entitled: Advances in Downy Mildew Research, Vol. 3 (A. Lebeda and P.T.N. Spencer-Phillips, Eds., Palacký University in Olomouc, 2007, 278 pp., ISBN 80-86636-19-4).

Knowledge of downy mildew pathogens, and the plant diseases they cause, has advanced substantially since publication of the book: The Downy Mildews, edited by D.M. Spencer (Academic Press, 1981). Over the last 25 years, many aspects of downy mildew biology have been investigated, and progress in some areas has been summarized in three volumes of the text: Advances in Downy Mildew Research (2002, 2004 and 2007). However, expansion of knowledge about downy mildews provided an opportunity to organize the Second International Symposium, and to publish this volume which contains 14 invited review and original research papers presented by keynote contributors. The result is a book comprehensively updating understanding across a wide spectrum of the most important advances, designed to provide a balanced overview of the subject area.

We hope that it will be of use to students, academics and researchers in plant pathology, mycology, crop protection, taxonomy, plant biology and ecology, as well as to others involved in the agricultural and horticultural advisory services and industry.

A. Lebeda (✉)
Faculty of Science, Department of Botany,
Palacký University in Olomouc,
Slechtitelu 11, 783 71 Olomouc-Holice,
Czech Republic
e-mail: ales.lebeda@upol.cz

P. T. N. Spencer-Phillips
School of Biosciences, University of the West of England,
Coldharbour Lane,
Bristol BS16 1QY, UK
e-mail: peter.spencer-phillips@uwe.ac

B. M. Cooke
School of Biology and Environmental Science,
College of Life Sciences, University College Dublin,
Belfield, Dublin 4, Ireland
e-mail: Mike.Cooke@ucd.ie

Eur J Plant Pathol (2008) 122:3–18
DOI 10.1007/s10658-008-9341-y

Progress and challenges in systematics of downy mildews and white blister rusts: new insights from genes and morphology

Hermann Voglmayr

Received: 31 August 2007 / Accepted: 29 May 2008

Abstract Recent advances in the classification of downy mildews and white blister rusts are presented from ordinal to species level. Using molecular data (mainly LSU of nuclear ribosomal DNA and ITS rDNA data, but also cox2, beta-tubulin and NADH genes), ordinal, family and generic circumscriptions have been reconsidered and changed during the last years; species circumscription and concepts are also changing. These rearrangements also lead to a reevaluation of the traditional morphological characters used for classification. The recent changes have various implications for applied sciences (phytopathology, molecular biology) mainly at the species level; besides name changes for some taxa, revised species circumscriptions and improved species identification using genetic markers have important consequences on host ranges, source inocula and risk assessment of phytopathologically important taxa. However, there are also some substantial unresolved problems which need to be addressed in the future with new data and methods. These include the systematic position of some rarely sampled taxa, the phylogenetic relationships of the main downy mildew lineages to each other, more detailed molecular studies on speciation processes to develop appropriate sound species concepts and circumscriptions, and the development of a molecular bar coding system for downy mildews enabling reliable species identification. Applying molecular methods has the potential to greatly enhance our knowledge on the overall biodiversity of downy mildews.

Keywords *Albugo* · Molecular phylogeny · Peronosporales · Peronosporaceae · *Phytophthora* · Species concept · White blister rusts

H. Voglmayr (✉)
Department of Systematic and Evolutionary Botany, Faculty Centre Botany, University of Vienna, Rennweg 14, 1030 Vienna, Austria
e-mail: hermann.voglmayr@univie.ac.at

Introduction

Downy mildews and white blister rusts are members of the class Oomycetes (Peronosporomycetes), a comparatively small lineage with estimates of <1,000 species (Kirk et al. 2001). Due to their morphological, physiological and ecological similarities to fungi, the Oomycetes are traditionally treated within mycology; however, ultrastructural, biochemical and molecular phylogenetic data confirm that they are not related to true fungi (kingdom Fungi), but belong to the kingdom Chromista (Straminipila) which also contains the chromistan (heterokont) algae (Dick 2001, 2002; Kirk et al. 2001). Downy mildews are an important group of obligate biotrophic plant parasites, which have a great economic impact on numerous crops (e.g. *Plasmopara viticola* on *Vitis vinifera*, *Pseudoperonospora humuli* on *Humulus lupulus*, *Perono-*

spora tabacina on *Nicotiana* spp.). With the commercial large-scale production of some crop and ornamental plants, some downy mildew diseases have recently become of major concern, e.g. downy mildew disease of *Ocimum basilicum* (basil; e.g. Belbahri et al. 2005), *Eruca sativa* (rocket; e.g. Larran et al. 2006 and references cited therein), or *Rubus* spp. (arctic bramble and boysenberry; e.g. Hukkanen et al. 2006 and references cited therein). In several of these cases, classification of the causal agent is still unclear, which demonstrates our ignorance of biodiversity of downy mildews. With the rapid development of new research techniques and questions in applied and theoretical plant pathology, the interest in phylogeny of downy mildews has increased over the last years. The availability of a substantial number of additional new characters less prone to subjective interpretations in general led to a change of paradigms in classification, as phylogenetic hypotheses could be vigorously tested for the first time, which led to a shift from a phenetic to a phylogenetic classification. In addition, phylogenetic analyses of DNA sequence data also enabled a re-evaluation of morphological features in an evolutionary background and a re-investigation of species boundaries and host specificity. Significant progress has been made towards a phylogenetic classification, which is presented here. However, several important questions are still unresolved, and these will be briefly discussed.

General considerations on systematics and taxonomy

Before considering the changes, progress and challenges in downy mildews systematics in detail, some general considerations will be briefly outlined. There are often conflicts between taxonomists and applied biologists, originating from different and sometimes incongruent expectations on taxonomy and systematics. Although there have been different methodological approaches in taxonomy during its history, it is nowadays commonly accepted that phylogenetic relationships should be the primary basis of a taxonomic system (Lecointre and Le Guyader 2006). Therefore, the taxonomist seeks consistency of a taxonomic system with theories on phylogeny and evolution. Consequently, classification should be in line with well-supported phylogenetic hypotheses. With the increase of knowledge, reinterpretation of phylogenetic relationships leads necessarily to name changes. In addition, taxonomy and classification has to be consistent with the current rules of the International Code of Botanical Nomenclature. Application of the latter sometimes necessitates the change of well-established names, which are often felt unnecessary, cumbersome or complicated.

Non-taxonomists are primarily interested in using scientific names without the need for a deeper knowledge of taxonomy itself. Therefore, stability of names is highly desirable. Consequently, taxonomy should be easily applicable outside the taxonomic community, e.g. for purposes of identification. As another important requirement, taxonomy should be also appropriate for legal measures such as quarantine lists of species.

Of course, ideally taxonomy should fit the needs of both taxonomists and non-taxonomists, and this is often possible if taxonomic decisions are made cautiously. In downy mildews, it has been possible up to now to avoid undesirable name changes for most phytopathologically important species. However, taxonomic changes are sometimes unavoidable to meet the standards of a consistent phylogenetic classification. This is no end in itself, but enables progress also in other research disciplines. Evidently, the non-taxonomist should be interested in reliable species concepts and boundaries, which are prerequisites for the development of reliable PCR-based identification systems, a record of the correct host ranges, and the application of pest control and quarantine measures.

Phylogenetic placement of Peronosporaceae (downy mildews) and Albuginaceae (white blister rusts)

Based on morphological and ultrastructural data, Oomycetes were subdivided into three subclasses, the Saprolegniomycetidae, Rhipidiomycetidae and Peronosporomycetidae, the latter including *Pythium*, *Phytophthora* and some other genera together with downy mildews and white blister rusts (Dick et al. 1984; Dick 1995). This subdivision was largely confirmed by subsequent molecular phylogenetic analyses (e.g. Dick et al. 1999; Riethmüller et al.

1999; Hudspeth et al. 2000; Petersen and Rosendahl 2000). However, the classification at lower ranks remained uncertain and changed quite substantially following molecular phylogenetic analyses (Table 1, Fig. 1). Whereas previously considered closely related to Peronosporaceae (Fig. 1a), the Albuginaceae were placed outside the Pythiaceae-Peronosporaceae lineage in cox2 (Fig. 1b), LSU/SSU (Fig. 1c) and ITS rDNA (Cooke et al. 2000) sequence analyses. Therefore, Albuginaceae should represent an ancient, evolutionarily–derived lineage of uncertain phylogenetic affinities (compare Fig. 1b,c). Early origin of Albuginaceae is in line with high sequence divergence (Riethmüller et al. 2002) and its unique conidial and oospore morphology (Riethmüller et al. 2002; Hudspeth et al. 2003; Thines and Spring 2005; Voglmayr and Riethmüller 2006). Consequently, Hudspeth et al. (2003) pleaded for exclusion of Albuginaceae from Peronosporales and for elevation to ordinal level, whereas Thines and Spring (2005) created even a new subclass, Albuginomycetidae.

As opposed to *Albugo*, DNA data confirmed a close phylogenetic relationship of downy mildews to the genus *Phytophthora* (e.g. Petersen and Rosendahl 2000; Cooke et al. 2000; Riethmüller et al. 1999, 2002; Hudspeth et al. 2003; Voglmayr 2003; Göker et al. 2007). There is strong molecular evidence that the genus *Peronophythora*, sometimes considered an

Table 1 Comparison of some ordinal, family and generic classifications of downy mildews, white blister rusts and relatives

Waterhouse (1973)	Kirk et al. (2001)	Riethmüller et al. (2002)	Göker et al. (2007)[a], Thines and Spring (2005)[b]
Peronosporales	Peronosporales	(no order name)	Peronosporales
Peronosporaceae	Peronosporaceae	Peronosporaceae	Peronosporaceae
Basidiophora	*Basidiophora*	*Basidiophora*	*Basidiophora*
Bremia	*Benua*	(*Benua*)	*Benua*
Bremiella	*Bremia*	*Bremia*	*Bremia*
Peronospora	*Bremiella*	*Paraperonospora*	*Graminivora*
Plasmopara	*Paraperonospora*	*Peronophythora*	*Hyaloperonospora*
Pseudoperonospora	*Peronospora*	(*Peronosclerospora*)	*Paraperonospora*
Sclerospora	*Plasmopara*	*Peronospora*	*Perofascia*
Albuginaceae	*Pseudoperonospora*	*Phytophthora*	*Peronosclerospora*
Albugo	Albuginaceae	*Peronospora*	*Peronospora*
Pythiaceae	*Albugo*	*Plasmopara*	*Peronospora*
Phytophthora	Pythiales	*Pseudoperonospora*	*Plasmopara*
Pythiogeton	Pythiaceae	*Sclerospora*	*Plasmoverna*
Pythium	*Halophytopthora*	Albuginaceae	*Protobremia*
Sclerophthora	*Peronophythora*	*Albugo*	*Pseudoperonospora*
Trachysphaera	*Phytophthora*	Pythiaceae	*Sclerospora*
	Pythium	*Lagenidium*	*Viennotia*
	Trachysphaera	*Pythium*	(family not formally classified)
	Pythiogetonaceae	(*Pythiogeton*)	*Phytophthora*
	Pythiogeton	(*Trachysphaera*)	Albuginales
	Sclerosporales	(*Sclerophthora*)	Albuginaceae
	Sclerosporaceae		*Albugo*
	Peronosclerospora		*Pustula*
	Sclerospora		*Wilsoniana*
	Verrucalvaceae		
	Sclerophthora		

For the Pythiaceae and Verrucalvaceae, only important genera (those with phytopathogenic and/or phylogenetic relevance for downy mildews) are listed (for complete genus lists, see the respective publications); from Peronosporaceae and Albuginaceae, all genera are considered. Taxa in parentheses: not included in the phylogenetic analyses. *Phytophthora* excluded from Peronosporaceae but placed in Peronosporales without formal family classification; other traditional Pythiaceae not classified.

[a] For downy mildews

[b] For white blister rusts

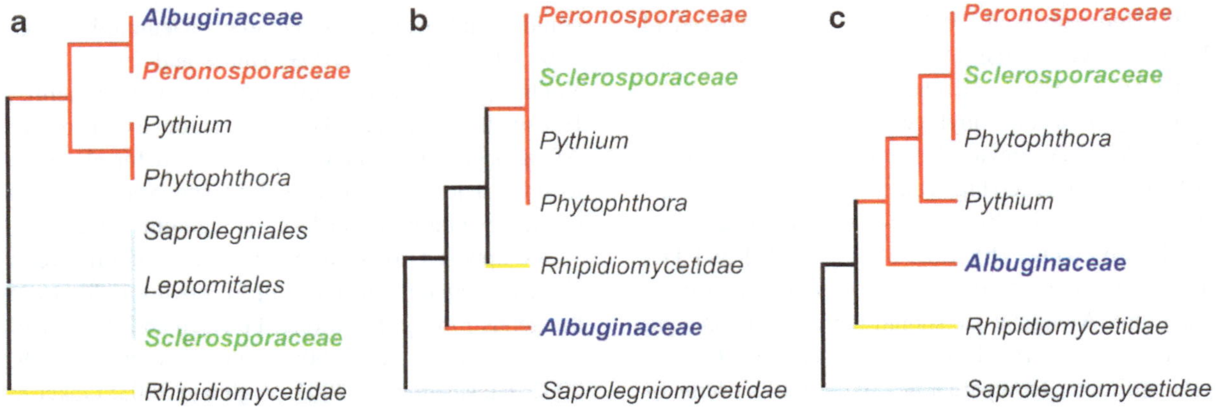

Fig. 1 Phylogenetic hypotheses among the Oomycetes, with special reference to obligate parasites and their closest relatives (*Pythium*, *Phytophthora*). **a** Topology reflecting the hierarchical classification of Dick (2001). **b** Topology obtained from cox2 sequence data (Hudspeth et al. 2003). **c** Topology obtained from SSU (Petersen and Rosendahl 2000) and LSU rDNA (Riethmüller et al. 2002) sequence data. *Red branches*: Peronosporomycetidae, *grey branches*: Saprolegniomycetidae, *yellow branches*: Rhipidiomycetidae. The three groups of obligate parasites (Albuginaceae = white blister rusts, Peronosporaceae = downy mildews, Sclerosporaceae = graminicolous downy mildews) are given in *colour* and *bold*; circumscription follows Kirk et al. (2001; see also Table 1). The molecular analyses are largely congruent except for the position of the Rhipidiomycetidae. Due to the isolated phylogenetic position of Albuginaceae, subclass Albuginomycetidae has been proposed (Thines and Spring 2005)

intermediate between *Phytophthora* and downy mildews, is not the closest relative to downy mildews and should rather be classified within *Phytophthora* (Riethmüller et al. 2002; Voglmayr 2003; Göker et al. 2007).

Monophyly versus polyphyly of downy mildews

The downy mildews (Peronosporaceae), in the traditional sense, are a morphologically diverse group, which is mainly united by obligate parasitism in combination with more or less complex conidio- or sporangiophores with determinate growth. In traditional morphological classifications it is generally assumed that obligate biotrophism of the downy mildews evolved only once.

In the first molecular phylogenetic analyses it was uncertain whether downy mildews are monophyletic or stem from different groups of *Phytophthora*, a situation which in fact has not yet been clarified with certainty (compare Riethmüller et al. 2002; Göker et al. 2003, 2007; Göker and Stamatakis 2006). In a recent multigene phylogeny involving five genes (Göker et al. 2007), monophyly of downy mildews was highly supported by various methods of phylogenetic reconstruction, which therefore seemed to be corroborated (Fig. 2a). However, in an analysis of the same dataset using different methods of phylogenetic reconstruction (Göker and Stamatakis 2006), downy mildews did not appear monophyletic, with the *Phytophthora infestans* group (*Phytophthora* 1) being embedded within the downy mildews clade (Fig. 2b). It is questionable whether this problem can be solved with comparably few DNA sequences alone; more detailed investigations on genome organisation and ultrastructure may provide better insights.

The paraphyly problem of *Phytophthora*

Phylogenetic classification using the monophyly criterion may raise severe problems for the genus *Phytophthora*, which is paraphyletic in most analyses in respect to downy mildews (e.g. Cooke et al. 2000, 2002; Göker et al. 2007; Fig. 2). Therefore, it remains open if a phylogenetic classification can be achieved without either merging downy mildews with *Phytophthora* or generic splitting of *Phytophthora* into several genera. It is foreseeable that these alternatives will not receive broad acceptance, representing a dilemma for classification. However, internal support for the tree backbone is usually low even in multigene analyses (e.g. Göker et al. 2007; Fig. 2), and additional data from other gene regions and taxa need to be collected before a robust phylogeny can be achieved.

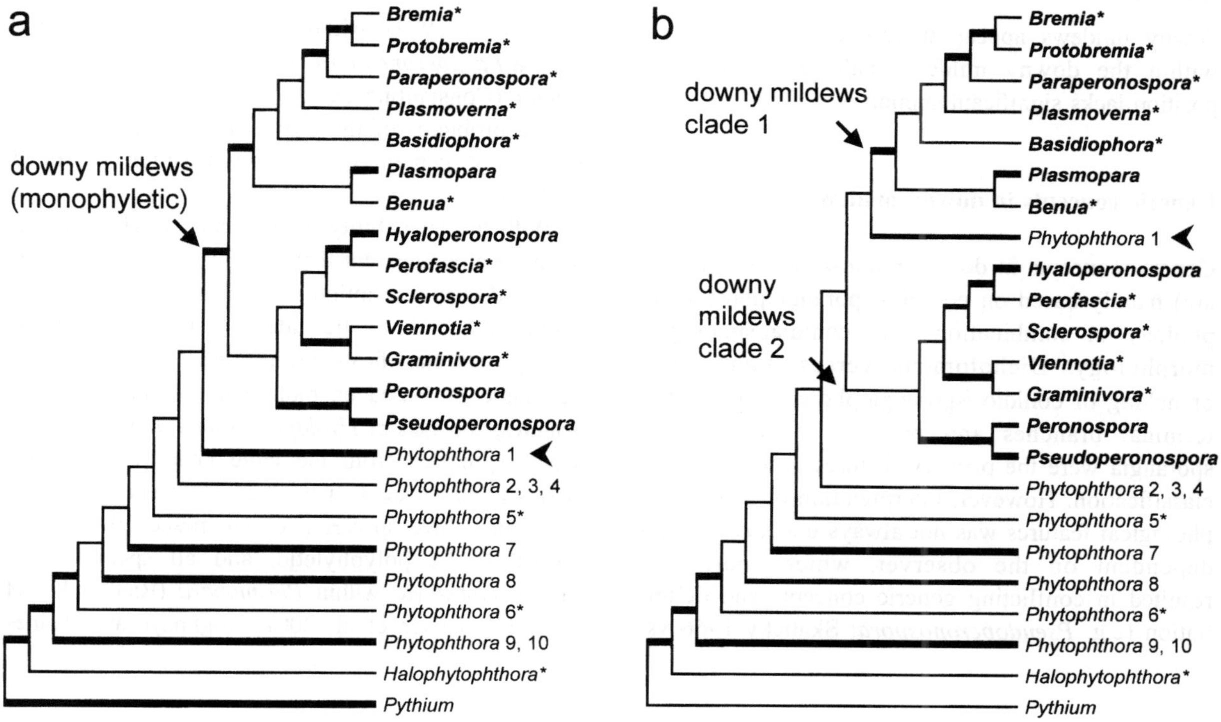

Fig. 2 Phylogenetic hypotheses among the downy mildews (genera in *bold*) inferred from DNA data. **a** Simplified tree illustrating the phylogenetic hypotheses of Göker et al. (2007) showing a single monophyletic downy mildews clade. **b** Simplified tree illustrating the phylogenetic hypotheses of Göker and Stamatakis (2006), showing polyphyly of two separate downy mildew clades; *Phytophthora* 1 (*P. infestans* group; arrowhead) is the closest relative of the downy mildews clade 1 containing, amongst others, *Plasmopara* and *Bremia*; the same tree topology was already largely revealed by Göker et al. (2003). *Bold* and medium branches indicate boostrap support equal or higher than 90% and 70%, respectively; genera marked with an *asterisk* are represented by a single taxon only. Note that in both trees the genus *Phytophthora* (closest relative of downy mildews) is paraphyletic. *Numbers* after *Phytophthora* correspond to the clade numbering in Cooke et al. (2000)

The status of the graminicolous downy mildews (Sclerosporaceae)

The controversial classification (Table 1, Fig. 1) of the graminicolous downy mildews (i.e. *Sclerospora* and *Peronosclerospora*) has been recently clarified with molecular phylogenetic data. Classification as a separate family and order (Dick et al. 1984), as well as the classification within the Saprolegniomycetidae (Dick et al. 1989), did not receive support from molecular data (Fig. 1). In all molecular phylogenetic investigations published so far (Riethmüller et al. 2002; Hudspeth et al. 2003; Göker et al. 2003, 2007; Thines et al. 2007, 2008), graminicolous downy mildews were unequivocally placed within Peronosporaceae. Therefore, separation of graminicolous downy mildews from downy mildews and classification within a separate family Sclerosporaceae has become obsolete (Figs. 1 and 2). The genus *Sclerophthora*, sometimes considered closely related to *Sclerospora* and *Peronosclerospora*, has been placed within *Phytophthora* in a recent molecular phylogenetic analysis (Thines et al. 2008), however, without significant internal support.

In addition to the graminicolous downy mildews in the strict sense (i.e. *Sclerospora*, *Peronosclerospora*), the species from the genera *Plasmopara* and *Bremia* parasitising Poaceae were also recently reclassified and transferred to newly described genera (Göker et al. 2003; Thines et al. 2006, 2007). Whether these are close relatives to *Sclerospora* and *Peronosclerospora* cannot be resolved with the current molecular data (Göker et al. 2007; Thines et al. 2008).

Phylogenetic relationship of the graminicolous downy mildews within the Peronosporaceae is much less clear. To date, only few accessions have been analysed. In some recent publications (Göker et al. 2007; Thines et al. 2007, 2008), graminicolous downy mildews appear at varying basal positions within the downy mildew clade. However, this position lacks significant support.

Generic concepts in downy mildews

Generic concepts in downy mildews were (and still are) mainly based on conidio-/sporangiophore morphology in combination with conidial/sporangial morphology. Dichotomous versus monopodial branching of conidio-/sporangiophore, shape of the terminal branches and presence of conidia or sporangia were the primary features used for genus classification. However, interpretation of these morphological features was not always unequivocal and dependent on the observer, which sometimes resulted in conflicting generic concepts and delimitation (e.g. *Pseudoperonospora*: Skalický 1966 vs. Waterhouse 1973). Although the segregation of the genera *Paraperonospora* (Constantinescu 1989) and *Benua* (Constantinescu 1998) resolved some taxonomic problems of the morphological classification scheme of genera, generic classification remained problematic.

With the availability of molecular phylogenies, it soon became apparent that current generic classification and circumscription contained numerous problems and had to be adapted if standards of phylogenetic classification were applied (for a summary, see Table 2). Based on molecular and morphological features, the genera *Hyaloperonspora* and *Perofascia* were segregated from the large genus *Peronospora* (Constantinescu and Faheti 2002). The genus *Bremiella*, containing three species (Constantinescu 1991a), was shown to be polyphyletic, and all species were clearly embedded within *Plasmopara* (Riethmüller et al. 2002; Göker et al. 2007; Voglmayr and Thines

Table 2 Recent taxonomic changes at the generic level of white blister rusts and downy mildews based on molecular and morphological data

Pre-2000	Recent (post-2000)	References
Albugo	*Albugo* (sensu stricto)	Thines and Spring (2005)
	Pustula	Thines and Spring (2005)
	Wilsoniana	Thines and Spring (2005)
Basidiophora	*Basidiophora*	Riethmüller et al. (2002), Voglmayr et al. (2004), Göker et al. (2007)
Benua	*Benua*	Göker et al. (2007)
Bremia	*Bremia* (*B. lactucae*)	Voglmayr et al. (2004)
	Graminivora (*B. graminicola*)	Thines et al. (2006)
Bremiella	*Plasmopara*	Riethmüller et al. (2002), Voglmayr et al. (2004), Göker et al. (2007), Voglmayr and Thines (2007)
Paraperonospora	*Paraperonospora*	Riethmüller et al. (2002), Voglmayr et al. (2004), Göker et al. (2007)
Peronosclerospora	*Peronosclerospora*	Hudspeth et al. (2003), Thines et al. (2008)
Peronospora	*Peronospora*	Göker et al. (2007)
	Hyaloperonospora (*P. parasitica* s.l.)	Constantinescu and Faheti (2002), Göker et al. (2003, 2004)
	Perofascia (*P. lepidii*)	Constantinescu and Faheti (2002)
Plasmopara	Core *Plasmopara*	Voglmayr et al. (2004)
	Novotelnova (*Pl. savulescui*)	Voglmayr and Constantinescu (2008)
	Plasmoverna (*Pl. pygmaea* s.l.)	Constantinescu et al. (2005)
	Poakatesthia (*Pl. penniseti*)	Thines et al. (2007)
	Protobremia (*Pl. sphaerosperma*)	Voglmayr et al. (2004)
	Viennotia (*Pl. oplismeni*)	Göker et al. (2003)
Pseudoperonospora	*Pseudoperonospora*	Riethmüller et al. (2002), Göker et al. (2007)
Sclerospora	*Sclerospora*	Riethmüller et al. (2002), Göker et al. (2007)

The pre-2000 classification is according to the Dictionary of the Fungi (Kirk et al. 2001); the references refer to the post-2000 classifications

2007). On the other hand, the genus *Plasmopara* itself was shown to be a non-monophyletic genus, and several genera were segregated (see Table 2). Fortunately enough, it was nomenclaturally possible to maintain the use of the name *Plasmopara* for the bulk of species including the phytopathologically important ones (for details, see Constantinescu et al. 2005). The genus *Bremia*, consisting of two species (one on Asteraceae, one on Poaceae), was also shown to be polyphyletic, and the genus *Graminivora* was segregated for the grass parasite (Thines et al. 2006). Therefore, the number of accepted genera has gradually increased during the last years (one merged versus eight new genera). However, except for *Hyaloperonospora*, as none of these new genera is phytopathologically important, these changes have had little effect for phytopathologists.

Morphology revisited: shortcomings, new features and interpretations

The results of the molecular phylogenetic investigations also stimulated reevaluation of the morphological features traditionally used for classification and the search for new, previously neglected characters (Spring and Thines 2004; Voglmayr et al. 2004; Constantinescu et al. 2005; Thines 2006; Voglmayr and Riethmüller 2006). Already Hall (1996) stressed the necessity of thorough morphological investigations, besides other methods, to obtain a more satisfactory classification. Downy mildews and white blister rusts show a low morphological complexity, and comparatively few characters are available for classification. However, taking the whole spectrum of these few available characters into account, it is astonishing that even fewer selected characters had been used in traditional morphological classifications (i.e. some features of conidio-/sporangiophore morphology and conidial/sporangial morphology; for a synopsis, see Waterhouse 1973). In addition, these few characters often had not been critically studied and evaluated for more than a few species, and often data were just taken non-critically from the original descriptions. The morphology of the few common and phytopathologically important downy mildew species was comparatively well known, but the vast majority of species were much less studied. Therefore, morphological classification and interpretation was mainly based on the few phytopathologically important species, neglecting much of the morphological diversity.

With the advent of molecular systematics, the morphological features used for classification were reinterpreted in a phylogenetic context, making previous interpretations of conidio-/sporangiophore morphology too simple, partly incorrect or unsuitable for generic classification (e.g., Riethmüller et al. 2002; Voglmayr et al. 2004; Thines et al. 2006). Nevertheless, importance of known but previously neglected features became apparent, as is the case for haustorial morphology. Presence of different haustorial types in downy mildews was recorded by Fraymouth (1956), but these were never used for classification. Recently, several haustorial types could be shown to be diagnostic for several lineages (e.g. *Hyaloperonospora*: Constantinescu and Faheti 2002; *Plasmopara*, *Bremia* and their close relatives: Riethmüller et al. 2002; Göker et al. 2003; Voglmayr et al. 2004; Thines et al. 2006; *Pseudoperonospora*: Voglmayr et al. 2004).

In addition, the importance of searching for additional features to be used in classification was emphasised by Spring (2004) and Spring and Thines (2004). Ideally, these should be preserved in herbarium specimens to enable investigation of as many representatives as possible, which would be difficult if living specimens were necessary. As herbarium specimens can be used, Spring (2004) and Spring and Thines (2004) argued for investigations of ultrastructural features by scanning electron microscopy (SEM) to reveal additional features. Thines (2006, 2007b) presented the results of detailed morphological investigations of downy mildews including SEM studies of conidio-/sporangiophores and sporangia. He produced a morphological character matrix and analysed it within the constraint of recently published molecular phylogenetic investigations to reveal and evaluate morphological synapomorphies. He concluded that the classical features used for classification like conidio-/sporangiophore branching or sporangial germination are not diagnostic for phylogenetic lineages above the generic level. However, the fine structure of the ultimate branchlets (revealed by SEM) and haustorial shape were considered to be phylogenetically informative.

The genus *Albugo* sensu lato has also recently been extensively re-investigated using light microscopy and SEM. Thines and Spring (2005) emphasised that different sporangial ornamentation types revealed by SEM were diagnostic for the *Albugo* clades revealed by molecular data, and segregated the two genera *Wilsoniana* (for species on Caryophyllidae) and *Pustula* (for species on Asteraceae). Voglmayr and Riethmüller (2006) confirmed the SEM data of Thines and Spring (2005) and added detailed light microscopical and SEM data of the oospores, which were also shown to be diagnostic for the different molecular phylogenetic lineages of *Albugo* sensu lato. Constantinescu and Thines (2006) investigated and clarified sporangiogenesis in Albuginaceae; they confirmed the presence of sporangial dimorphism (primary and secondary sporangia) for all species investigated.

Biochemical characters

Apart from DNA sequence data, few characters at the biochemical level have been used for classification of downy mildews. Recently, fatty acids were recorded as potentially promising for the characterisation of downy mildew species (e.g., Spring et al. 2003; Spring, 2004; Spring and Thines 2004; Spring et al. 2005). In the case of *Albugo* sensu lato, Spring et al. (2005) recorded significant differences between three species and considered fatty acid pattern characteristic for the species. Investigations on closely related species are still missing, and therefore additional data are necessary to evaluate applicability and resolution limits of fatty acid composition for taxonomic differentiation.

Species concepts in downy mildews

The species concept is probably the most controversial issue in downy mildew systematics, partly due to experimental difficulties to test it and partly due to its profound implications for researchers outside the systematics research community. A well-written account of the history and implications of the different species concepts is given in Hall (1996). In downy mildews, several species concepts were applied, which resulted in highly different numbers of accepted species depending on the criteria used. The main problem in species delimitation in downy mildews is that there are numerous indications that, due to their obligate parasitism, they often have narrow host ranges and therefore represent genetically distinct species. On the other hand, host specificity is not always paralleled by morphological distinctness. Therefore, if morphology is used as a primary criterion for species definition, only a few species can be defined and accepted in many lineages, resulting in genetically heterogeneous species.

Historically, two approaches were commonly followed, which were both mainly based on host ranges: the splitting approach of Gäumann (1918, 1923) versus the lumping approach of Yerkes and Shaw (1959). Gäumann (1918, 1923) advocated a narrow species concept in *Peronospora*, based on his results of cross-inoculation studies and minute morphological differences. Each species was usually assumed to be confined to one host genus or even a few host species (one host-one species concept; see Hall 1996). Conversely, Yerkes and Shaw (1959) argued that host specificity is neither sufficient nor suitable for the recognition of a species without clear-cut morphological differences. As a consequence, they lumped the numerous *Peronospora* species on Brassicaceae and Chenopodiaceae each into a single species (*Peronospora parasitica* and *P. farinosa*, respectively), resulting in a wide one host family-one species concept.

Both the splitting as well as the lumping approach have sincere shortcomings. Using the narrow species concept, identification of morphologically similar species is often difficult or impossible without correct identification of the host. In addition, high host specificity has rarely been conclusively demonstrated, weakening the primary underlying assumption of the narrow species concept.

In a wide species concept, there is the problem that genetically distinct or even unrelated entities may be classified in the same species, raising incorrect assumptions on biology and host ranges. This is especially problematic if host jumps are common and parasitism on the same host family has evolved multiple times, resulting in polyphyletic species. However, due to its easier applicability, the approach to classify all accessions of a given host family within a single species was widely followed by phytopathologists and molecular biologists.

Impact of molecular data on downy mildew species concept and circumscription

Recently, molecular phylogenetic investigations have enabled the evaluation of the species problem using new perspectives and have led to the shift from a morphological to a phylogenetic species concept. A biological species concept directly addressing mating barriers has never been applied to downy mildews due to sincere methodological difficulties, and it is unlikely that these can be overcome. Therefore, reproductive isolation can only be indirectly assessed, e. g. by genetic distance of sequence data. The impact of molecular data is manifold: (1) numerous additional characters are available for recognition and distinction; (2) presence and amount of reproductive isolation can be assessed; (3) presence and amount of genetic distances provide indirect but strong evidence for host specificity and host ranges; (4) molecular data are less variable and prone to subjective interpretation than morphological data; (5) molecular data provide a sound basis for species identification even if morphological data are missing or incomplete; (6) pathotypes or races, the basic entities for experiments in applied sciences, can be properly attributed to a species and their phylogenetic relationships can be assessed. Therefore, in the absence of sound morphological characters, the species concept is increasingly based on molecular evidence of reproductive isolation, which is a general tendency within mycology. Consequently, morphologically similar cryptic species are often recognised as distinct species if reproductive isolation and genetic distinctness can be demonstrated. However, evaluation of species boundaries by molecular data require thorough sampling throughout the distribution area to assess genetic variability as well as reproductive isolation, and at best several molecular markers should be used for corroboration of species boundaries.

Due to easy amplification and variability, the ITS rDNA region has been used in most investigations addressing the species concept in downy mildews and white blister rusts (e.g. Choi et al. 2003, 2005, 2006, 2007a, b, c, d; Voglmayr 2003; Göker et al. 2004; Scott et al. 2004; Cunnington 2006; Spring et al. 2006; Voglmayr et al. 2006; Landa et al. 2007; García Blázquez et al. 2008). However, the mitochondrial cox2 region may also be a promising candidate to resolve species boundaries and for species identification (e.g. Choi et al. 2006, 2007d). Interestingly, the current evidence from molecular phylogenetic investigations often supports a narrow species concept as advocated by Gäumann (1918, 1923), although there are sometimes marked differences in detail. In the following, the results of recent molecular investigations will be briefly summarised for the different genera.

Hyaloperonospora

According to Constantinescu and Faheti (2002), about 140 species names were published attributable to this genus. In their separation of *Hyaloperonospora* from *Peronospora*, Constantinescu and Faheti (2002) only recognised six morphologically distinct species, and accessions from most hosts of Brassicaceae were placed in *Hyaloperonospora parasitica*. However, subsequent molecular phylogenetic investigations demonstrated that the latter was a paraphyletic assemblage with respect to the other five *Hyaloperonospora* species, and that many more species should be accepted based on the high genetic distances observed (Choi et al. 2003; Göker et al. 2003, 2004; Voglmayr 2003). Usually, these genetically distinct entities deserving species rank have a narrow host range and are confined to host genera or even species; however, in some cases accessions from the same host do not form a monophylum (e.g. from *Armoracia rusticana*; see Göker et al. 2004). Therefore, it is problematic when species are determined solely on host association, as this is often but not always conclusive. The case study of *Hyaloperonospora* is also relevant for investigations at the molecular level of plant–pathogen interactions, as numerous studies are performed with the plant model organism *Arabidopsis thaliana* and its *Hyaloperonospora* parasite. The parasite is usually named *H. parasitica*, but it is genetically quite distinct from *H. parasitica* sensu stricto which is confined to *Capsella bursa-pastoris* (Göker et al. 2004); therefore, the name *H. arabidopsidis* should be used for the *Arabidopsis* parasite.

Peronospora

The molecular phylogenetic analyses dealing with the genus *Peronospora* (e. g. Voglmayr 2003; Choi et al.

2007c; García Blázquez et al. 2008) also provide evidence for a narrow species circumscription. Species parasitising e.g. Chenopodiaceae do not form a monophyletic lineage (Voglmayr 2003; Choi et al. 2007c), but are interspersed with species from other host families and can often also be separated morphologically (Choi et al. 2007c). Interestingly, the type species of *Peronospora*, *P. rumicis*, a parasite of *Rumex* spp. (Polygonaceae), is embedded within a group of species infecting Chenopodiaceae (Choi et al. 2007c), which may indicate a recent host jump. Choi et al. (2007c) demonstrated that *Peronospora effusa* (spinach downy mildew) is genetically homogeneous world-wide, but distinct from *P. farinosa* sensu stricto. Reasons for this genetic homogeneity may include the recent introduction to most of its present growth area as well as pathogen transmission by seeds, enabling rapid dispersal from a small geographic source area via international seed trade. Choi et al. (2008a), investigating five *Peronospora* species from different species of *Chenopodium*, recorded significant molecular and morphological differences, and they concluded that these are well-distinct species and should not be merged with *P. farinosa*. Therefore, the approach of Yerkes and Shaw (1959) to merge all species on Chenopodiaceae under *P. farinosa* appears to be inappropriate, as *P. farinosa* according to Yerkes and Shaw (1959) is evidently a polyphyletic assemblage. However, further studies using additional variable genes are required to reveal and evaluate the phylogenetic species, as the ITS region does not give significant support or resolution for many nodes of the backbone (Voglmayr 2003; Choi et al. 2007c).

Traditionally, from de Bary (1863) onwards, only two *Peronospora* species have been recognised on Fabaceae, *P. trifoliorum* and *P. viciae*, which were observed to differ in their oospore ornamentation. Gäumann (1923) and subsequent authors again described numerous new species from different host species, resulting in >100 *Peronospora* binomials described from 25 host genera of Fabaceae (Constantinescu 1991b). The results of Voglmayr (2003) indicated that *Peronospora* on Fabaceae does not form a monophyletic lineage, although most accessions including those from *Vicia* and *Trifolium* were united in a single monophyletic clade. However, the clades did not correspond to the classical two species recognised from Fabaceae and showed that more than two species are involved. Cunnington (2006) confirmed high genetic distances between accessions from different hosts, which were traditionally included in *P. viciae*. In the most extensive study on *Peronospora* parasites of Fabaceae, García-Blázquez et al. (2008) showed that numerous host-specific lineages are present on Fabaceae. Although their study did not include *Peronospora* species from other host families, there is evidence that they do not even form a monophyletic lineage (Voglmayr 2003), a situation comparable to the *Peronospora* species on Chenopodiaceae. Although there are numerous nomenclatural problems left in the Peronosporas from Fabaceae which need additional studies (García-Blázquez et al. 2008), the results are roughly concordant with the classification of Gäumann (1923).

Molecular phylogenies using ITS data not only often confirmed a rather narrow species concept in *Peronospora*, but also helped to clarify species attribution. Scott et al. (2004) identified the parasite of oilseed poppy (*Papaver somniferum*) from Tasmania, which was previously listed as *Peronospora arborescens*, as *P. cristata*, and both species were shown to be genetically distinct. Conversely, Landa et al. (2007) reported *Peronospora arborescens* as the causal agent of downy mildew of *P. somniferum* from Spain. Therefore, apparently both *P. arborescens* and *P. cristata* can infect *P. somniferum* and are of different importance in various regions of the world.

Pseudoperonospora

Conversely to the other examples listed above, in *Pseudoperonospora* molecular data provided evidence for conspecificity of species from different host families. Choi et al. (2005) performed a study including *Pseudoperonospora humuli* (from *Humulus* spp., Cannabinaceae) and *P. cubensis* (from *Citrullus vulgaris*, *Cucumis* spp. and *Cucurbita* spp., Cucurbitaceae). Distinction as separate species was mainly based on their occurrence on two non-related host families, but almost identical ITS sequences and absence of significant morphological differences indicate conspecificity. Therefore, Choi et al. (2005) synonymised *Pseudoperonospora humuli* with *P. cubensis*. These data confirm recent and possibly multiple host shift from Cannabinaceae to Cucurbitaceae. However, for detailed insights into the evolutionary processes involved, the ITS data offer too little resolution, and molecular markers with higher resolution should be applied.

ITS length differences as potential markers for species

In addition to phylogenetic analyses of ITS data, the structure of the ITS itself has also recently received increasing attention. This appears to be especially promising for the *Plasmopara-Bremia* clade, which was shown to have an ITS region of variable length up to about 3,200 bp (Choi et al. 2007b; Komjáti et al. 2007; Thines et al. 2005; Thines 2007a), which is remarkable compared to the usually about 800 bp in downy mildews (Voglmayr 2003; Göker et al. 2004). Size increase mainly concerns the ITS2 region and originates from repetitive elements, the number and length of which appear to be taxon-specific (Choi et al. 2007b; Thines 2007a, c). The number and sequence of repetitive elements is usually conserved within different lineages. Choi et al. (2007b) investigated *Bremia* accessions from different hosts and identified nine repetitive elements showing high sequence heterogeneity between the accessions from different hosts, which indicates that they may represent distinct species. Thines (2007a) investigated repetitive elements for several species and recorded a highly variable number. He concluded that these repetitive elements could be useful for investigation of speciation and radiation processes. Based on sequence variability observed in the ITS2 region, Spring et al. (2006) could separate and characterise two groups of pathotypes of *Plasmopara halstedii* from sunflower.

Repetitive elements are also present in the ITS2 region of some *Hyaloperonospora* species (Voglmayr 2003; Göker et al. 2004); however, they are confined to a few species for which they may be diagnostic. Thines (2007a) recorded similarities to the repetitive elements of the *Plasmopara-Bremia* clade and suggested that this may be indication of a closer relationship for these two lineages.

Repetitive elements are also observed within *Peronospora* sensu stricto, where they have been reported within the ITS1 region (Voglmayr 2003). The number of repetitive elements has recently been investigated and analysed in detail for *Peronospora* on *Trifolium* (García Blázquez et al. 2008). The number of additional copies of a region about 70 bp long ranged from one to 11 and was, with few exceptions, diagnostic for different host-specific lineages. Therefore, different lengths of ITS may, with limits, be suitable for species identification. However, little is yet known on the mechanisms governing the copy number of repetitive elements, and taxon specificity needs additional detailed investigations.

Discovery of morphologically distinct species by molecular data

Molecular phylogenetic data also stimulated closer morphological examinations at the species level, which showed that the non-critical but widely applied species identification solely on host association can be misleading. Several new species could be described which were morphologically well distinct but remained remarkably unnoticed (e.g., Voglmayr et al. 2006; Choi et al. 2007a). *Plasmopara* on Geraniaceae provides an excellent example for this. Traditionally, two morphologically distinct species, *Plasmopara pusilla* on European and *P. geranii* on North American *Geranium* species, were accepted and recognised (Constantinescu 2004). Detailed molecular and morphological investigations confirmed distinctness of another previously described but neglected species and revealed two additional undescribed species which are morphologically distinct, widespread and rather common (Voglmayr et al. 2006). This indicates that biodiversity of downy mildews is still imperfectly known even in regions which are considered well-studied.

Unresolved questions and future perspectives

Detailed phylogenetic relationships and evolutionary scenario within downy mildews

It remains still unresolved whether downy mildews are monophyletic and how the major groups of downy mildews are related to each other. Clarification of phylogenetic relationships will necessitate additional molecular markers, new methods of phylogenetic inference as well as improved taxon sampling. The graminicolous downy mildews and representatives of remote, little sampled areas (Africa, South America, East and Southern Asia) are especially underrepresented in the present studies, and additional taxa need to be sampled. Albuginales are also little-investigated, and recent investigations on the *Albugo candida* complex indicate a high level of, as yet, undescribed biodiversity (Choi et al. 2007d, 2008a, b, which could also be true in other lineages. In other cases such as the downy mildews with pyriform haustoria *Plasmopara*, *Bremia* and relatives, additional sequence data are needed to confirm more robust phylogenetic relationships, a precondition for a sound delimitation of genera. In addition, the investigations need to be embedded in a wider taxonomic context. *Phytophthora* especially needs to be included, which is a key genus for the evolution of downy mildews. In the recent multigene phylogeny of *Phytophthora* by Blair et al. 2008), obtained from a data matrix of 8,700 bp from seven genes, various *Phytophthora* groups were highly supported; however, internal support for the tree backbone was still only moderate to low in maximum parsimony and likelihood analyses, despite the large data matrix, which may indicate that it may be difficult to obtain highly supported gene phylogenies for *Phytophthora* and its relatives. To test this, these sequence data should be complemented and analysed with corresponding sequence data of representative downy mildews as well as from the genus *Pythium*. Sound phylogenetic hypotheses are a precondition for detailed insights into the processes of character evolution, adaptive radiation and speciation of downy mildews on different host groups. The species-rich genera *Peronospora* and *Plasmopara* especially require extended, representative taxon sampling as well as additional molecular markers to provide a sound phylogenetic framework. It is still little investigated whether polyploidy is involved in speciation, and detailed studies on nuclear genome size may provide information on genome evolution (Voglmayr and Greilhuber 1998).

Whole genome analysis in an evolutionary context

For a better understanding of the evolutionary processes and of phylogenetic relationships, the integration of whole genome data is also promising, especially where phylogenetic relationships remain unsettled with the current molecular phylogenetic analyses (e.g. monophyly of downy mildews, detailed relationships of graminicolous downy mildews, detailed relationships of genera). The availability of whole genome sequences of several species of *Phytophthora* should accelerate our understanding of molecular evolution in plant pathogenic oomycetes (Lamour et al. 2007). However, for progress in comparative genome analysis, it is critical that high quality genome sequence assemblies and gene models are developed, which is a next, urgently needed step (Lamour et al. 2007). When such high-quality data become available for *Phytophthora*, they could provide the basis for new evolutionary and phylogenetic investigations on downy mildews. As *Phytophthora* is the closest relative of downy mildews, the genetic models of host specificity and host jumps could be adapted to and tested in downy mildews. The *Hyaloperonospora-Arabidopsis* pathosystem would be an ideal candidate, as both host and pathogen are genetically well studied, and as the genus *Hyaloperonospora* contains numerous host-specific, genetically distinct entities commonly recognised as separate species (Choi et al. 2003; Göker et al. 2004). In addition, comparative analysis of genome organisation could provide detailed evolutionary insights. However, it should be mentioned that these investigations are methodologically difficult and require the development of sophisticated techniques as downy mildews cannot be cultivated on artificial media. It is evident that sequencing of the whole genome or even large quantities of the genome can practically only be done for a few selected representatives. In addition, suitable computational methods of phylogenetic analysis of such large data sets need to be developed. Despite these limitations, whole genome data could contribute to more robust phylogenetic hypotheses than investigations based on one or a few sequences regions only, and could provide help for the selection of promising proper genome regions for phylogenetic analyses on an extended taxon selection.

Applicable species definitions and the need for taxonomic and nomenclatural revisions

The most important issues of downy mildew systematics outside the taxonomic community concern the species concept. Nomenclatural stability as well as sound applicable species circumscription are eagerly anticipated. For this, more detailed and conclusive molecular studies are required to resolve the species boundaries, especially in the species-rich genera *Peronospora*, *Hyaloperonospora* and *Plasmopara*. In addition, to achieve nomenclatural stability, numerous taxonomic and nomenclatural problems involving host range and correct typification need to be solved. Although molecular data provide strong evidence that a narrow species concept as advocated by Gäumann (1918, 1923) may in many cases be appropriate, there are numerous problems in his classification in detail. Whereas some of his species are apparently conspecific, others are heterogeneous. As he did not designate type collections, lectotypification is necessary which for heterogeneous assemblages has great impact on species nomenclature. This has been recently discussed for *Peronospora* on Fabaceae (García Blázquez et al. 2008), but is true also for other lineages of *Peronospora* and *Plasmopara*. Therefore, thorough investigations of types and proper typification whenever necessary are tedious but unavoidable prerequisites to receive correct and stable circumscription for many species. In addition, as shown above, numerous distinct species remained undetected up to now, which require additional thorough studies involving morphological as well as molecular data. Even lineages containing important plant pathogens such as the *Plasmopara halstedii* group are in need of critical taxonomic and nomenclatural revision to reveal an appropriate species delimitation.

Molecular bar coding systems for improved and reliable species identification

Another important issue is the development of methods for reliable and easy species identification. As discussed above, morphology is often not the best basis for identification due to a lack of morphological distinctness of numerous genetically well distinct lineages. In addition, all morphological structures necessary for identification are not always present on a specimen to be identified. Especially for the plant pathology community, molecular methods are much easier to use, and are nowadays routine procedures, and provide reliable identification even with the lack of morphological structures required for identification (e.g. sporangiophores, oospores). Therefore, a species identification system based on sequence data is highly desirable. Numerous sequence data are already available in sequence databases like Genbank; however, these data usually suffer from the lack of standards concerning correct identification, documentation, nomenclature but also the sequence data quality. Therefore, molecular bar coding initials have been recently proposed and started for many taxonomic groups of organisms to facilitate identification (see the publications of the themed issue of the Philosophical Transactions of the Royal Society, Biological Sciences 360, Number 1462, 2005). Molecular bar coding requires strict quality standards for the laboratory routine, the sequence data as well as identification, documentation, specimen deposition and nomenclature; for details see the homepage of the Consortium for the Barcode of Life (http://barcoding.si.edu/index.htm). Evidently, a bar coding approach needs to be accompanied by thorough taxonomic revisions to provide a proper taxonomic framework.

Crucial for the resolution of a molecular bar coding system is the selection of the sequence region used. Ideally, species identification should be possible with a single and the same sequence over a wide range of organisms, at best with the same primers, which is problematic for downy mildews. Therefore, a compromise between applicability for as many different taxa as possible and sufficient taxonomic resolution needs to be found. One possible candidate is the ITS rDNA: it is a multicopy region easy to amplify in most downy mildews, can detect pathogens with high sensitivity, specific as well as universal primers are available, and it is also a region of choice in *Phytophthora*, the closest relative of downy mildews. In addition, numerous data on ITS are already available for downy mildews which is important for testing the discriminatory power of the data. However, amplification and sequencing can be troublesome in lineages affected by substantial size increases of the ITS region (see above), limiting universal applicability. Alternatively, sequences from the mitochondrial DNA should be considered. It is usually much easier to amplify from historic material than nuclear DNA due to a higher copy number, and therefore herbarium collections can be used. A promising candidate for bar coding is cox2, which offers high resolution in some closely related species groups (e.g. Choi et al. 2006, 2007d), also mtDNA sequence stretches less variable than cox2 should be considered, e.g. ribosomal mtDNA and cox1 or cox3. Additional detailed, comparative studies are necessary before the most appropriate sequence region can be selected. However, the development of a molecular bar coding system is now within reach.

References

Belbahri, L., Calmin, G., Pawlowski, J., & Lefort, F. (2005). Phylogenetic analysis and real time PCR detection of a presumably undescribed Peronospora species on sweet basil and sage. *Mycological Research, 109*, 1276–1287.

Blair, J. E., Coffey, M. D., Park, S.-Y., Geiser, D. M., & Kang, S. (2008). A multi-locus phylogeny for *Phytophthora* utilizing markers derived from complete genome sequences. *Fungal Genetics and Biology, 45*, 266–277.

Choi, Y.-J., Hong, S.-B., & Shin, H.-D. (2003). Diversity of the *Hyaloperonospora parasitica* complex from core brassicaceous hosts based on ITS rDNA sequences. *Mycological Research, 107*, 1314–1322.

Choi, Y.-J., Hong, S.-B., & Shin, H.-D. (2005). A reconsideration of *Pseudoperonospora cubensis* and *P. humuli* based on molecular and morphological data. *Mycological Research, 109*, 841–848.

Choi, Y.-J., Hong, S.-B., & Shin, H.-D. (2006). Genetic diversity within the *Albugo candida* complex (Peronosporales, Oomycota) inferred from phylogenetic analysis of ITS rDNA and COX2 mt DNA sequences. *Molecular Phylogenetics and Evolution, 40*, 400–409.

Choi, Y.-J., Constantinescu, O., & Shin, H.-D. (2007a). A new downy-mildew of the Rosaceae: *Peronospora oblongispora* sp. nov. (Chromista, Peronosporales). *Nova Hedwigia, 85*, 93–101.

Choi, Y.-J., Hong, S.-B., & Shin, H.-D. (2007b). Extreme size and sequence variation in the ITS rDNA of *Bremia lactucae*. *Mycopathologia, 163*, 91–95.

Choi, Y.-J., Hong, S.-B., & Shin, H.-D. (2007c). Re-consideration of *Peronospora farinosa* infecting *Spinacia oleracea* as distinct species, *Peronospora effusa*. *Mycological Research, 110*, 381–391.

Choi, Y.-J., Shin, H.-D., Hong, S.-B., & Thines, M. (2007d). Morphological and molecular discrimination among *Albugo candida* materials infecting *Capsella bursa-pastoris* world-wide. *Fungal Diversity, 27*, 11–34.

Choi, Y.-J., Denchev, C. M., & Shin, H.-D. (2008a). Morphological and molecular analyses support the existence of host-specific *Peronospora* species infecting *Chenopodium*. *Mycopathologia, 168*, 155–164.

Choi, Y.-J., Shin, H.-D., Ploch, S., & Thines, M. (2008b). Evidence for uncharted biodiversity in the *Albugo candida* complex, with the description of a new species. *Mycological Research* (in press).

Constantinescu, O. (1989). *Peronospora* complex on Compositae. *Sydowia, 41*, 79–107.

Constantinescu, O. (1991a). *Bremiella sphaerosperma* sp. nov. and *Plasmopara borreriae* comb. nov. *Mycologia, 83*, 473–479.

Constantinescu, O. (1991b). An annotated list of *Peronospora* names. *Thunbergia, 15*, 1–110.

Constantinescu, O. (1998). A revision of *Basidiophora* (Chromista, Peronosporales). *Nova Hedwigia, 66*, 251–265.

Constantinescu, O. (2004). The nomenclature of *Plasmopara* (Chromista, Peronosporales) parasitic on Geraniaceae. *Taxon, 53*, 523–525.

Constantinescu, O., & Faheti, J. (2002). *Peronospora*-like fungi (Chromista, Peronosporales) parasitic on Brassicaceae and related hosts. *Nova Hedwigia, 74*, 291–338.

Constantinescu, O., Voglmayr, H., Fatehi, J., & Thines, M. (2005). *Plasmoverna* gen. nov., and the taxonomy and nomenclature of *Plasmopara* (Chromista, Peronosporales). *Taxon, 54*, 813–821.

Constantinescu, O., & Thines, M. (2006). Dimorphism of sporangia in Albuginaceae (Chromista, Peronosporomycetes). *Sydowia*, 58, 178–190.

Cooke, D. E. L., Drenth, A., Duncan, J. M., Wagels, G., & Brasier, C. M. (2000). A molecular phylogeny of *Phytophthora* and related Oomycetes. *Fungal Genetics and Biology, 30*, 17–32.

Cooke, D. E. L., Williams, N. A., Williamson, B., & Duncan, J. M. (2002). An ITS-based phylogenetic analysis of the relationships between *Peronospora* and *Phytophthora*. In P. T. N. Spencer-Phillips, U. Gisi, & A. Lebeda (Eds.), *Advances in downy mildew research, Vol 1* (pp. 161–165). Dordrecht: Kluwer.

Cunnington, J. H. (2006). DNA sequence variation supports multiple host-specialised taxa in the *Peronospora viciae* complex (Chromista: Peronosporales). *Nova Hedwigia, 82*, 23–29.

de Bary, A. (1863). Recherches sur le développement de quelques champignons parasites. *Annales Des Sciences Naturelles. Botanique, Sér.*, *4*(20), 5–148.

Dick, M. W. (1995). Sexual reproduction in the Peronosporomycetes (chromistan fungi). *Canadian Journal of Botany, 73*(Suppl. 1), S712–S724.

Dick, M. W. (2001). The Peronosporomycetes. In D. J. McLaughlin, E. G. McLaughlin, & P. A. Lemke (Eds.), *The Mycota, Vol VII A* (pp. 39–72). Berlin: Springer.

Dick, M. W. (2002). Towards an understanding of the evolution of the downy mildews. In P. T. N. Spencer-Phillips, U. Gisi, & A. Lebeda (Eds.), *Advances in downy mildew research, Vol 1* (pp. 1–57). Dordrecht: Kluwer.

Dick, M. W., Wong, P. T. W., & Clark, G. (1984). The identity of the oomycete causing 'Kikuyu Yellows', with a reclassification of the downy mildews.. *Botanical Journal of the Linnean Society, 89*, 171–197.

Dick, M. W., Croft, B. J., Margarey, R. C., DeCock, A. W. A. M., & Clark, G. (1989). A new genus of the Verrucalvaceae (Oomycetes). *Botanical Journal of the Linnean Society, 99*, 97–113.

Dick, M. W., Vick, M. C., Gibbings, J. G., Hedderson, T. A., & Lopez Lastra, C. C. (1999). 18S rDNA for species of *Leptolegnia* and other Peronosporomycetes: Justification of the subclass taxa Saprolegniomycetidae and Peronosporomycetidae and division of the Saprolegniaceae sensu lato into the Leptolegniaceae and Saprolegniaceae. *Mycological Research, 103*, 1119–1125.

Fraymouth, J. (1956). Haustoria of the *Peronosporales*. *Transactions of the British Mycological Society, 39*, 79–107.

García-Blázquez, G., Göker, M., Voglmayr, H., Martín, M. P., Tellería, M. T., & Oberwinkler, F. (2008). Phylogeny of *Peronospora*, parasitic on Fabaceae, based on ITS sequences. *Mycological Research, 112*, 502–512.

Gäumann, E. (1918). Über die Formen der *Peronospora parasitica* (Pers.) Fries. *Beihefte zum Botanischen Centralblatt, 35*, 395–533.

Gäumann, E. (1923). Beiträge zu einer Monographie der Gattung *Peronospora* Corda. *Beiträge zur Kryptogamenflora der Schweiz, 5*, 1–360.

Göker, M., Riethmüller, A., Voglmayr, H., Weiß, M., & Oberwinkler, F. (2004). Phylogeny of *Hyaloperonospora* based on nuclear ribosomal internal transcribed spacer sequences. *Mycological Progress, 3*, 83–94.

Göker, M., & Stamatakis, A. (2006). Maximum likelihood phylogenetic inference: An empirical comparison on a multi-locus dataset. (Poster presented at the German Conference on Bioinformatics 2006, Tübingen, Germany.) Retrieved June 12, 2007 from http://diwww.epfl.ch/~stamatak/index-Dateien/publications/GCB2006_Poster.pdf.

Göker, M., Voglmayr, H., Riethmüller, A., & Oberwinkler, F. (2007). How do obligate parasites evolve? A multi-gene phylogenetic analysis of downy mildews. *Fungal Genetics and Biology, 44*, 105–122.

Göker, M., Voglmayr, H., Riethmüller, A., Weiß, M., & Oberwinkler, F. (2003). Taxonomic aspects of Peronosporaceae inferred from Bayesian molecular phylogenetics. *Canadian Journal of Botany*, 81, 672–683.

Hall, G. S. (1996). Modern approaches to species concepts in downy mildews. *Plant Pathology, 45*, 1009–1026.

Hudspeth, D. S. S., Nadler, S. A., & Hudspeth, M. E. S. (2000). A cox2 molecular phylogeny of the Peronosporomycetes. *Mycologia, 92*, 674–684.

Hudspeth, D. S. S., Stenger, D. C., & Hudspeth, M. E. S. (2003). A cox2 phylogenetic hypothesis for the downy mildews and white rusts. *Fungal Diversity, 13*, 47–57.

Hukkanen, A., Pietikäinen, L., Kärenlampi, S., & Kokko, H. (2006). Quantification of downy mildew (*Peronospora sparsa*) in *Rubus* species using real-time PCR. *European Journal of Plant Pathology, 116*, 225–235.

Kirk, P. M., Cannon, P. F., David, J. C., & Stalpers, J. A. (2001). *Ainsworth & Bisby's dictionary of the fungi* (9th ed.). Wallingford: CAB International.

Komjáti, H., Walcz, I., Virányi, F., Zipper, R., Thines, M., & Spring, O. (2007). Characteristics of a *Plasmopara angustiterminalis* isolate from *Xanthium strumarium*. *European Journal of Plant Pathology, 119*, 421–428.

Lamour, K. H., Win, J., & Kamoun, S. (2007). Oomycete genomics: New insights and future directions. *FEMS Microbiology Letters, 274*, 1–8.

Landa, B. B., Montes-Borrego, M., Muñoz-Ledesma, F. J., & Jiménez-Díaz, R. M. (2007). Phylogenetic analysis of downy mildew pathogens of opium poppy and PCR-based in planta and seed detection of *Peronospora arborescens*. *Phytopathology, 97*, 1380–1390.

Larran, S., Ronco, L., Mónaco, C., & Andreau, R. H. (2006). First report of *Peronospora parasitica* on rocket (*Eruca sativa*) in Argentina. *Australasian Plant Pathology, 35*, 377–378.

Lecointre, G., & Le Guyader, H. (2006). *The tree of life: A phylogenetic classification*. Cambridge, MA: The Belknap Press of Harvard University Press.

Petersen, A. B., & Rosendahl, S. (2000). Phylogeny of the Peronosporomycetes (Oomycota) based on partial sequences of the large ribosomal subunit (LSU rDNA). *Mycological Research, 104*, 1295–1303.

Riethmüller, A., Voglmayr, H., Göker, M., Weiß, M., & Oberwinkler, F. (2002). Phylogenetic relationships of the downy mildews (Peronosporales) and related groups based on nuclear large subunit ribosomal DNA sequences. *Mycologia, 94*, 834–849.

Riethmüller, A., Weiβ, M., & Oberwinkler, F. (1999). Phylogenetic studies of Saprolegniomycetidae and related groups based on nuclear large subunit DNA sequences. *Canadian Journal of Botany, 77*, 1790–1800.

Scott, J. B., Hay, F. S., & Wilson, C. R. (2004). Phylogenetic analysis of the downy mildew pathogen of oilseed poppy in Tasmania, and its detection by PCR. *Mycological Research, 108*, 198–205.

Skalický, V. (1966). Taxonomie der Gattungen der Familie *Peronosporaceae*. *Preslia (Praha), 38*, 117–129.

Spring, O. (2004). Potential and limits for the use of new characters in the systematics of biotrophic oomycetes. In P. Spencer-Phillips, & M. Jeger (Eds.), *Advances in downy mildew research, Vol 2* (pp. 211–231). Dordrecht: Kluwer.

Spring, O., Bachofer, M., Thines, M., Riethmüller, A., Göker, M., & Oberwinkler, F. (2006). Intraspecific relationship of *Plasmopara halstedii* isolates differing in pathogenicity and geographic origin based on ITS sequence data. *European Journal of Plant Pathology, 114*, 309–315.

Spring, O., Haas, K., Lamla, I., Thurnhofer, S., & Vetter, W. (2005). The composition and taxonomic significance of fatty acid patterns in three white rust species: *Albugo amaranthi*, *A. candida* and *A. tragopogonis* (Peronosporales, Albuginaceae). *Mycological Progress, 4*, 179–184.

Spring, O., & Thines, M. (2004). On the necessity of new characters for classification and systematics of biotrophic Peronosporomycetes. *Planta, 219*, 910–914.

Spring, O., Voglmayr, H., Riethmüller, A., & Oberwinkler, F. (2003). Characterization of a *Plasmopara* isolate from *Helianthus* x *laetiflorus* based on cross infection, morphological, fatty acids, and molecular phylogenetic data. *Mycological Progress, 2*, 163–170.

Thines, M. (2006). Evaluation of characters available from herbarium vouchers for the phylogeny of the downy mildew genera (Chromista, Peronosporales), with focus on scanning electron microscopy. *Mycotaxon, 97*, 195–218.

Thines, M. (2007a). Characterisation and phylogeny of repeated elements giving rise to exceptional length of ITS2 in several downy mildew genera (Peronosporaceae). *Fungal Genetics and Biology, 44*, 199–207.

Thines, M. (2007b). Use of scanning electron microscopy in downy mildew systematics. In A. Lebeda, & P. Spencer-Phillips (Eds.), *Advances in downy mildew research, Vol 3* (pp. 17–23). Olomouc: Palacký University and Kostelec na Hané, JOLA, v.o.s.

Thines, M. (2007c). Repeats of the ITS2 of Plasmopara species and their relevance for phylogenetic studies. In A. Lebeda, & P. Spencer-Phillips (Eds.), *Advances in downy mildew research, Vol 3* (pp. 31–35). Olomouc: Palacký University and Kostelec na Hané, JOLA, v.o.s.

Thines, M., Göker, M., & Oberwinkler, F. (2006). A revision of *Bremia graminicola*. *Mycological Research, 110*, 646–656.

Thines, M., Göker, M., Oberwinkler, F., & Spring, O. (2007). A revision of *Plasmopara penniseti*, with implications for the host range of the downy mildews with pyriform haustoria. *Mycological Research, 111*, 1377–1385.

Thines, M., Göker, M., Telle, S., Ryley, M., Mathur, K., Narayana, Y. D., et al. (2008). Phylogenetic relationships of graminicolous downy mildews based on cox2 sequence data. *Mycological Research, 112*, 345–351.

Thines, M., Komjáti, H., & Spring, O. (2005). Exceptional length of ITS in *Plasmopara halstedii* is due to multiple repetitions in the ITS-2 region. *European Journal of Plant Pathology, 112*, 395–398.

Thines, M., & Spring, O. (2005). A revision of *Albugo* (Chromista, Peronosporomycetes). *Mycotaxon, 92*, 443–458.

Voglmayr, H. (2003). Phylogenetic study of *Peronospora* and related genera based on nuclear ribosomal ITS sequences. *Mycological Research, 107*, 1132–1142.

Voglmayr, H., & Constantinescu, O. (2008). Revision and reclassification of three *Plasmopara* species based on morphological and molecular phylogenetic data. *Mycological Research, 112*, 487–501.

Voglmayr, H., Fatehi, J., & Constantinescu, O. (2006). Revision of *Plasmopara* (Chromista, Peronosporales) parasitic on Geraniaceae. *Mycological Research, 110*, 633–645.

Voglmayr, H., & Greilhuber, J. (1998). Genome size determination in *Peronosporales* (Oomycota) by Feulgen image analysis. *Fungal Genetics and Biology, 25*, 181–195.

Voglmayr, H., & Riethmüller, A. (2006). Phylogenetic relationships of *Albugo* species (white blister rusts) based on LSU rDNA sequence and oospore data. *Mycological Research, 110*, 75–85.

Voglmayr, H., Riethmüller, A., Göker, M., Weiß, M., & Oberwinkler, F. (2004). Phylogenetic relationships of *Plasmopara*, *Bremia* and other genera of downy mildews with pyriform haustoria based on Bayesian analysis of partial LSU rDNA sequence data. *Mycological Research, 108*, 1011–1024.

Voglmayr, H., & Thines, M. (2007). Phylogenetic relationships and nomenclature of *Bremiella sphaerosperma* (Chromista, Peronosporales). *Mycotaxon, 100*, 11–20.

Waterhouse, G. M. (1973). 10. Peronosporales. In G. C. Ainsworth, F. K. Sparrow, & A. S. Sussman (Eds.), *The fungi—An advanced treatise* (pp. 165–183). New York: Academic.

Yerkes, W. D., & Shaw, C. G. (1959). Taxonomy of *Peronospora* species on Cruciferae and Chenopodiaceae. *Phytopathology, 49*, 499–507.

Eur J Plant Pathol (2008) 122:19–30
DOI 10.1007/s10658-008-9305-2

Classical and molecular genetics of *Bremia lactucae*, cause of lettuce downy mildew

Richard Michelmore · Joan Wong

Received: 30 November 2007 / Accepted: 3 March 2008 / Published online: 3 April 2008

Abstract Lettuce downy mildew caused by *Bremia lactucae* has long been a model for understanding biotrophic oomycete–plant interactions. Initial research involved physiological and cytological studies that have been reviewed earlier. This review provides an overview of the genetic and molecular analyses that have occurred in the past 25 years as well as perspectives on future directions. The interaction between *B. lactucae* and lettuce (*Lactuca sativa*) is determined by an extensively characterized gene-for-gene relationship. Resistance genes have been cloned from *L. sativa* that encode proteins similar to resistance proteins isolated from other plant species. Avirulence genes have yet to be cloned from *B. lactucae*, although candidate sequences have been identified on the basis of motifs present in secreted avirulence proteins characterized from other oomycetes. *Bremia lactucae* has a minimum of 7 or 8 chromosome pairs ranging in size from 3 to at least 8 Mb and a set of linear polymorphic molecules that range in size between 0.3 and 1.6 Mb and are inherited in a non-Mendelian manner. Several methods indicated the genome size of *B. lactucae* to be ca. 50 Mb, although this is probably an underestimate, comprising approximately equal fractions of highly repeated sequences, intermediate repeats, and low-copy sequences. The genome of *B. lactucae* still awaits sequencing. To date, several EST libraries have been sequenced to provide an incomplete view of the gene space. *Bremia lactucae* has yet to be transformed, but regulatory sequences from it form components of transformation vectors used for other oomycetes. Molecular technology has now advanced to the point where rapid progress is likely in determining the molecular basis of specificity, mating type, and fungicide insensitivity.

Keywords *Bremia lactucae* · Lettuce · Virulence · Resistance · Oomycete

R. Michelmore (✉) · J. Wong
The Genome Center and Department of Plant Sciences, University of California,
Davis, CA 95616, USA
e-mail: rwmichelmore@ucdavis.edu

Introduction

Bremia lactucae is an obligate oomycete pathogen belonging to the Peronosporales. Members of the Peronosporales exhibit a gradient in modes of parasitism from saprotrophy through necrotrophy and varying degrees of biotrophy (Ingram 1981; Göker et al. 2007). *Bremia lactucae* represents one of the most highly specialized downy mildews at the biotrophic end of this spectrum. Like all members of the Peronosporaceae, it is an obligate biotroph (i.e. it can only currently be cultured in association with its host). However, the asexual spore germinates directly rather than via zoospores that are used by most other members of the Peronosporaceae. *Bremia lactucae* also directly penetrates through the plant cuticle and epidermal cells rather than entering the leaf through stomata. Both of these attributes indicate that it is one of the most highly evolved downy mildews. Recent molecular phylogenetics supports the advanced taxonomic position of *B. lactucae* (Voglmayr et al. 2004).

Bremia lactucae has a long history as a model for understanding biotrophy in the Oomycetes (Maclean et al. 1974; Andrews 1975; Ingram et al. 1976; Maclean and Tommerup 1979; Ingram 1981; Woods et al. 1988). Its biotrophic mode of nutrition involves a close interaction with its host, in which the plant plasmalemma is invaginated around simply lobed haustoria. Compatible interactions result in minimal macroscopic disturbance until sporulation. Although it is an obligate pathogen, *B. lactucae* can readily be cultured in the laboratory on lettuce seedlings; it is a tractable genetic system and many of the necessary tools for manipulating it in the laboratory in conjunction with its host (*Lactuca* spp.) have been developed. The classical genetics of specificity in lettuce downy mildew is one of the best understood of any gene-for-gene plant–pathogen interaction. Simultaneous studies of host and pathogen showed that specificity is determined by numerous gene-for-gene interactions (Crute and Johnson 1976; Farrara and Michelmore 1987). The molecular determinants on the host side are becoming increasingly well worked out (Meyers et al. 1998a, b; Shen et al. 2002; Kuang et al. 2004). However, the molecular biology of *B. lactucae* has lagged behind.

Bremia lactucae causes lettuce downy mildew, the most important disease affecting lettuce worldwide. Lettuce ranks as one of the top ten most valuable crops in the USA with an annual value of over $2.26 billion (US Department of Agriculture 2003, 2006). Lettuce is grown as extensive monocultures, often with several crops per year. Such intensive production makes the crop susceptible to major epidemics and lettuce suffers from several economically important pests and diseases, particularly downy mildew. These are currently controlled by a combination of genetic resistance, cultural practices, and chemical protection including the application of over 1.5 million pounds of insecticides and fungicides per year (US Department of Agriculture 2003). Several of these compounds are being withdrawn from agricultural use due to environmental concerns over their safety or have been rendered ineffective by changes in *B. lactucae* (Crute et al. 1987; Schettini et al. 1991; Brown et al. 2004). Breeding for resistance to *B. lactucae* is a major activity of most lettuce improvement programmes, and there is an increasing need for information and methods to accelerate the development of new disease-resistant cultivars. Downy mildew resistance (*Dm*) genes provide high levels of resistance but have only remained effective for limited periods of time due to changes in pathogen virulence. Much of the breeding effort is currently focused on introgressing new genes from wild species in response to pathogen changes. New strategies are needed to provide more durable forms of resistance.

The purpose of this review is to summarize what is now known of the classical and molecular genetics of *B. lactucae* and its interaction with lettuce, as well as to consider future developments that are imminent due to the application of genomics approaches.

Classical genetics of resistance

The interaction between lettuce and *B. lactucae* is one of the most extensively characterized gene-for-gene plant–pathogen relationships (Crute and Johnson 1976; Farrara et al. 1987; Hulbert and Michelmore 1985; Michelmore et al. 1984; Norwood and Crute 1984; Norwood et al. 1983; Ilott et al. 1987, 1989). The genetics of resistance has been facilitated by simultaneous studies of avirulence. At least 27 major *Dm* genes or resistance (R) factors are now known that provide resistance against specific isolates of *B. lactucae* in a gene-for-gene manner (Farrara et al. 1987; Bonnier et al. 1994; Maisonneuve et al. 1994; Jeuken and Lindhout 2002). Many other sources of resistance have been identified but have not yet been extensively characterized genetically (e.g. Farrara and Michelmore 1987; Gustafsson 1989; Bonnier et al. 1994; Lebeda and Zinkernagel 2003; Beharav et al. 2006). As more *Dm* genes are characterized from these and other sources, it is likely that many hundred *Dm* genes with specificity to *B. lactucae* will be identified.

Most of the currently identified *Dm* genes confer high levels of resistance. This may be a consequence of these genes being the ones identified and used by breeders. Some *Dm* genes, e.g. *Dm6*, confer incomplete resistance phenotypes (Crute and Norwood 1978). Partial phenotypes do not necessarily imply quantitative inheritance or more durable resistance. The phenotype of the interaction depends on the gene and environment. Heterozygotes of some *Dm* genes, e.g. *Dm18*, also confer incomplete resistance (Maisonneuve et al. 1994). In addition, different isolates of *B. lactucae* can exhibit different levels of incompatibility to the same *Dm* gene (Ilott et al. 1989). At lower temperatures, resistance conferred by several *Dm* genes becomes less effective; temperature shift experiments suggested that the determinants of specificity are present in most host cells and expressed throughout pathogen development (Judelson and Michelmore 1992). There are also resistance genes of minor effect that confer incomplete or field resistance (Eenink et al. 1983; Jeuken and Lindhout 2002). Many genes of minor effect will probably be identified in the future by quantitative trait locus (QTL) analysis using molecular markers.

The known *Dm* resistance phenotypes are located in at least five clusters in the lettuce genome (Hulbert and Michelmore 1985; Farrara et al. 1987; Bonnier et al. 1994). The major cluster contains over nine genetically separable *Dm* specificities, as well as resistance to root aphid. Another large cluster contains several *Dm* genes, resistance to the root-infecting downy mildew *Plasmopara lactucae-radicis*, and the hypersensitive reaction to *Turnip mosaic virus* (Witsenboer et al. 1995).

Molecular genetics of resistance

One downy mildew resistance gene, *Dm3*, has been cloned through a combination of map-based cloning and candidate gene approaches (Shen et al. 1998, 2002; Meyers et al. 1998a). *Dm3* encodes a nucleotide binding site and leucine-rich repeat (NBS-LRR) protein, similar to genes cloned from other species for resistance to downy mildews and other pathogens (McHale et al. 2006). *Dm3* is large, containing nearly double the number of LRRs compared to proteins characterized in other species. *Dm3* is a member of the large *RGC2* (*Resistance Gene Candidate2*) multigene family that can vary in copy number from 12 to over 30 (Meyers et al. 1998a, b; Kuang et al. 2004). Sequence analysis of paralogues from several species indicated that this large cluster evolves by a birth-and-death mechanism (Michelmore and Meyers 1998; Kuang et al. 2004). Genes in the *RGC2* family exhibit two distinct patterns of evolution. Type I genes are extensive chimeras resulting from frequent sequence exchange between paralogues, and individual genes are rare in nature. *Dm3* is a Type I gene and only rarely present in nature (Kuang et al. 2006). Type II genes occur more frequently in nature, and sequence exchanges only rarely occur between individual lineages (Kuang et al. 2004). Trans-specific polymorphism was observed for different groups of Type II orthologues, suggesting balancing selection. Different evolutionary forces have impacted different parts of *RGC2* genes. The *RGC2* cluster is not highly recombinogenic; it exhibits a recombination frequency 18 times lower than the genome-wide average (Chin et al. 2001). This is consistent with reduced pairing during meiosis between haplotypes due to structural heterozygosity.

The meiotic spontaneous mutation rates differ between the *Dm* genes (Chin et al. 2001). Spontaneous mutations in *Dm1*, *Dm3* and *Dm7* occurred at the rate of 10^{-3} to 10^{-4} per generation. No spontaneous mutations were detected for *Dm5/8*. Spontaneous mutations at the *Dm3* locus but not the *Dm7* locus were frequently associated with large deletions resulting from unequal crossing-over. One spontaneous loss of *Dm3* resistance was observed to be the result of a gene conversion event between the LRR-encoding regions of similar paralogues (Chin et al. 2001). Given that a lettuce plant is capable of producing several thousand seeds per generation, such mutation rates suggest that in every generation an average of one progeny with a novel haplotype at a resistance locus is produced per plant.

PCR using degenerate oligonucleotides designed to sequences encoding conserved NBS domains has resulted in the identification of over 20 distinct families of resistance gene candidates (*RGCs*; Shen et al. 1998; McHale and Michelmore, unpublished). These are being mapped relative to phenotypic resistances to provide a comprehensive view of the genomic distribution of resistance genes, including many *Dm* genes.

The clustered genomic distribution of *Dm* genes suggests that they are similar genes. This has been confirmed for the major cluster of *Dm* genes. An interfering hairpin RNA (ihpRNA) construct containing fragments encoding the LRR of *Dm3* was used to induce post-transcriptional gene silencing of the

RGC2 family (Wroblewski et al. 2007). This showed that the resistance specificity encoded by the genetically defined *Dm18* locus is the combination of two resistance specificities, only one of which was silenced by ihpRNA derived from *Dm3*. Analysis of progeny from crosses between transgenic, silenced tester stocks and lettuce accessions carrying other resistance genes previously mapped to the *RGC2* locus indicated that two additional resistance specificities *to B. lactucae*, *Dm14* and *Dm16*, as well as resistance to lettuce root aphid (*Pemphigus bursarius*), *Ra*, are encoded by *RGC2* family members. This strategy is now being extended to other clusters of resistance genes for which *RGC* sequences and phenotypic resistances co-segregate.

Numerous haplotypes and homologues at the major cluster of resistance genes that contains *Dm3* have been identified. Fifty-one different haplotypes were identified in 74 accessions studied using molecular markers diagnostic of the *RGC2* cluster (Sicard et al. 1999). The copy number of *RGC2* paralogues at a locus can vary from 12 to >30 (Kuang et al. 2004). No accessions have been observed that completely lack *RGC2* genes even though they do not carry detectable *Dm* specificities. The large number of different haplotypes is consistent with there being a minimum of several hundred distinct *Dm* genes in *Lactuca* species and indicates that wild germplasm will be a rich source of new resistance genes that can be introgressed and pyramided using molecular markers.

There is also a growing understanding of the signalling pathways and downstream genes and proteins that are involved in plant resistance (Jones and Dangl 2006). However, there are little specific data on genes involved downstream of *Dm* genes in the interaction with *B. lactucae*. Homologues of genes from other species known to be involved in pathogen interactions are present in ESTs from *Lactuca* spp. (http://compgenomics.ucdavis.edu), and therefore it is likely that similar processes are involved in lettuce as in other plants. Ultrastructural and biochemical studies indicate that the hypersensitive response is typical but includes the induction of phytoalexins characteristic of the Compositae (Maclean and Tommerup 1979; Bennett et al. 1996; Bestwick et al. 1998; Lebeda et al. 2008). As the molecular understanding of *B. lactucae* develops, it will be interesting to determine how the pathogen has adapted to deal with these defences.

Mating system

Bremia lactucae is diploid for the majority of its life-cycle and predominantly heterothallic (Michelmore and Ingram 1980; Michelmore and Sansome 1982). Both the asexual life-cycle of 1 to 3 weeks and the sexual cycle of several months' to many years' duration can be readily induced in the laboratory. The asexual cycle allows the facile clonal propagation of individual phenotypes. Its heterothallic nature allows controlled crosses between isolates of known phenotypes for the investigation of the genetics of (a)virulence.

When hyphae of opposite mating type come into physical contact, asexual sporulation is suppressed, clusters of gametangia are elaborated at the point of contact, synchronous meioses occur in the oogonium and periclinal antheridium, and haploid gametes are transferred from the antheridium to the oogonium to effect fertilization (Michelmore and Ingram 1981; Michelmore and Sansome 1982). Each mating type can probably produce both oogonia and antheridia, as do *Phytophthora* species. Differences in maleness and femaleness have not been investigated.

Heterothallism seems to be determined by two haplotypes at a single locus, with the B_1 compatibility type being conferred by a homozygous recessive condition and the B_2 mating type by a heterozygous condition. The two mating types segregate in approximately 1:1 ratios in sexual progeny (Michelmore and Ingram 1981; Norwood et al. 1983; Michelmore et al. 1984; Sicard et al. 2003). However, the current data do not preclude a more complicated situation such as double heterozygotes and balanced lethals, as has been proposed for *Phytophthora infestans* (Fabritius and Judelson 1997). The molecular determinants of mating type for *B. lactucae* await characterization as they do for all oomycetes.

Some isolates exhibit secondary homothallism (Michelmore and Ingram 1982). These isolates behave predominantly as B_2 types in that they usually reproduce asexually except when cultured in combination with B_1 isolates, whereupon they produce abundant oospores. However, they also produce oospores at low frequency when cultured alone, particularly at high inoculum densities. This is due to the generation of B_1 components at low frequency, as shown by the isolation of stable B_1 and B_2 as well as self-fertile derivatives by single–spore analysis (Michelmore and Ingram 1982). This self-fertility may be due to trisomy of the determinants of mating type (Michelmore and Sansome 1982). Somatic segregation of self-sterile lines from self-fertile progenitors involves at least transitory heterokaryosis.

The prevalence of each mating type varies in nature. Isolates of both mating types have been frequently identified in Europe and New York State, although the B_2 type sometimes predominated (Michelmore and Ingram 1980; Lebeda and Blok 1990; Gustafsson et al. 1985; Yuen and Lorbeer 1987; Petrželová and Lebeda 2003). This is consistent with a sexually reproducing population and the high diversity of virulence phenotypes observed. In contrast, the B_2 mating type predominates in isolates from cultivated lettuce in California; B_1 isolates are identified extremely rarely. In addition, the one B_1 isolate analyzed from California had reduced fertility (Ilott et al. 1987). The data for California isolates are indicative of an asexual population that propagates clonally. This is consistent with the more restricted spectrum of virulence phenotypes observed and widespread pathotypes that are stable from year to year. However, even in the apparent absence of the sexual cycle and the oospore as a survival stage, *B. lactucae* has been able to change virulence phenotype in response to the deployment of new *Dm* genes and it is unclear how the pathogen survives crop-free periods in California.

Genetics of avirulence

Several initial studies established that avirulence to specific *Dm* genes was inherited as single dominant unlinked loci (Michelmore and Ingram 1981; Norwood et al. 1983; Norwood and Crute 1984; Michelmore et al. 1984; Ilott et al. 1987). The gene-for-gene interaction between lettuce and *B. lactucae* was subsequently analyzed critically, involving extensive crosses between 20 isolates of diverse worldwide geographical origins to complement the simultaneous genetic analysis of resistance (Farrara et al. 1987; Ilott et al. 1989). The majority of the data were consistent with the underlying tenets of a gene-for-gene interaction. Avirulence was usually determined by dominant alleles at unlinked loci, although their expression could be modified depending on the genetic background of the host and pathogen. Some segregation anomalies could be explained by hyperploidy and gene dosage effects. In order to test for complementation between *Avr* loci, 125 tests involving 19 crosses were analyzed. In no case were all progeny avirulent to a specific *Dm* gene when both parental isolates had been virulent; therefore, there was no evidence for complementation, indicating that avirulence to individual *Dm* genes was conferred at the same locus. To investigate the presence of dominant inhibitors of avirulence, crosses were made between avirulent and virulent isolates. The data for an inhibitor locus epistatic to *Avr5/8* were good but not unequivocal; there was no evidence for inhibitors of other *Avr* loci (Ilott et al. 1989). Therefore, unlike the situation in phytopathogenic bacteria (Abramovitch et al. 2003; Espinosa et al. 2003; Jamir et al. 2004; Fu et al. 2007), inhibitor loci do not seem to be common in *B. lactucae*.

Genetic mapping

A preliminary genetic linkage map of *B. lactucae* was constructed using the segregation of 53 RFLP loci, 8 *Avr* loci, and the mating type locus in a total of 70 F_1 individuals from two crosses (Hulbert et al. 1988). This map consisted of 13 small linkage groups, including 35 RFLP loci and one *Avr* gene. However, construction of a more detailed genetic map was hindered by the ambiguous phase of the alleles in the parents and an insufficient number of markers due to the type of marker technology available at the time.

A more comprehensive genetic map of *B. lactucae* was subsequently constructed using PCR-based markers as well as additional RFLP loci (Sicard et al. 2003). The more heterozygous of the two crosses that had been used previously was expanded to 97 F_1 progeny to facilitate the identification of the phase of the parental alleles and to improve the detection of linkage. Two parental maps and a consensus map were constructed using a total of 347 AFLP and 83 RFLP markers, six *Avr* genes, and the mating-type locus. One parental map contained 24 linkage groups distributed over 835 cM; the second map contained 21 linkage groups distributed over 606 cM. The consensus map contained 12 linkage groups with markers from both maps and 24 parent-specific groups.

There was no evidence for clustering of *Avr* genes. All six mapped to different linkage groups. This is consistent with the lack of linkage observed in classical segregation analysis of 12 *Avr* loci (Ilott et al. 1989). Also, the genetic data provided no evidence for pathogenicity islands that have been identified in bacteria (Alfano et al. 2000; Guttman et al. 2002; Jackson et al. 1999; Sugio et al. 2005). Four *Avr* loci were located at the ends of linkage groups. Telomeric locations of *Avr* genes would be consistent with the high instability of the avirulence phenotype in *B. lactucae*. In the fungal pathogen *Magnaporthe grisea*, four out of eight known *Avr* genes are close to a telomere, and losses in avirulence were associated with deletions (Mandel et al. 1997; Dioh et al. 2000). Linkage of three *Avr* genes with distorted markers in *B. lactucae* may be indicative of other mechanisms of instability of *Avr* genes, such as high frequencies of mitotic gene conversion as observed in *P. sojae* (Chamnanpunt et al. 2001).

The current genetic map of *B. lactucae* is far from saturated. Over 20% of the markers remain unlinked. It is difficult to estimate the total number of chromosomal groups and genetic genome size because of the possible redundancy between the parent-specific linkage groups. The mating type locus and two *Avr* loci are flanked by molecular markers; however, no close linkages have been identified. The closest marker is 1 cM, and only loose linkages have been identified for the majority of *Avr* genes. Whether this represents a dearth of polymorphic low-copy sequences or high rates of recombination close to avirulence genes is unknown. We attempted bulked segregant analysis (Michelmore et al. 1991) to identify markers closely linked to several avirulence genes; however, this was unsuccessful (Zungri and Michelmore, unpublished).

Karyotype and chromosomal assignment of markers

Cytological analysis of *B. lactucae* resolved at least 7 or 8 chromosome pairs at meiosis (Michelmore and Sansome 1982). However, these chromosomes are too small to be resolved clearly using conventional light microscopy. Examination of isolates of diverse geographical origins as well as progeny from sexual crosses by pulsed-field gel electrophoresis (PFGE) revealed a minimum of seven chromosomes ranging in size from 3 to at least 8 Mb and a set of linear polymorphic molecules from 0.3 to 1.6 Mb (Francis and Michelmore 1993). Genetic and hybridization analyses confirmed the existence of two classes of molecules.

The class of smaller molecules is sequence-related, non-ribosomal, nuclear, highly polymorphic, variable in number, and inherited in a non-Mendelian manner. These small polymorphic molecules are therefore B chromosomes or large liner plasmids. No RFLP markers, and consequently none of the *Avr* genes, were assigned to the small polymorphic 0.3–1.6 Mb molecules. Therefore, there was no evidence that these small variable molecules are involved in variation in specificity of *B. lactucae*.

The second class of molecules is larger than 2 Mb, is more constant in size and number and represents the true chromosomes. A total of 25 probes were successfully hybridized to these chromosomes (Sicard et al. 2003). Of these, 23 had been mapped and represented 16 of the linkage groups in the consensus map; two were unlinked. This resulted in two consensus linkage groups and seven parent-specific linkage groups being assigned to chromosomes. Linkage to RFLP markers allowed three *Avr* loci also to be assigned to chromosomes. The mating–type locus could not be assigned to any chromosome-sized molecule. Together the genetic and physical data suggest that there are at least 10 chromosomes in *B. lactucae*.

Somatic variation

Bremia lactucae can exhibit somatic variation in addition to the segregation of phenotypes following sexual reproduction. RFLP analysis of 25 isolates from diverse worldwide geographical origins revealed different ploidy levels and somatic variants (Hulbert and Michelmore 1987). Most European isolates were clearly diploid. They were heterozygous at approximately 44% of their loci and had highly variable genotypes consistent with the frequent occurrence of the sexual cycle. In contrast, many of the isolates from Australia, Japan, Wisconsin and Australia had more than two alleles at multiple loci, indicating that they were either polyploids or stable heterokaryons (hyperploid). Variation between similar sympatric isolates indicated that they had arisen by the somatic loss of alleles. One hyperploid California isolate had resulted from the fusion of two diploid California isolates of the same mating type, providing the first evidence for natural somatic fusion in the Oomycetes.

Several phenotypic changes in *B. lactucae* seem to have resulted from somatic changes. The segregation of self-sterile lines in secondary homothallic isolates is one example (Michelmore and Ingram 1982). Fungicide insensitivity seems to have arisen in the most common virulence phenotype, rather than involving sexual progeny (Crute et al. 1987; Schettini et al. 1991; Brown et al. 2004). Recent changes in virulence phenotype in California seem also to be somatic (Ilott et al. 1987; Ochoa and Michelmore, unpublished). The molecular genetic changes underlying these changes are unknown, but they are becoming amenable to analysis with the advent of technologies for whole genome analysis.

Genome size and complexity

The physical genome size of *B. lactucae* has been estimated using several methods: comparisons of hybridizations between cloned DNA fragments and genomic DNA in dot blot reconstructions, DNA–DNA reassociation kinetics assayed by hydroxyapatite chromatography, and summation of chromosomal sizes determined by CHEF gel electrophoresis (Francis et al. 1990; Francis and Michelmore 1993). All three methods gave similar estimates of 50 Mb; however, this may be an underestimate. *Aspergillus nidulans* and *Arabidopsis thaliana* were used as controls in the genomic reconstruction experiments and their sizes were estimated to be 17 and 52 Mb, respectively; genomic sequencing has now shown their genome sizes to be 30 and 125 Mb, respectively (Galagan et al. 2005; The *Arabidopsis* Genome Initiative 2000). Therefore the estimate for the genome size of *B. lactucae* should probably be revised upward to approximately 100 Mb. This is consistent with estimates of 70 to 144 Mb, depending on the isolate, measured by Feulgen absorbance cytophotometry (Voglmayr and Greilhuber 1998). This size is comparable to that of *Phytophthora* species that range from 65 Mb for *P. capsici* to 240 Mb for *P. infestans* as well as similar to *Hyaloperonospora parasitica* (75 Mb; Govers and Gijzen 2006). Only sequencing the entire genome of *B. lactucae* will provide an accurate determination of its genome size.

DNA reassociation kinetics indicated that the nuclear DNA of *B. lactucae* is comprised of approximately 65% repeated sequences and 35% low-copy sequences (Francis et al. 1990). The repeat fraction is made up of approximately 21% high-copy sequences and 38% intermediate-copy sequences. Hybridization analysis of random genomic λ clones demonstrated that the low-copy-number sequences are interspersed with repeated sequences.

Regulatory sequences for transformation of *B. lactucae* and other oomycetes

Transformation of *B. lactucae* has yet to be achieved. Early work towards this goal involved the isolation of regulatory sequences from *B. lactucae*. These included the promoters and terminators from *Hsp70* and a constitutively highly expressed single-copy gene, *HAM34* (Judelson and Michelmore 1989, 1990). Although there was evidence for transient expression, no stable transformants of *B. lactucae* were obtained. Efforts were therefore directed towards transformation of culturable oomycetes including *P. infestans* (Judelson and Michelmore 1991). These studies ultimately resulted in the stable transformation of several *Phytophthora* species using vectors originally developed for *B. lactucae* (Judelson et al. 1991, 1993). The function of *HAM34* is still unknown; it is present in *P. infestans* (Win et al. 2005) but not yet evident in the sequence of *H. parasitica*.

These experiments indicated that the transcriptional machinery of oomycetes differs significantly from that of higher fungi but that sufficient similarity exists so vectors developed using regulatory sequences from one oomycete will likely function in other oomycetes (Judelson et al. 1992). It is now time to reinitiate experiments on the transformation of *B. lactucae* using better selectable markers and reporter genes that have become available, as well as novel methods for introducing the transgenes.

(A)virulence effectors

Pathogens have evolved sophisticated mechanisms to alter their hosts' metabolism and interfere with host defences (Jones and Dangl 2006). This is best understood for Gram-negative bacteria that secrete virulence effector proteins into host cells and the extracellular space (Nomura et al. 2005). Some effectors can trigger defences dependent on specific resistance genes. Some can also block the resistance response elicited by the activities of other effectors (Abramovitch et al. 2003; Espinosa et al. 2003; Jamir et al. 2004; Fu et al. 2007). Such effectors exhibit a dominant inhibitor of avirulence phenotype. The recent availabilities of sequenced genomes of phytopathogenic bacteria, bioinformatic tools, and efficient functional screens have resulted in the identification of numerous genes encoding candidate effectors (e.g. Guttman et al. 2002; Petnicki-Ocwieja et al. 2002; Greenberg and Vinatzer 2003; Chang et al. 2005). It is now recognized that individual strains of phytopathogenic bacteria secrete ~40 effectors into their hosts. Functional studies and the sequences of several effectors suggest that they alter plant defence signalling (reviewed in Grant et al. 2006).

There is increasing evidence that fungi and oomycete pathogens also secrete diverse effector proteins into their hosts (Torto et al. 2003; Birch et al. 2006; Kamoun 2006). Initially avirulence genes have been cloned from *Phytophthora* spp. and *H. parasitica* on a gene-by-gene basis (Tyler 2002; MacGregor et al. 2002; Shan et al. 2004; Allen et al. 2004; Rehmany et al. 2005; Armstrong et al. 2005). Recent studies of avirulence and secreted proteins from *H. parasitica* and *Phytophthora* spp. revealed a novel, highly conserved RXLR amino acid motif (Rehmany et al. 2005). This motif is predicted to be required for translocation from the pathogen to the host (Bhattacharjee et al. 2006) and it was recently shown to be required for translocation of the avirulence protein *Avr3a* by *P. infestans* (Whisson et al. 2007). Bioinformatic analyses have identified hundreds of genes encoding other potentially secreted proteins in the genome sequences of *Phytophthora* spp. (Birch et al. 2006; Tyler et al. 2006).

In order to identify (a)virulence effector proteins in *B. lactucae*, we have generated several cDNA libraries of *B. lactucae* from a variety of sources including conidia, germlings and infected tissue. One subtraction library was made by subtracting mock-inoculated leaf material against heavily infected leaf material. The resulting sequences had a bimodal distribution of GC contents. On the basis of GC content and (dis)similarity to plant or oomycete sequences, sequences were categorized as most likely to be of *B. lactucae* origin (38%), lettuce origin (35%), or uncertain origin (27%). Many of the putative *B. lactucae* unigenes had an average GC content of 50%. In order to obtain more full-length clones, we generated and sequenced a new library that was enriched for *B. lactucae* sequences by hybridizing cDNA from heavily infected leaves to *B. lactucae* genomic DNA using a protocol developed by J. Jones (Sainsbury Laboratory, Norwich, UK). Sequences from all libraries are being analyzed for candidate effectors using several bioinformatics approaches. We are searching for sequence similarity to genes encoding known avirulence proteins and putative secreted proteins from other oomycetes. Candidate effector sequences have yet to be identified; however, this is not surprising as effector proteins may be evolving rapidly. We are also searching for the presence of a secretion signal peptide and the RXLR amino acid motif. These analyses have so far yielded over 15 candidate sequences that satisfied one or more of these criteria. These are currently being assayed for function in lettuce using *Agrobacterium*-mediated transient assays (Wroblewski et al. 2005).

The impact of genomic sequencing

Although *B. lactucae* was ranked as one of the high-priority plant pathogens targeted for sequencing since 2002 (American Phytopathological Society 2006), this has yet to occur. The latest generation of sequencing technologies combined with conventional Sanger sequencing will provide large amounts of sequence information for *B. lactucae*. Sequencing the whole genome will provide an expedient and cost-efficient approach to the identification of effector proteins and other types of molecules involved in determining specificity and mating type. It will also provide targets for disease control strategies as well as provide an important reference genome.

Sequencing of multiple isolates will provide large numbers of single nucleotide polymorphisms that, combined with the new generation of marker technologies, will allow large-scale population analyses for variation in both effector genes and genes involved in other aspects of the pathogen's biology. It is likely that these whole-genome analyses will reveal a variety of mechanisms of variation. It will be particularly interesting to determine the basis of insensitivity to the fungicides metalaxyl (Ridomil) and fosetyl A (Alliette), as well as the bases for changes in virulence phenotype.

Sequencing the genome will also provide insights into what extent the genome of *B. lactucae* has become streamlined in parallel with its total dependence on its host. Also, it will facilitate the identification of which biosynthetic capabilities appear to be lacking and therefore can be supplemented in media for axenic culture.

Acknowledgements The work described here has been the result of many people's efforts spread over the past 25 years. We thank them all for their contributions. Financial support has come from numerous sources including sustained support from the California Lettuce Research Board and the USDA CREES National Research Initiative.

References

Abramovitch, R. B., Kim, Y. J., Chen, S., Dickman, M. B., & Martin, G. B. (2003). *Pseudomonas* type III effector AvrPtoB induces plant disease susceptibility by inhibition of host programmed cell death. *EMBO Journal, 22*, 60–69.

Alfano, J. R., Charkowski, A. O., Deng, W. L., Badel, J. L., Petnicki-Ocwieja, T., van Dijk, K., et al. (2000). The *Pseudomonas syringae* Hrp pathogenicity island has a tripartite mosaic structure composed of a cluster of type III secretion genes bounded by exchangeable effector and conserved effector loci that contribute to parasitic fitness and pathogenicity in plants. *Proceedings of the National Academy of Sciences of the United States of America, 97*, 4856–4861.

Allen, R. L., Bittner-Eddy, P. D., Grenville-Briggs, L. J., Meitz, J. C., Rehmany, A. P., Rose, L. E., et al. (2004). Host–parasite co-evolutionary conflict between *Arabidopsis* and downy mildew. *Science, 306*, 1957–1960.

American Phytopathological Society (2006). *Microbial genomic sequencing. Perspectives of the American Phytopathological Society (Revised 2006)*. http://199.86.26.56/members/ppb/PDFs/MicrobialGenomicsSeq06.pdf.

Andrews, J. H. (1975). Distribution of label from ^{3}H-glucose and ^{3}H-leucine in lettuce cotyledons during early stages of infection with *Bremia lactucae*. *Canadian Journal of Botany, 53*, 1103–1115.

Armstrong, M. R., Whisson, S. C., Pritchard, L., Bos, J. L., Venter, E., Avrova, A. O., et al. (2005). An ancestral oomycete locus contains late blight avirulence gene *Avr3a*, encoding a protein that is recognized in the host cytoplasm. *Proceedings of the National Academy of Sciences of the United States of America, 102*, 7766–7771.

Beharav, A., Lewinsohn, D., Lebeda, A., & Nevo, E. (2006). New wild *Lactuca* genetic resources with resistance against *Bremia lactucae*. *Genetic Resources and Crop Evolution, 53*, 467–474.

Bennett, M., Gallagher, M., Fagg, J., Bestwick, C., Paul, T., Beale, M., et al. (1996). The hypersensitive reaction, membrane damage and accumulation of autofluorescent phenolics in lettuce cells challenged by *Bremia lactucae*. *Plant Journal, 9*, 851–865.

Bestwick, C. S., Brown, I. R., & Mansfield, J. W. (1998). Localized changes in peroxidase activity accompany hydrogen peroxide generation during the development of a nonhost hypersensitive reaction in lettuce. *Plant Physiology, 118*, 1067–1078.

Bhattacharjee, S., Hiller, L. N., Liolios, K., Win, J., Kanneganti, T.-D., Young, C., et al. (2006). The malarial host-targeting signal is conserved in the Irish potato famine pathogen. *PLoS Pathogens, 2*, 50.

Birch, P. R., Rehmany, A. P., Pritchard, L., Kamoun, S., & Beynon, J. L. (2006). Trafficking arms: Oomycete effectors enter host plant cells. *Trends in Microbiology, 14*, 8–11.

Bonnier, F. J. K., Reinink, K., & Groenwold, R. (1994). Genetic analysis of *Lactuca* accessions with new major gene resistance to lettuce downy mildew. *Phytopathology, 84*, 462–468.

Brown, S., Koike, S., Ochoa, O., Laemmlen, F., & Michelmore, R. W. (2004). Insensitivity to the fungicide, fosetyl-aluminum, in California isolates of lettuce downy mildew, *Bremia lactucae*. *Plant Disease, 46*, 1059–1069.

Chamnanpunt, J., Shan, W.-X., & Tyler, B. M. (2001). High frequency mitotic gene conversion in genetic hybrids of the oomycete *Phytophthora sojae*. *Proceedings of the National Academy of Sciences of the United States of America, 98*, 14530–14535.

Chang, J. H., Urbach, J. M., Law, T. F., Arnold, L. W., Hu, A., Gombar, S., et al. (2005). A high-throughput, near-saturating screen for type III effector genes from *Pseudomonas syringae*. *Proceedings of the National Academy of Sciences of the United States of America, 102*, 2549–2554.

Chin, D. B., Arroyo-Garcia, R., Ochoa, O., Kesseli, R. V., Lavelle, D. O., & Michelmore, R. W. (2001). Recombination and spontaneous mutation at the major cluster of resistance genes in lettuce (*Lactuca sativa*). *Genetics, 157*, 831–849.

Crute, I. R., & Johnson, A. G. (1976). The genetic relationship between races of *Bremia lactucae* and cultivars of *Lactuca sativa*. *Annals of Applied Biology, 83*, 125–137.

Crute, I. R., & Norwood, J. M. (1978). Incomplete specific resistance to *Bremia lactucae* in lettuce. *Annals of Applied Biology, 89*, 467–474.

Crute, I. R., Norwood, J. M., & Gordon, P. L. (1987). The occurrence, characteristics, and distribution in the United Kingdom of resistance to phenylamide fungicides in *Bremia lactucae* (lettuce downy mildew). *Plant Pathology, 36*, 297–315.

Dioh, W., Tharreau, D., Notteghem, J.-L., Orbach, M., & Lebrun, M.-H. (2000). Mapping of avirulence genes in the rice blast fungus, *Magnaporthe grisea*, with RFLP and RAPD markers. *Molecular Plant–Microbe Interactions, 13*, 217–227.

Eenink, A. H., Groenwold, R., & Bijker, W. (1983). Partial resistance in lettuce downy mildew (*Bremia lactucae*). 4. Resistance after natural, semi-artificial and artificial infestation and examples of mutual interference of resistance levels. *Euphytica, 32*, 139–149.

Espinosa, A., Guo, M., Tam, V. C., Fu, Z. Q., & Alfano, J. R. (2003). The *Pseudomonas syringae* type III-secreted protein HopPtoD2 possesses protein tyrosine phosphatase activity and suppresses programmed cell death in plants. *Molecular Microbiology, 49*, 377–387.

Fabritius, A. L., & Judelson, H. S. (1997). Mating-type loci segregate aberrantly in *Phytophthora infestans* but normally in *Phytophthora parasitica*: Implications for models of mating-type determination. *Current Genetics, 32*, 60–65.

Farrara, B., Ilott, T. W., & Michelmore, R. W. (1987). Genetic analysis of factors for resistance to downy mildew (*Bremia lactucae*) in species of lettuce (*Lactuca sativa* and *L. serriola*). *Plant Pathology, 36*, 499–514.

Farrara, B., & Michelmore, R. W. (1987). Identification of new sources of resistance to downy mildew in *Lactuca* germplasm. *HortScience, 22*, 647–649.

Francis, D. M., Hulbert, S. H., & Michelmore, R. W. (1990). Genome size and complexity of the obligate fungal pathogen, *Bremia lactucae*. *Experimental Mycology, 14*, 299–309.

Francis, D. M., & Michelmore, R. W. (1993). Two classes of chromosome-sized molecules are present in *Bremia lactucae*. *Experimental Mycology, 17*, 284–300.

Fu, Z. Q., Guo, M., Jeong, B.-R., Tian, F., Elthon, T. E., Cerny, R. L., et al. (2007). A type III effector ADP-ribosylates RNA-binding proteins and quells plant immunity. *Nature, 447*, 284–288.

Galagan, J. E., Calvo, S. E., Cuomo, C., Ma, L.-J., Wortman, J. R., Batzoglou, S., et al. (2005). Sequencing of *Aspergillus nidulans* and comparative analysis with *A. fumigatus* and *A. oryzae*. *Nature, 438*, 1105–1115.

Göker, M., Voglmayer, H., Reithmüller, A., & Oberwinkler, F. (2007). How do obligate parasties evolve? A multi-gene phylogenetic analysis of the downy mildews. *Fungal Genetics and Biology, 44*, 105–122.

Govers, F., & Gijzen, M. (2006). *Phytophthora* genomics: The plant destroyers' genome decoded. *Molecular Plant–Microbe Interactions, 19*, 1295–1301.

Grant, S. R., Fisher, E. J., Chang, J. H., Mole, B. M., & Dangl, J. L. (2006). Subterfuge and manipulation: Type III effector proteins of phytopathogenic bacteria. *Annual Review of Microbiology, 60*, 425–429.

Greenberg, J., & Vinatzer, B. A. (2003). Identifying type III effectors of plant pathogens and analyzing their interaction with plant cells. *Current Opinion in Microbiology, 6*, 20–28.

Gustafsson, I. (1989). Potential sources of resistance to lettuce downy mildew (*Bremia lactucae*) in different *Lactuca* species. *Euphytica, 40*, 227–232.

Gustafsson, M., Liljeroth, E., & Gustafsson, I. (1985). Pathogenic variation and sexual reproduction in Swedish populations of *Bremia lactucae*. *Theoretical and Applied Genetics, 70*, 643–649.

Guttman, D. S., Vinatzer, B. A., Sarkar, S. F., Ranall, M. V., Kettler, G., & Greenberg, J. T. (2002). A functional screen for the type III (Hrp) secretome of the plant pathogen *Pseudomonas syringae*. *Science, 295*, 1722–1726.

Hulbert, S. H., Hott, T. W., Legg, E. J., Lincoln, S. E., Lander, E. S., & Michelmore, R. W. (1988). Genetic analysis of the fungus, *Bremia lactucae*, using restriction fragment length polymorphisms. *Genetics, 120*, 947–958.

Hulbert, S. H., & Michelmore, R. W. (1985). Linkage analysis of genes for resistance to downy mildew (*Bremia lactucae*) in lettuce (*Lactuca sativa*). *Theoretical and Applied Genetics, 70*, 520–528.

Hulbert, S. H., & Michelmore, R. W. (1987). DNA restriction fragment length polymorphism and somatic variation in the lettuce downy mildew fungus, *Bremia lactucae*. *Molecular Plant–Microbe Interactions, 1*, 17–24.

Ilott, T. W., Durgan, M. E., & Michelmore, R. W. (1987). Genetics of virulence in California populations of *Bremia lactucae* (lettuce downy mildew). *Phytopathology, 77*, 1381–1386.

Ilott, T. W., Hulbert, S. H., & Michelmore, R. W. (1989). Genetic analysis of the gene-for-gene interaction between lettuce (*Lactuca sativa*) and *Bremia lactucae*. *Phytopathology, 79*, 888–897.

Ingram, D. S. (1981). Physiology and biochemistry of host–parasite interaction. In D. M. Spencer (Ed.), *The downy mildews* (pp. 143–163). London: Academic.

Ingram, D. S., Sargent, J. A., & Tommerup, I. C. (1976). Structural aspects of infection by biotrophic fungi. In J. Friend, & D. R. Threlfall (Eds.), *Biochemical aspects of plant–parasite relationships* (pp. 43–78). New York: Academic.

Jackson, R. W., Athanassopoulos, E., Tsiamis, G., Mansfield, J. W., Sesma, A., Arnold, D. L., et al. (1999). Identification of a pathogenicity island, which contains genes for virulence and avirulence, on a large native plasmid in the bean pathogen *Pseudomonas syringae* pathovar *phaseolicola*. *Proceedings of the National Academy of Sciences of the United States of America, 96*, 10875–10880.

Jamir, Y., Guo, M., Oh, H. S., Petnicki-Ocwieja, T., Chen, S., Tang, X., et al. (2004). Identification of *Pseudomonas syringae* type III effectors that can suppress programmed cell death in plants and yeast. *Plant Journal, 37*, 554–565.

Jeuken, M., & Lindhout, P. (2002). *Lactuca saligna*, a non-host for lettuce downy mildew (*Bremia lactucae*), harbors a new race-specific *Dm* gene and three QTLs for resistance. *Theoretical and Applied Genetics, 105*, 384–391.

Jones, J. D., & Dangl, J. L. (2006). The plant immune system. *Nature, 444*, 323–329.

Judelson, H. S., Dudler, R., Pieterse, C. M. J., Unkles, S. E., & Michelmore, R. W. (1993). Expression and antisense inhibition of transgenes in *Phytophthora infestans* is modulated by choice of promoter and position effects. *Gene, 133*, 63–69.

Judelson, H. S., & Michelmore, R. W. (1989). Structure and expression of a gene encoding heat-shock protein Hsp70 from the oomycete fungus *Bremia lactucae*. *Gene, 79*, 207–217.

Judelson, H. S., & Michelmore, R. W. (1990). Highly abundant and stage-specific mRNAs in the obligate pathogen *Bremia lactucae*. *Molecular Plant–Microbe Interactions, 3*, 225–232.

Judelson, H. S., & Michelmore, R. W. (1991). Transient expression of genes in the oomycete *Phytophthora infestans* using *Bremia lactucae* regulatory sequences. *Current Genetics, 19*, 453–459.

Judelson, H. S., & Michelmore, R. W. (1992). Temperature and genotype interactions in the expression of host resistance in lettuce downy mildew. *Physiological and Molecular Plant Pathology, 40*, 233–245.

Judelson, H. S., Tyler, B., & Michelmore, R. W. (1991). Stable transformation of the potato late blight fungus, *Phytophthora infestans. Molecular Plant–Microbe Interactions, 4*, 602–607.

Judelson, H. S., Tyler, B. M., & Michelmore, R. W. (1992). Regulatory sequences for expressing genes in oomycete fungi. *Molecular and General Genetics, 234*, 138–146.

Kamoun, S. (2006). A catalogue of the effector secretome of plant pathogenic oomycetes. *Annual Review on Phytopathology, 44*, 41–60.

Kuang, H., Ochoa, O. E., Nevo, E., & Michelmore, R. W. (2006). The disease resistance gene *Dm3* is infrequent in natural populations of *Lactuca serriola* due to deletions and frequent gene conversions. *Plant Journal, 47*, 38–48.

Kuang, H., Woo, S.-S., Meyers, B. C., Nevo, E., & Michelmore, R. W. (2004). Multiple genetic processes result in heterogeneous rates of evolution within the major cluster disease resistance genes in lettuce. *Plant Cell, 16*, 2870–2894.

Lebeda, A., & Blok, I. (1990). Sexual compatibility types of *Bremia lactucae* originating from *Lactuca serriola. Netherlands Journal of Plant Pathology, 96*, 51–54.

Lebeda, A., Sedlarova, M., Petrivalsky, M., & Prokopova, J. (2008). Diversity of defense mechanisms in plant–pathogen interactions: A case study of *Lactuca* spp.–*Bremia lactucae. European Journal of Plant Pathology* (this issue).

Lebeda, A., & Zinkernagel, V. (2003). Characterization of new highly virulent German isolates of *Bremia lactucae* and efficiency of resistance in wild *Lactuca* germplasm. *Journal of Phytopathology, 151*, 274–282.

MacGregor, T., Bhattacharyya, M., Tyler, B., Bhat, B., Schmitthenner, A. F., & Gijzen, M. (2002). Genetic and physical mapping of *Avr1a* in *Phytophthora sojae. Genetics, 160*, 949–959.

Maclean, D. J., Sargent, J. A., Tommerup, I. C., & Ingram, D. S. (1974). Hypersensitivity as a primary event in resistance to fungal parasites. *Nature, 249*, 186–187.

Maclean, D. J., & Tommerup, I. C. (1979). Histology and physiology of compatibility and incompatibility between lettuce and the downy mildew fungus, *Bremia lactucae* Regel. *Physiological Plant Pathology, 14*, 291–312.

Maisonneuve, B., Anderson, P., & Michelmore, R. W. (1994). Rapid mapping of two genes for resistance to downy mildew derived from *Lactuca serriola* to existing clusters of resistance genes. *Theoretical and Applied Genetics, 89*, 96–104.

Mandel, M. A., Crouch, V. W., Gunawardena, U. P., Harper, T. M., & Orbach, M. J. (1997). Physical mapping of the *Magnaporthe grisea Avr1-MARA* locus reveals the virulent allele contains two deletions. *Molecular Plant–Microbe Interactions, 10*, 1102–1105.

McHale, L., Tan, X., Koehl, P., & Michelmore, R. W. (2006). Plant NBS-LRR proteins: Adaptable guards. *Genome Biology, 7*, 212.

Meyers, B. C., Chin, D. B., Shen, K. A., Sivaramakrishnan, S., Lavelle, D. O., Zhang, Z., et al. (1998a). The major resistance gene cluster in lettuce is highly duplicated and spans several megabases. *Plant Cell, 10*, 1817–1832.

Meyers, B. C., Shen, K. A., Rohani, P., Gaut, B. S., & Michelmore, R. W. (1998b). Receptor-like genes in the major resistance locus of lettuce are subject to divergent selection. *Plant Cell, 10*, 1833–1846.

Michelmore, R. W., & Ingram, D. S. (1980). Heterothallism in *Bremia lactucae. Transactions of the British Mycological Society, 75*, 47–56.

Michelmore, R. W., & Ingram, D. S. (1981). The origin of gametangia in the heterothallic isolates of *Bremia lactucae* Regel. *Transactions of the British Mycological Society, 76*, 425–432.

Michelmore, R. W., & Ingram, D. S. (1982). Secondary homothallism in *Bremia lactucae. Transactions of the British Mycological Society, 78*, 1–9.

Michelmore, R. W., & Meyers, B. C. (1998). Clusters of resistance genes in plants evolve by divergent selection and a birth-and-death process. *Genome Research, 8*, 1113–1130.

Michelmore, R. W., Norwood, J. M., Ingram, D. S., Crute, I. R., & Nicholson, P. (1984). The inheritance of virulence in *Bremia lactucae* to match resistant factors 3, 4, 5, 6, 8, 9, 10 and 11 in lettuce (*Lactuca sativa*). *Plant Pathology, 33*, 301–315.

Michelmore, R. W., Paran, I., & Kesseli, R. V. (1991). Identification of markers linked to disease resistance genes by bulked segregant analysis: A rapid method to detect markers in specific genomic regions by using segregating populations. *Proceedings of the National Academy of Sciences of the United States of America, 88*, 9828–9832.

Michelmore, R. W., & Sansome, E. R. (1982). Cytological studies of heterothallism and secondary homothallism in *Bremia lactucae. Transactions of the British Mycological Society, 79*, 291–297.

Nomura, K., Melotto, M., & He, S.-Y. (2005). Suppression of host defense in compatible plant–*Pseudomonas syringae* interactions. *Current Opinion in Plant Biology, 8*, 361–368.

Norwood, J. M., & Crute, I. R. (1984). The genetic control and expression of specificity in *Bremia lactucae* (lettuce downy mildew). *Plant Pathology, 33*, 385–400.

Norwood, J. M., Michelmore, R. W., Crute, I. R., & Ingram, D. S. (1983). The inheritance of specific virulence of *Bremia lactucae* (downy mildew) to match resistance factors 1, 2, 4, 6 and 11 in *Lactuca sativa* (lettuce). *Plant Pathology, 32*, 176–177.

Petnicki-Ocwieja, T., Schneider, D. J., Tam, V. C., Chancey, S. T., Shan, L., Jamir, Y., et al. (2002). Genomewide identification of proteins secreted by the Hrp type III protein secretion system of *Pseudomonas syringae* pv. *tomato* DC3000. *Proceedings of the National Academy of Sciences of the United States of America, 99*, 7652–7657.

Petrželová, I., & Lebeda, A. (2003). Distribution of compatibility types and occurrence of sexual reproduction in natural populations of *Bremia lactucae* on wild *Lactuca serriola* plants. *Acta Phytopathologica et Entomologica Hungarica, 38*, 43–52.

Rehmany, A. P., Gordon, A., Rose, L. E., Allen, R. L., Armstrong, M. R., Whisson, S. C., et al. (2005). Differential recognition of highly divergent downy mildew avirulence gene alleles by *RPP1* resistance genes from two *Arabidopsis* lines. *Plant Cell, 17*, 1839–1850.

Schettini, T. M., Legg, E. J., & Michelmore, R. W. (1991). Insensitivity to metalaxyl in California populations of *Bremia lactucae* and resistance of California lettuce cultivars to downy mildew. *Phytopathology, 81*, 64–69.

Shan, W., Cao, M., Leung, D., & Tyler, B. M. (2004). The Avr1b locus of *Phytophthora sojae* encodes an elicitor and a regulator required for avirulence on soybean plants carrying resistance gene *Rps*1b. *Molecular Plant–Microbe Interactions, 17*, 394–403.

Shen, K. A., Chin, D. B., Arroyo-Garcia, R., Ochoa, O. E., Lavelle, D. O., Wroblewski, T., et al. (2002). *Dm3* is one member of a large constitutively-expressed family of NBS-LRR encoding genes. *Molecular Plant–Microbe Interactions, 15*, 251–261.

Shen, K. A., Meyers, B. C., Islam-Faridi, M. N., Chin, D. B., Stelly, D. M., & Michelmore, R. W. (1998). Resistance gene candidates identified using PCR with degenerate primers map to resistance genes clusters in lettuce. *Molecular Plant–Microbe Interactions, 11*, 815–823.

Sicard, D., Legg, E. J., Brown, S., Babu, N., Ochoa, O., & Michelmore, R. W. (2003). A genetic map of the lettuce downy mildew fungus, *Bremia lactucae*, constructed from molecular markers and avirulence genes. *Fungal Genetics and Biology, 39*, 16–30.

Sicard, D., Woo, S.-S., Arroyo-Garcia, R., Ochoa, O., Nguyen, D., Korol, A., et al. (1999). Molecular diversity at the major cluster of disease resistance genes in cultivated and wild *Lactuca* spp. *Theoretical and Applied Genetics, 99*, 405–418.

Sugio, A., Yang, B., & White, F. F. (2005). Characterization of the *hrpF* pathogenicity peninsula of *Xanthomonas oryzae* pv. *oryzae*. *Molecular Plant–Microbe Interactions, 18*, 546–554.

The Arabidopsis Genome Initiative (2000). Analysis of the genome sequence of the flowering plant *Arabidopsis thaliana*. *Nature, 408*, 796–815.

Torto, T. A., Li, S., Styer, A., Huitema, E., Testa, A., Gow, N. A., et al. (2003). EST mining and functional expression assays identify extracellular effector proteins from the plant pathogen *Phytophthora*. *Genome Research, 13*, 1675–1685.

Tyler, B. M. (2002). Molecular basis of recognition between Phytophthora pathogens and their hosts. *Annual Review of Phytopathology, 40*, 137–167.

Tyler, B. M., Tripathy, S., Zhang, X., Dehal, P., Jiang, R. H. Y., Aerts, A., et al. (2006). *Phytophthora* genome sequences uncover evolutionary origins and mechanisms of pathogenesis. *Science, 313*, 1197–1313.

US Department of Agriculture (2003). *Agricultural chemical usage: 2002 vegetables summary*. http://usda.mannlib.cornell.edu/reports/nassr/other/pcu-bb/agcv0703.pdf.

US Department of Agriculture (2006). *USDA Economics, Statistics and Market Information System (ESMIS)*. http://usda.mannlib.cornell.edu/MannUsda/homepage.do.

Voglmayr, H., & Greilhuber, J. (1998). Genome size determination in Peronosporales (Oomycota) by Feulgen image analysis. *Fungal Genetics and Biology, 25*, 181–195.

Voglmayr, H., Riethmüller, A., Göker, M., Weiss, M., & Oberwinkler, F. (2004). Phylogenetic relationships of *Plasmopara*, *Bremia*, and other genera of downy mildews with pyriform haustoria based on Bayesian analysis of partial LSU rDNA sequence. *Mycological Research, 108*, 1011–1024.

Whisson, S. C., Boevink, P. C., Moleleki, L., Avrova, A. O., Morales, J. G., Gilroy, E. M., et al. (2007). A translocation signal for delivery of oomycete effector proteins into host plant cells. *Nature, 450*, 115–118.

Win, J., Kanneganti, T. D., Torto-Alalibo, T., & Kamoun, S. (2005). Computational and comparative analyses of 150 full-length cDNA sequences from the oomycete plant pathogen *Phytophthora infestans*. *Fungal Genetics and Biology, 43*, 20–33.

Witsenboer, H., Kesseli, R. V., Fortin, M., Stangellini, M., & Michelmore, R. W. (1995). Sources and genetic structure of a cluster of genes for resistance to three pathogens in lettuce. *Theoretical and Applied Genetics, 91*, 178–188.

Woods, A. M., Didehvar, F., Gay, J. L., & Mansfield, J. W. (1988). Modification of the host plasmalemma in haustorial interactions of *Lactuca sativa* by *Bremia lactucae*. *Physiological and Molecular Plant Pathology, 33*, 299–310.

Wroblewski, T., Piskurewicz, U., Tomczak, A., Ochoa, O., & Michelmore, R. W. (2007). Silencing of the major family of NBS-LRR-encoding genes in lettuce results in the loss of multiple resistance specificities. *Plant Journal, 51*, 803–818.

Wroblewski, T., Tomczak, A., & Michelmore, R. W. (2005). Optimization of *Agrobacterium*-mediated transient assays of gene expression in lettuce, tomato and *Arabidopsis*. *Plant Biotechnology Journal, 3*, 259–273.

Yuen, J. E., & Lorbeer, J. W. (1987). Natural and experimental production of oospores of *Bremia lactucae* in lettuce in New York. (1987). *Plant Disease, 71*, 63–64.

Eur J Plant Pathol (2008) 122:31–39
DOI 10.1007/s10658-008-9296-z

Species hybrids in the genus *Phytophthora* with emphasis on the alder pathogen *Phytophthora alni*: a review

Tibor Érsek · Zoltán Á. Nagy

Received: 7 August 2007 / Accepted: 22 February 2008

Abstract This review provides a summary of recent examples of interspecific hybridisation within the oomycetous genus *Phytophthora*. Species hybrids either created in the laboratory or evolved in natural environments are discussed in association with evolutionary issues and possible threats they may pose to agriculture, horticulture and forestry. It is suggested that sustainable control of such hybrids will depend on the better understanding of temporal and spatial aspects of genetic mechanisms and environmental factors that lead to the hybridisation process and thus the genetic diversity in *Phytophthora* populations.

Keywords Evolution · Interspecific hybridisation · Oomycetes

Introduction

Hybridisation between individuals from two populations has long been known in the plant kingdom, but its occurrence among populations of eukaryotic microorganisms has received delayed recognition. Indeed, eukaryotic microorganisms, such as fungi and oomycetes, possess a variety of reproductive mechanisms whereby they might undergo interspecific genetic exchange. As a consequence of the combination of two distinct genomes via sexual or parasexual processes, new allopolyploid hybrid species may evolve.

The possibility of hybridisation between two closely related species of plant pathogenic fungi or oomycetes has been considered for many years. For instance, seven decades ago Flor (1932) pointed out the potential for hybridisation among fungi based on the appearance of isolates of *Tilletia* with atypical morphological phenotypes. Decades later, Burdon et al. (1981) used isozyme analysis to confirm that a rust virulent on rough wheat grass and barley evolved via somatic hybridisation between rye stem rust and wheat stem rust i.e., between two formae speciales (f.sp.) of *Puccinia graminis*. Conclusive proof of species hybridisation has been provided only recently with the arrival of modern molecular genetic tools. Since the mid 1990s, a limited number of phytopathogenic species hybrids have been detected in the fungal phyla, Ascomycota and Basidiomycota (Brasier 2000; Olson and Stenlid 2002; Schardl and Craven 2003).

Within the Oomycota in the Kingdom Staminipila (Dick 2002), efforts to establish the occurrence of species hybridisation have focused on the genus *Phytophthora*. This genus contains approximately 80

T. Érsek (✉)
Faculty of Agricultural and Food Sciences,
Department of Plant Protection,
University of West Hungary,
Vár 2, 9200 Mosonmagyaróvár, Hungary
e-mail: ters@mtk.nyme.hu

Z. Á. Nagy
Department of Plant Pathology, Plant Protection Institute,
Hungarian Academy of Sciences,
P.O. Box 102, 1525 Budapest, Hungary

species, including the infamous late blight pathogen, *P. infestans*. These species are primarily soil-borne pathogens that affect a wide range of crop plants, shrubs and trees throughout the world. Although fungus-like in appearance, phytophthoras differ fundamentally from true fungi in terms of cell wall composition, reproductive biology and genetics (Erwin and Ribeiro 1996). In this review we summarise the evidence for species hybridisation in this important genus. Evidence for hybridisation as a source of genetic variability has derived from both laboratory attempts to create hybrids, and from studies of *Phytophthora* hybrids found in either natural or agro-ecosystems.

Hybrids created in the laboratory

Species hybrid formation in nature is likely to be a rare event and, consequently, it is difficult to detect and study. A tractable approach to the study of the possibility of genetic exchange and evolution derived from species hybridisation would be to create such organisms artificially.

Sexual crosses

An early report of this approach was by Boccas (1981), who induced sexual crosses among isolates of numerous heterothallic (outcrossing) *Phytophthora* species. This first effort was minimally successful and produced only one putative species hybrid among 220 progeny derived from several species crosses. Subsequent attempts to induce and confirm species hybridisation in other laboratories were more successful. For instance, Goodwin and Fry (1994) induced sexual crosses of the sympatric, heterothallic species, *P. mirabilis* and *P. infestans*. They confirmed that 79 out of 86 progeny were species hybrids based on DNA fingerprinting and isozyme analyses. Notably, mitochondrial DNA was uniparentally inherited, predominantly from the *P. infestans* parental isolates. Interestingly, most of these hybrids lost their ability to attack hosts of either parental species, including *Mirabilis jalapa*, the host for *P. mirabilis*, and potato or tomato, common hosts for *P. infestans*. May et al. (2003) more recently induced sexual crosses between the homothallic (self-fertile) species, *P. sojae* and *P. vignae*. They confirmed the hybrid nature of offspring by RAPD and AFLP analyses. They also noted that both of the tested F1 hybrids were pathogenic to soybean, the host for *P. sojae*, and cowpea, the host for *P. vignae*. However, the aggressiveness of these hybrids was reduced and was substantially more variable when compared to the parental isolates on their respective hosts.

Somatic hybridisations

Somatic fusion has also been suggested as a mechanism for hybridisation in nature among *Phytophthora* species that are heterothallic and that temporarily or spatially lack compatible mating types (Brasier 1992; Érsek et al. 1995). Although protoplast fusion between strains of a *Phytophthora* species was relatively easy to induce, the same method appeared to be insufficient for the creation of interspecific hybrids between *P. sojae* (syn.: *P. megasperma* f.sp. *glycinea*) and *P. medicaginis* (syn.: *P. megasperma* f. sp. *medicaginis*) (Layton and Kuhn 1988) or between *P. nicotianae* (syn.: *P. parasitica*) and *P. capsici* (Gu and Ko 2000). The first evidence of the formation of such hybrids was obtained by the induced fusion of uninucleate zoospores (Fig. 1) derived from non-compatible mating-type isolates of the closely related heterothallic species, *P. capsici* and *P. nicotianae* (Érsek et al. 1995). In these laboratory experiments, the morphologies of the four resultant hybrid isolates resembled *P. capsici* more closely than *P. nicotianae*. All of the hybrids were pathogenic to tomato, a plant

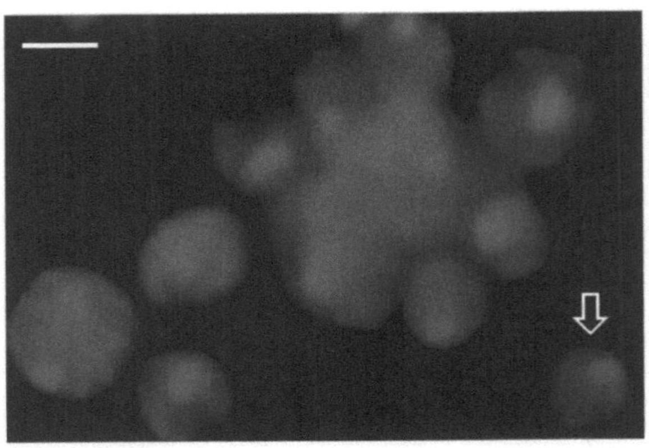

Fig. 1 Nuclear status of regenerating cells stained with DAPI following induced fusion of zoospores of *Phytophthora capsici* and *P. nicotianae*, as viewed by epifluorescence microscopy. Note enlarged cells and/or nuclei and multiple nuclei as compared to a uninucleate cell of normal size (*arrow*). Bar: 10 μm

susceptible to both parental species. However, two of the hybrid isolates exhibited an expanded host range that included both radish and lemon, hosts that are susceptible to *P. capsici* or *P. nicotianae*, respectively. The hybrid nature of the fusion offspring was confirmed by detection of DNA sequences specific to each parental species. In these hybrids repetitive DNA of *P. capsici* was detected readily by hybridisation with a species-specific DNA probe, whereas *P. nicotianae*-specific DNA was revealed after PCR amplification of DNA from hybrids using *P. nicotianae*-specific primers or random primers (Érsek et al. 1995; English et al. 1999). Bakonyi et al. (2002) also induced zoospore fusion to create hybrids from the morphologically distinct species, *P. nicotianae* and *P. infestans*. Resultant fusion offspring were more similar to *P. nicotianae* than to *P. infestans* on the basis of morphological and molecular evidence. Again, these hybrids expressed modified pathogenicity traits compared to parental species.

As a final example of the tractability of somatic fusion, Érsek et al. (1997) created tri-parental hybrids derived from *P. capsici*, *P. nicotiane* and *P. citrophthora*. In these studies, zoospore fusion offspring contained DNA from each parental species; however, all offspring failed to express pathogenicity to any of the hosts susceptible to the parental species. Notwithstanding the failure of protoplast fusion to create hybrids between *P. nicotianae* and *P. capsici*, Gu and Ko (2000) successfully generated hybrids by transfer of isolated nuclei from one species to the other. Analysis of zoospore progeny of these nuclear hybrids suggested the completion of events leading to a parasexual cycle.

Studies on interspecific zoospore fusion and nuclear transfer support the suggestion of Brasier (1992) that non-sexual genetic exchange might generate variability in pathogenicity or virulence within pathogen populations, particularly when complementary mating types needed for sexual reproduction are lacking. Artificially induced hybrids also suggest that it may be difficult to predict the effects of hybridisation on pathogen survival and dominance among populations in nature. With the development of molecular and biochemical markers, however, there have recently been noteworthy findings of naturally occurring *Phytophthora* species hybrids that may provide further insight into the mechanism of interspecific genetic exchange.

Naturally formed hybrids

Phytophthora alni

The potential for hybrid formation among *Phytophthora* species that has been established in laboratory studies has, in recent years, been confirmed by the detection of true or putative hybrids in natural and agro-ecosystems. One of the best-studied examples of *Phytophthora* species hybridisation in nature is that of *P. alni*, a newly recognised pathogen of alder (*Alnus* spp.). This pathogen was first discovered on dying alders in southern Britain in the beginning of the 1990s, and it has since been found throughout Europe, including Hungary (Brasier et al. 1995; Szabó et al. 2000; Streito 2003). *Phytophthora alni* killed approximately 10% of the alders in southern Britain within a few years of its initial discovery (Brasier et al. 1995; Gibbs et al. 1999). Recent surveys indicate that the disease is even more severe in riparian ecosystems in north-eastern France (Streito et al. 2002) and in Bavaria (Jung and Blaschke 2004).

Initial studies showed that certain isolates of this new pathogen of alder resembled *P. cambivora*. The similarity to *P. cambivora* was notable through the morphology of the gametangia (Brasier et al. 1995). However, the new pathogen differed from *P. cambivora* in several other traits, for instance, in being homothallic rather than heterothallic and in exhibiting an extremely high level of zygotic abortion. These properties, in addition to assessments of internal transcribed spacer (ITS) sequences and genomic polymorphisms, ultimately suggested that the alder pathogen might be a hybrid of two species, the heterothallic *P. cambivora* and a homothallic *P. fragariae*-like species (Brasier et al. 1999). Neither of these organisms is a known pathogen of alder. Ultimately, the alder *Phytophthora* was formally designated by Brasier et al. (2004) as a new species, *P. alni* Brasier & S.A. Kirk.

Because the newly defined species comprises a range of phenotypically diverse allopolyploid genotypes, *P. alni* was split into three subspecies, *P. alni* ssp. *alni (Paa)*, *P. alni* ssp. *uniformis (Pau)* and *P. alni* ssp. *multiformis (Pam)* (Brasier et al. 2004). In addition to the three subspecies, a series of emerging variant types of *P. alni* have recently been recovered (Brasier et al. 2004; Jung and Blaschke 2004). Prior to their designation as subspecies, isolates of *Paa* and

Pau were termed standard types and Swedish variants, respectively, whereas *Pam*, including divergent hybrid types, were considered to be Dutch, German and UK variants of the pathogen (Brasier et al. 1999).

Paa occurs more commonly across much of Europe, and isolates of this form are generally more aggressive than those of the other two subspecies that are also present in several countries (Brasier and Kirk 2001). Furthermore, *Paa* produces *P. cambivora*-like ornamented oogonia and elongated two-celled antheridia. In contrast, the other two subspecies exhibit unique reproductive structures. Isolates of *Pam* produce oogonia that are typically ornamented, but antheridia and gametangial fusions may vary in morphology. *Pau* uniformly forms oogonia with a smooth surface under ordinary conditions but develops *Paa*-like ornamented female organs when grown at sub-optimal temperatures, i.e. $\leq$15°C, thus indicating that morphology-based differentiation of the two subspecies might fail under variable conditions (Fig. 2).

As opposed to typical *Phytophthora* species, which are diploid (2n) organisms, the standard type isolate of *Paa* was determined by acetoorcein staining to be an approximate tetraploid (~4n), having a chromosome number of ca. 18–22 at the first metaphase division. This isolate, however, is unable to complete meiosis (Brasier et al. 1999). In contrast, ploidy levels of *Pau* and *Pam* are intermediate between diploid and tetraploid and range from 2n+2 to 2n+4–7, respectively.

In addition to differences in ploidy, subspecies of *P. alni* differ in details of molecular features. For instance, *Paa* has dimorphic sites in the ITS region of its rDNA genes, in which DNA sequences are representative of two species, *P. cambivora* and a *P. fragariae*-like species. In contrast with *Paa*, ITS sequences in both *Pam* and *Pau* are homogeneous and resemble the ITS sequences of either *P. fragariae* or *P. cambivora*, respectively (Brasier et al. 1999, 2004). Subspecies of *P. alni* also differ on the basis of AFLP profiles (Brasier et al. 1999), RAPD or isozyme patterns (Nagy et al. 2003; Brasier et al. 2004) and diagnostic *P. alni*-specific PCR primer sets (Table 1). A firm correspondence, as shown in Table 2, has also been established between PCR markers and expression patterns for glucose-phosphate isomerase (Gpi) and malate dehydrogenase (Mdh) (Bakonyi et al. 2007). Since both isozymes are known to be nuclear-encoded, this suggests that the above-mentioned PCR-targeted DNAs that differentiate the subspecies are likely to be of nuclear origin.

Mitochondrial (mt) genomic variability within *P. alni* has been examined to only a limited extent. In studies by Nagy et al. (2003), RFLP analyses of mtDNA showed several bands that co-migrated between *P. alni* isolates and either *P. cambivora* or *P. fragariae*. However, it was not clear whether the appearance of such bands refers to the presence of biparental mtDNA fragments in hybrid isolates or whether it reflects intraspecific variation in the mt genome of either parental species. Based on sexual crosses within individual *Phytophthora* species, it has been suggested that the mitochondrial genome is transmitted uniparentally through the maternal line, whereas the nuclear genome is inherited from both the maternal and paternal lines (Förster and Coffey 1991; Whittaker et al. 1994).

The comparatively meagre knowledge about the mt genome of *P. alni* has been broadened recently. Ioos et al. (2006) performed phylogenetic analysis of the mt genes, *cox*1 and *nadh*1, that revealed that mtDNA sequences from either *P. cambivora* or *P. fragariae*

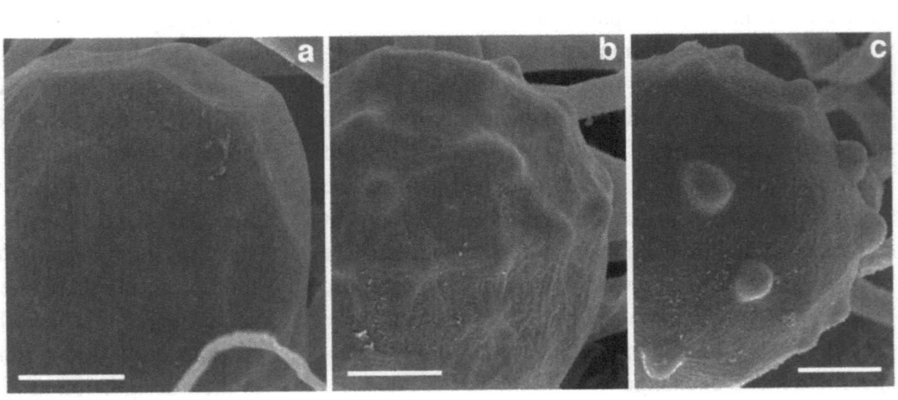

Fig. 2 Scanning electron micrographs of the surface of oogonia of *Phytophthora alni* subsp. *uniformis* grown at optimal temperature (**a**), sub-optimal temperature (**b**) and of *P. alni* subsp. *alni* (**c**). Note the environment-dependent change in morphology of *P. alni* subsp. *uniformis*. Bars: 20 µm

Table 1 PCR primer pairs developed for specific detection of *Phytophthora alni*

Primer name	Sequence (5′–3′)	Amplicon size (bo)	Specificity[a]	Reference
SAP1	GGC ACT GAG GGT TCC TC	930	*Paa, Pam*	Bakonyi et al. (2006)
SAP2	GGC ACT GAG GTC TAG ATT			
SWAP1	TGG CCC TCA CAT TAA AAC TGC TGC	1130	*Pau, Paa*	
SWAP2	GGC CCT CAC CAA ATG CGA AAT GA			
PA-F	GGT GAT CAG GGG AAT ATG TG	450	*Paa, Pau, Pam*	Ioos et al. (2005)
PA-R	ATG TCG GAG TGT TTC CCA AG			
PAM-F	CTG ACC AGC CCC TTA TTG GC	590	*Paa, Pam*	
PAM-R	CTG ACC AGC CAT CCC ACA TG			
PAU-F	GAG GAT CCC TAA CAC TGA ATG G	750	*Pau, Paa*	
PAU-R	GAT CCC TGG TTG AAG CTG AG			
D16F	AGG GCG TAA GGG TGC GAA ATA	366	*Paa, Pau*	De Merlier et al. (2005)
D16R	AGG GCG TAA GCC TGG ACC G			

[a] *Paa*, *Pau* and *Pam* are for *Phytophthora alni* subsp. *alni* subsp. *uniformis* subsp. *multiformis*, respectively.

did not cluster with those from the hybrid isolates, likely to be the result of uniparental inheritance of the mt genome. Furthermore, mtDNA sequences of *P. alni* isolates from the three subspecies clustered into only two groups, one that included *Paa* and *Pam*, and the other, *Pau*. Surprisingly, the mtDNA profiles of certain isolates that had been identified as *Paa* using morphological and nuclear markers were identical to those of *Pau* isolates (Ioos et al. 2006; Bakonyi et al. 2007). Such isolates may represent additional hybrid forms that encompass the nuclear type of *Paa* and a mitotype represented by *Pau* (Table 2).

The complexity of the nuclear and mitochondrial genomes of *Paa*, *Pau*, and *Pam* suggests that *P. alni* may be a species in a state of continuing evolution. As compared to its putative parental species, *P. alni* has exploited a new host (Brasier and Kirk 2001). The source of the parental species is not clear, since *P. cambivora* and *P. fragariare* are believed to be exotic to Europe. The evolutionary mechanism leading to the formation of subspecies of *P. alni* is also uncertain. Based on cytological and molecular analyses, Brasier et al. (1999) favoured the view that *Paa* could have arisen via somatic fusion followed by further segregation, rather than via a sexual cross between *P. cambivora* and a *P. fragariae*-like species. These authors further suggested that *Pau* and *Pam* might have then evolved through subsequent recombination events and chromosome losses in *Paa* that led to reversions towards the *P. cambivora*-like or *P. fragariae*-like parental genotypes.

Recently, Ioos et al. (2006) proposed an alternative evolutionary model by which *Paa* might have arisen via hybridisation of *Pam* and *Pau*. They suggested

Table 2 Patterns of nuclear and mithochondrial traits in subspecies of *Phytophthora alni* according to the results of Bakonyi et al. (2006, 2007)

Marker type	*Pam* (M)[a]	*Pau* (U)	*Paa* (A)
PCR with primer set SAP1/SAP2	+[b]	–	+
PCR with primer set SWAP1/SWAP2	–	+	+
RAPDs with primer OPG-02 or OPG-05	M	U	M + U
Isozyme locus *Mdh-1*	91/100	83/83	83/91/100
Isozyme locus *Mdh-2*	94/94	100/100	94/100
Isozyme locus *Gpi*	85/100	93/93	85/93/100
mtDNA-RFLP with *Msp*I or *Hae*III	M (=A)	U	A (=M) or U

[a] *Pam* (M) *Pau* (U) *Paa* (A) are for *Phytophthora alni* subsp. *multiformis*, subsp. *uniformis* and subsp. *alni*, respectively.

[b] Specific amplicon is produced (+) or not produced (–).

that *Pau* might have evolved from *P. cambivora*, whereas *Pam* might have either been generated itself by an ancient reticulation or by autopolyploidisation. Their hypotheses were based on analyses of a large European-wide collection of *P. alni* isolates showing that *Paa* possessed three alleles for each of four nuclear genes studied, two of which were also present in *Pam*, and a third one that matched a single allele in *Pau*. Furthermore, the *Paa* isolates displayed a mtDNA RFLP pattern identical to isolates of either *Pam* or *Pau*, implying uniparental inheritance of the mt genome in the suspected hybridisation process. These results are supported by data of Bakonyi et al. (2007) who found that the studied nuclear-encoded traits expressed in *Paa* included combined expression profiles of *Pam* and *Pau*, whereas mtDNA restriction profiles of *Paa* matched that of either *Pam* or *Pau* (Table 2).

Isolates of *Pam* and *Pau* have been recovered from alder lesions far less frequently than have isolates of *Paa*, and they have also proven to be significantly less aggressive in colonising alder bark (Brasier and Kirk 2001). On the basis of these observations, Bakonyi et al. (2007) suggested that the emergence of atypical *Paa* isolates with a *Pau* mitotype might have occurred in bark tissue co-colonised by *Paa* and *Pau*. In this niche, *Paa* and *Pau* isolates might have hybridised by either somatic or gametangial interaction. In this scenario, Bakonyi et al. (2007) also suggested the possibility that these atypical *Paa* isolates may have arisen through the introgression of mitochondria from *Pau* into the nuclear background of *Paa*. Interactions like these must be very rare in nature, and indeed, there has been only one such report, in association with the causal agents of Dutch elm disease. According to Bates et al. (1993), certain *Ophiostoma novo-ulmi* isolates exhibited typical *O. novo-ulmi* nuclear DNA profiles, but they also exhibited *O. ulmi*-like mtDNA patterns. They attributed these patterns to somatic fusion between the two related species.

The hybridisation event that led to the emergence of *P. alni* is believed to be recent, and it may have occurred in a European nursery, perhaps on raspberry or another host that is common to the putative parental species (Brasier et al. 1999; Brasier and Jung 2003). It is assumed that *P. alni* arrived in Britain, the country of first record, as a result of commercial trade of colonised plant material. Its subsequent spread over long distances is likely to have occurred via distribution and planting of infested nursery stock (Brasier et al. 1999).

Local spread from points of *P. alni* introduction is not likely to be related to the movement of oospores, since these structures have poor survival ability in soil (Delcan and Brasier 2001). More likely, zoospores and plant debris containing mycelium contribute to pathogen movement at this scale. Alders are key trees in wetlands and riparian environments, where they stabilise river and stream banks. In these habitats, the presence of saturated or flooded soils, and water movement, would enhance spread of zoospores and debris. This scenario is supported by observations of higher disease incidence among alders growing near rivers than those some distance away (Gibbs et al. 1999).

Phytophthora cactorum × *P. nicotianae*

Species hybridisation has also been reported in hydroponic greenhouse systems in The Netherlands. Under such circumstances novel *Phytophthora* diseases have appeared on diverse ornamental species. For instance, Man in't Veld et al. (1998) reported the isolation of *Phytophthora* that differed morphologically from known pathogenic species from *Spathiphyllum* and *Primula* plants. Isozyme and RAPD analyses revealed that the unusual isolates represented hybrids of *P. nicotianae* and *P. cactorum*. In addition, mtDNA restriction patterns of the hybrid isolates were identical to those of *P. nicotianae* (Man in't Veld et al. 1998). *Phytophthora nicotianae* is an introduced species in The Netherlands, and it can infect both *Spathiphyllum* and *Primula*. In contrast, *P. cactorum* is a resident species, but it does not cause disease on these host plants. Additional hybrid isolates were obtained from a *Cyclamen* sp., which is not known to be a host of either of the parental species (Bonants et al. 2000). Subsequent analysis of the ITS region of rDNA and AFLP analyses provided further evidence of the biparental origin of the recovered isolates.

Similar hybrids have been characterised recently from loquat trees (*Eriobotrya japonica*) grown in orchards in central Taiwan (Man in't Veld 2001). The unlikely movement of hybrid isolates between such distinctly different and separated agricultural and horticultural systems suggests a potential for hybridisation between *P. cactorum* and *P. nicotianae* when both species occupy the same habitat.

Phytophthora cactorum × *P. hedraiandra*

Man in't Veld et al. (2007) recently reported the involvement of *P. cactorum* in the formation of yet another hybrid species after hybridisation with *P. hedraiandra*. These new hybrids were shown to be heterozygous at the dimeric malic enzyme (*Mdhp*) locus, possessing the MDHP alleles of the two parental species. They also contained dimorphic sites in the ITS region, exactly at those positions where the parental sequences differ. Consistent with the hybrid hypothesis, most hybrid isolates contained the mitochondrial-encoded cytochrome oxidase I (Cox I) gene sequences that were identical to those of *P. hedraiandra*, and one isolate had the gene sequences of the other putative parent.

Phytophthora cactorum has been isolated from numerous hosts, including *Rhododendron* spp. During the past decade, however, only these novel hybrids have been found on *Rhododendron* in the Netherlands, suggesting that they are replacing the resident *P. cactorum* population on this host. While *P. cactorum* is an indigenous species in Europe, *P. hedraiandra* is believed to be a recent introduction from North America, where it infects *Rhododendron* spp. In comparison with the parental species, the hybrid isolates exhibit expanded host ranges, including monocots (*Allium* spp.) as well as dicots (*Idesia* and *Penstemon* spp.). These isolates are known to be proliferating in the environment in the Netherlands and in Germany (Man in't Veld et al. 2007).

Concluding remarks

Laboratory and field studies suggest that interspecific hybridisation in *Phytophthora* populations occurs rarely. However, such rare events may prove to be an important source of genetic diversity, in addition to the more commonly recognised processes of mutation and sexual or parasexual reproduction within individual species. Studies summarised in this review suggest that *Phytophthora* species hybridisation may produce unique offspring capable of exploiting an expanded range of host plants. In addition, hybrid offspring with increased aggressiveness may be selected to such an extent that they become a dominant component of *Phytophthora* populations within a region.

The studies summarised here also suggest that species hybridisation occurs readily between allopatric species that have not co-evolved in the same location. To date, known *Phytophthora* species hybrids represent the offspring of a native and an exotic or two exotic species that occupy the same habitat and niche. No reports have described similar hybridisation among indigenous, sympatric *Phytophthora* species populations, even though such hybrids can be generated in the laboratory. The reason for this limitation is uncertain, although it is believed that strong genetic barriers have evolved to restrict hybridisation among sympatric oomycete and fungal species (Brasier 2000; Olson and Stenlid 2002; Schardl and Craven 2003).

The mechanisms of species hybridisation in *Phytophthora* populations are not known, but studies of several cases have provided evidence for hybrid populations in various states of genomic evolution. The role of diverse ploidy levels and genomic reorganisation in determining host range, aggressiveness, and population dynamics bears further investigation. In addition, it is noteworthy that most *Phytophthora* species hybrids have acquired the mitochondrial genome of the exotic, introduced parental species (Man in't Veld et al. 2007). Since mitochondrial control of virulence was reported for artificially made hybrids of the basidiomycete fungus *Heterobasidion annosum* (Olson and Stenlid 2001), further research is needed to examine the influence of the acquired mitochondrial genome on host selection by species hybrids of *Phytophthora*.

Interspecific hybridisation among *Phytophthora* species is likely to increase with expanding world trade that introduces plants and associated pathogens into new regions with uniquely different environmental conditions (Brasier 2000). Opportunities for interactions between *Phytophthora* species are also enhanced as plants are managed under hydroponic and other non-traditional agricultural conditions (Man in't Veld et al. 1998, 2007; Bonants et al. 2000). Finally, over longer periods of time, human disturbance factors such as pollution, climate change, and land use may accentuate the emergence of newly adapted hybrids with unique pathogenicity attributes. Limited evidence for these possibilities was provided by Gibbs et al. (1999), who showed that pollution of water with oxidised nitrogen can sensitise alder trees to *P. alni* infection. Consequently, interspecific hybrids appear to be the products of recent evolutionary events. However, some of them might have existed for a long period of time without being identified as hybrids, due to the lack of appropriate tools. Natural hybridisation has been suspected but never proven with *P. meadii* (Sansome et al. 1991). As for fungi, a particular poplar rust identified recently as a hybrid of *Melampsora medusae* and *M. occidentalis* is represented in specimens from nearly a century ago (Newcombe et al. 2000).

Although species hybridisation as a source of new epidemic outbreaks should not be exaggerated, it is of interest to regulatory officials to monitor the emergence of new hybrid genotypes. Unfortunately, emerging hybrid populations are unlikely to be detectable by conventional, morphology-based approaches. Consequently, population sampling methods and molecular techniques for characterising pathogen genetic structure require further development for effective detection and management.

Acknowledgements Part of this study was supported by the Hungarian Scientific Research Fund (OTKA) grant K-61107. Special thanks are due to James T. English for critically reviewing the manuscript.

References

Bakonyi, J., Láday, M., & Érsek, T. (2002). Characterization of parental traits in somatic fusion progeny of *Phytophthora infestans* and *Phytophthora nicotianae*. *Acta Phytopathologica et Entomologica Hungarica, 37*, 33–46.

Bakonyi, J., Nagy, Z. Á., & Érsek, T. (2006). PCR-based DNA markers for identifying hybrids within *Phytophthora alni*. *Journal of Phytopathology, 154*, 168–177.

Bakonyi, J., Nagy, Z. Á., & Érsek, T. (2007). A novel hybrid with nuclear background of *Phytophthora alni* subsp. *alni* exhibits a mitochondrial DNA profile characteristic of *P. alni* subsp. *uniformis*. *Acta Phytopathologica et Entomologica Hungarica, 42*, 1–7.

Bates, M. R., Buck, K. W., & Brasier, C. M. (1993). Molecular relationship of the mitochondrial DNA of *Ophiostoma ulmi* and the NAN and EAN races of *O. novo-ulmi* determined by restriction fragment length polymorphisms. *Mycological Research, 97*, 1093–1100.

Boccas, B. R. (1981). Interspecific crosses between closely related heterothallic *Phytophthora* species. *Phytopathology, 71*, 60–65.

Bonants, P. J. M., Hagenaar-de Weerdt, M., Man in't Veld, W., & Baayen, R. P. (2000). Molecular characterization of natural hybrids of *Phytophthora nicotianae* and *P. cactorum*. *Phytopathology, 90*, 867–874.

Brasier, C. M. (1992). Evolutionary biology of *Phytophthora*. Part I: Genetic system, sexuality and the generation of variation. *Annual Review of Phytopathology, 30*, 153–171.

Brasier, C. M. (2000). The rise of the hybrid fungi. *Nature, 405*, 134–135.

Brasier, C. M., Cooke, D. E. L., & Duncan, J. M. (1999). Origin of a new *Phytophthora* pathogen through interspecific hybridization. *Proceedings of the National Academy of Sciences USA, 96*, 5878–5883.

Brasier, C. M., & Jung, T. (2003). Progress in understanding *Phytophthora* diseases of trees in Europe. In J. McComb, G. Hardy, & I. Tommerup (Eds.), *Proceedings, 2nd International Meeting of IUFRO Working Party 7.02.09, Albany, W. Australia 2001: Phytophthora in Forest and Natural Ecosystems* (pp. 4–18). Perth, Australia: Murdoch University.

Brasier, C. M., & Kirk, S. A. (2001). Comparative aggressiveness of standard and variant hybrid alder phytophthoras, *Phytophthora cambivora* and other *Phytophthora* species on bark of *Alnus, Quercus* and other woody plants. *Plant Pathology, 50*, 218–229.

Brasier, C. M., Kirk, S. A., Delcan, J., Cooke, D. E. L., Jung, T., & Man in't Veld, W. A. (2004). *Phytophthora alni* sp. nov. and its variants: Designation of emerging heteroploid hybrid pathogens spreading on *Alnus* trees. *Mycological Research, 108*, 1172–1184.

Brasier, C. M., Rose, J., & Gibbs, J. M. (1995). An unusual *Phytophthora* associated with widespread alder mortality in Britain. *Plant Pathology, 44*, 999–1007.

Burdon, J. J., Marshall, D. R., & Luig, N. H. (1981). Isozyme analysis indicates that a virulent cereal rust pathogen is a somatic hybrid. *Nature, 293*, 565–566.

Delcan, J., & Brasier, C. M. (2001). Oospore viability and variation in zoospore and hyphal tip derivatives of the hybrid alder Phytophthoras. *Forest Pathology, 31*, 65–83.

De Merlier, D., Chandelier, A., Debruxelles, N., Noldus, M., Laurent, F., Dufays, E., et al. (2005). Characterization of alder *Phytophthora* isolates from Wallonia and development of SCAR primers for their specific detection. *Journal of Phytopathology, 153*, 99–107.

Dick, M. W. (2002). Towards an understanding of the evolution of the downy mildews. In P. T. N. Spencer-Phillips, U. Gisi, & A. Lebeda (Eds.), *Advances in Downy Mildew Research* (pp. 1–57). Dordrecht: Kluwer.

English, J. T., Láday, M., Bakonyi, J., Schoelz, J. E., & Érsek, T. (1999). Phenotypic and molecular characterization of species hybrids derived from induced fusion of zoospores of *Phytophthora capsici* and *Phytophthora nicotianae*. *Mycological Research, 103*, 1003–1008.

Érsek, T., English, J. T., & Schoelz, J. E. (1995). Creation of species hybrids of *Phytophthora* with modifyied host ranges using zoospore fusion. *Phytopathology, 85*, 1343–1347.

Érsek, T., English, J. T., & Schoelz, J. E. (1997). Triparental species hybrids from fused zoospores of *Phytophthora*. *Czech Mycology, 50*, 13–20.

Erwin, D. C., & Ribeiro, O. K. (1996). *Phytophthora diseases worldwide*. St. Paul, MN: American Phytopathological Society Press.

Flor, H. H. (1932). Heterothallism and hybridization in *Tilletia tritici* and *T. laevis*. *Journal of Agricultural Research, 44*, 49–58.

Förster, H., & Coffey, M. D. (1991). Approaches to the taxonomy of *Phytophthora* using polymorphisms in mitochondrial and nuclear DNA. In J. A. Lucas, R. C.

Shattock, D. S. Shaw, & L. R. Cooke (Eds.), *Phytophthora* (pp. 164–183). Cambridge: Cambridge University Press.

Gibbs, J. N., Lipscombe, M. A., & Peace, A. J. (1999). The impact of *Phytophthora* disease on riparian populations of common alder (*Alnus glutinosa*) in southern Britain. *European Journal of Forest Pathology, 29*, 39–50.

Goodwin, S. B., & Fry, W. E. (1994). Genetic analysis of interspecific hybrids between *Phytophthora infestans* and *Phytophthora mirabilis*. *Experimental Mycology, 18*, 20–32.

Gu, Y. H., & Ko, W. H. (2000). Segregation following interspecific transfer of isolated nuclei between *Phytophthora parasitica* and *P. capsici*. *Canadian Journal of Microbiology, 46*, 410–416.

Ioos, R., Andrieux, A., Marcais, B., & Frey, P. (2006). Genetic characterization of the natural hybrid species *Phytophthora alni* as inferred from nuclear and mitochondrial DNA analyses. *Fungal Genetics and Biology, 43*, 511–529.

Ioos, R., Husson, C., Andrieux, A., & Frey, P. (2005). SCAR-based PCR primers to detect the hybrid pathogen *Phytophthora alni* and its subspecies causing alder disease in Europe. *European Journal of Plant Pathology, 112*, 323–335.

Jung, T., & Blaschke, M. (2004). *Phytophthora* root and collar rot of alders in Bavaria: distribution, modes of spread and possible management strategies. *Plant Pathology, 53*, 197–208.

Layton, A. C., & Kuhn, D. N. (1988). Heterokaryon formation by protoplast fusion of drug-resistant mutants in *Phytophthora megasperma* f.sp. *glycinea*. *Experimental Mycology, 12*, 180–194.

Man in't Veld, W. A. (2001). First report of natural hybrids of *Phytophthora nicotianae* and *P. cactorum* on loquat in Taiwan. *Plant Disease, 85*, 98.

Man in't Veld, W. A., de Cock, A. W. A. M., & Summerbell, R. C. (2007). Natural hybrids of resident and introduced *Phytophthora* species proliferating on multiple new hosts. *European Journal of Plant Pathology, 117*, 25–33.

Man in't Veld, W. A., Veenbaas-Rijks, W. J., Ilieva, E., de Cock, A. W. A. M., Bonants, P. J. M., & Pieters, R. (1998). Natural hybrids of *Phytophthora nicotianae* and *Phytophthora cactorum* demonstrated by isozyme analysis and random amplified polymorphic DNA. *Phytopathology, 88*, 922–929.

May, K. J., Drent, A., & Irwin, J. A. G. (2003). Interspecific hybrids between the homothallic *Phytophthora sojae* and *Phytophthora vignae*. *Australasian Plant Pathology, 32*, 353–359.

Nagy, Z. Á., Bakonyi, J., & Érsek, T. (2003). Standard and Swedish variant types of the hybrid alder *Phytophthora* attacking alder in Hungary. *Pest Management Science, 59*, 484–492.

Newcombe, G., Stirling, B., McDonald, S., & Bradshaw, H. D., Jr. (2000). *Melampsora ×columbiana*, a natural hybrid of *M. medusae* and *M. occidentalis*. *Mycological Research, 104*, 261–274.

Olson, Å., & Stenlid, J. (2001). Mitochondrial control of fungal hybrid virulence. *Nature, 411*, 438.

Olson, Å., & Stenlid, J. (2002). Pathogenic fungal species hybrids infecting plants. *Microbes and Infection, 4*, 1353–1359.

Sansome, E., Brasier, C. M., & Hamm, P. B. (1991). *Phytophthora meadii* may be a species hybrid. *Mycological Research, 95*, 273–277.

Schardl, C. L., & Craven, K. D. (2003). Interspecific hybridization in plant-associated fungi and oomycetes: A review. *Molecular Ecology, 12*, 2861–2873.

Streito, J.-C. (2003). *Phytophthora* disease of alder: Identification and distribution. In J. N. Gibbs, C. Van Dijk, & J. F. Webber (Eds.), *Phytophthora* disease of alder in Europe. *Forestry Commission Bulletin, 126*, 25–38.

Streito, J.-C., Legrand, P., Tabary, F., & Jarnouen de Villartay, G. (2002). *Phytophthora* disease of alder (*Alnus glutinosa*) in France: Investigations between 1995 and 1999. *Forest Pathology, 32*, 179–191.

Szabó, I., Nagy, Z., Bakonyi, J., & Érsek, T. (2000). First report of *Phytophthora* root and collar rot of alder in Hungary. *Plant Disease, 84*, 1251.

Whittaker, S. L., Assinder, S. J., & Shaw, D. S. (1994). Inheritance of mitochondrial DNA in *Phytophthora infestans*. *Mycological Research, 98*, 569–575.

Eur J Plant Pathol (2008) 122:41–55
DOI 10.1007/s10658-008-9313-2

Proteomic analysis of a compatible interaction between *Pisum sativum* (pea) and the downy mildew pathogen *Peronospora viciae*

R. C. Amey · T. Schleicher · J. Slinn · M. Lewis · H. Macdonald · S. J. Neill · P. T. N. Spencer-Phillips

Received: 11 October 2007 / Accepted: 27 March 2008 / Published online: 9 May 2008

Abstract A proteomic approach was used to identify host proteins altering in abundance during *Peronospora viciae* infection of a susceptible cultivar of pea (*Pisum sativum* cv. Livioletta). Proteins were extracted from fully developed pea leaflets at 4 days post-inoculation, before visible symptoms were apparent. Cytoplasmic proteins and membrane- and nucleic acid-associated proteins from infected and control leaves were examined using two-dimensional difference gel electrophoresis. The majority of proteins had a similar abundance in control and infected leaves; however, several proteins were altered in abundance and twelve were found to have increased significantly in the latter. These proteins were selected for either matrix-assisted laser desorption/ionisation time-of-flight mass spectrometry or electro-spray ionisation quadrupole time-of-flight tandem mass spectrometry analysis following trypsin digestion, with sequence identity being assigned to eight of the proteins. These included the ABR17 stress-response protein, the pathogen-induced PI176 protein, three photosynthetic proteins, a glycine-rich RNA binding protein and two glyceraldehyde 3-phosphate dehydrogenases (cytosolic and chloroplastic) which can be induced by a range of abiotic and biotic stresses in many plant species. The possible roles of these proteins in the response of the pea plant during *P. viciae* infection are discussed. This study represents the first proteomic analysis of downy mildew infection of pea leaves, and provides the basis for further work to elucidate molecular mechanisms of compatibility in *P. viciae* infections.

Keywords Electrophoresis · DIGE · MALDI-TOF · Mass spectrometry · Oomycete · Protein

Abbreviations

2-D DIGE	two-dimensional difference gel electrophoresis
dpi	days post-inoculation
ESI Q-TOF MS/MS	electro-spray ionisation quadrupole time-of-flight tandem mass spectrometry
GAPDH	glyceraldehyde 3-phosphate dehydrogenase
MALDI-TOF MS	matrix-assisted laser desorption/ionisation time-of-flight mass spectrometry

R. C. Amey · T. Schleicher · J. Slinn · M. Lewis · H. Macdonald · S. J. Neill · P. T. N. Spencer-Phillips (✉)
Centre for Research in Plant Science,
Faculty of Health and Life Sciences,
University of the West of England,
Coldharbour Lane,
Bristol BS16 1QY, UK
e-mail: peter.spencer-phillips@uwe.ac.uk

Introduction

Downy mildew is the most common foliar disease of the pea crop (*Pisum sativum*) in the UK, with up to 55% losses in yield observed where plant resistance is ineffective (Clark and Spencer-Phillips 2000). Downy mildew is also a significant problem in other parts of the world where peas are grown (Amey and Spencer-Phillips 2006).

Production of conidia by *Peronospora viciae* results in a substantial loss of photosynthate from the host to the pathogen, contributing to symptoms such as stunted growth, distortion and early death of the infected plant (Mence and Pegg 1971). This re-direction of photosynthates and the other effects are likely to be accompanied by changes in the abundance of certain host proteins. Proteomics not only has the potential to identify these proteins, but also to provide quantitative data which signify their relative importance to the process.

Proteomic technologies such as 2D-DIGE (Ünlü et al. 1997), mass spectrometry and bioinformatics are an effective and accurate way of identifying and measuring protein differences between cell types (Beranova-Giorgianni 2003). Typically, the proteome of a control cell type or tissue is compared to a treated or diseased cell type or tissue. Protein differences observed between the two samples are investigated further to identify protein function and origin. 2D-DIGE and mass spectrometry are being used increasingly to identify proteins that increase or decrease in abundance during plant-microbe interactions (Corbett et al. 2005; Coulthurst et al. 2006). At present, little is known about mechanisms of pathogenesis in *P. viciae* infections of pea, with very few host and pathogen factors explained at a biochemical and molecular level (Clark and Spencer-Phillips 2004). Proteomics provides a global approach to explore changes in abundance of specific components of the pathosystem proteome, and hence to identify specific proteins and processes likely to be central to the outcome of infection.

To date, few proteomic studies of oomycete pathogens of plants have been performed. Most have focused on *Phytophthora* species such as *P. infestans*, a devastating pathogen of solanaceous plants (Ebstrup et al. 2005; Grenville-Briggs et al. 2005), *P. nicotianae* which has a wide host range (Mitchell et al. 2002) and *P. palmivora*, a serious pathogen of tropical crops including cocoa (Shepherd et al. 2003). The latter study examined the proteome of the asexual spores at various stages of development and germination, and identified a number of proteins that may be specific to different phases of the asexual life-cycle. The study by Mitchell et al. (2002) examined proteins from zoospores and cysts, whereas Grenville-Briggs et al. (2005) examined the proteomes of mycelium, zoospores and germinating cysts with appressoria, and Ebstrup et al. (2005) compared cysts, germinated cysts and appressoria. This work, however, only provided information on pathogen proteins during pre-invasion stages of infection. In contrast, Colditz et al. (2004) examined the proteome of the roots of the legume *Medicago truncatula* during infection by the oomycete *Aphanomyces euteiches*, with the majority of induced proteins belonging to the pathogenesis-related (PR)-10 group of pathogenesis-related (PR) proteins.

Proteomics has been applied previously to the study of pea proteins, including two host genotypes inoculated with the powdery mildew pathogen *Erysiphe pisi* (Curto et al. 2006). These authors compared the proteomes of *E. pisi*-infected and non-inoculated control leaves, and identified seven and 16 proteins with increased abundance following infection in resistant and susceptible interactions, respectively. The proteins functioned in photosynthesis, carbon metabolism, energy production, stress and defence, protein synthesis, and degradation, and signal transduction. Other published data relate to abiotic stresses such as salinity (Kav et al. 2004), biotic stresses such as infection by the parasitic plant *Orobanche crenata* (Castillejo et al. 2004) or a combination of interactions such as in *Glomus mosseae*-inoculated pea roots treated with calcium (Repetto et al. 2003). Numerous studies have examined the proteomes of other legumes. These include the symbiotic relationships of *Medicago* with *G. mosseae* (Bestel-Corre et al. 2002) and *Sinorhizobium meliloti* (Djordjevic et al. 2003), and soybean infected by the bacterium *Bradyrhizobium japonicum* to gain further information about the processes involved in nodulation (Wan et al. 2005).

The aim of the present work was to generate fundamental information on the most abundant proteins specifically involved in a compatible *P. viciae*–pea interaction, and to compare this with data on proteins in compatible and resistant interactions between pea and *E. pisi* (Curto et al. 2006).

Materials and methods

Inoculation of pea with *P. viciae*

P. viciae isolate Nitouche (kindly provided by Dr David Kenyon, National Institute of Agricultural

Botany, Cambridge, UK) was maintained on a mixture of seven pea cultivars (*P. sativum* cvs Livioletta, Kelvedon Wonder, Maro, Krupp Pelushka, Early Onward, Solara and Progreta). For protein extractions, plants of *P. sativum* cv. Livioletta were cultivated from seed in compost (Levington F2S) in growth chambers (Sanyo; 16 h light at 20°C, 8 h dark at 14°C). At 10 days after sowing, the surfaces of fully developed leaflets were rubbed gently to flatten waxes before being inoculated with *P. viciae* conidia by the method of El-Gariani and Spencer-Phillips (2004). Control plants were inoculated with sterile distilled water only.

Protein extraction and preparation

Proteins were extracted from fully developed leaflets of healthy pea plants and from *P. viciae*-infected plants at 4 dpi according to the method of Giavalisco et al. (2003). Their method is claimed to result in three fractions, comprising enriched preparations of (I) cytosolic proteins, (II) membrane-bound proteins and (III) nucleic acid-associated proteins. Protein fractions II and III were pooled for analysis in the present investigation. In brief, leaves were ground in liquid nitrogen before addition of 0.125 (*v/w*) inhibitor mixture 1 (100 mM KCl; 20% *v/v* glycerol; 50 mM Tris, pH 7.1), including Complete Protease Inhibitor Cocktail Tablet (Roche, Germany) used according to the manufacturer's instructions, and 0.05 (*w/w*) of inhibitor mixture 2 (1 mM Pepstatin A, 1.4 mM PMSF). Samples were centrifuged for 60 min at 22,000 g at 4°C. The supernatant containing the soluble cytosolic protein (fraction I) was removed and stored at −80°C. The pellet was ground further in liquid nitrogen before addition of 0.125 (*v/w*) of inhibitor mixture 3 (200 mM KCl; 20% *v/v* glycerol; 100 mM phosphate buffer, pH 7.1; Complete Protease Inhibitor Cocktail, as before), one volume of buffer A (100 mM phosphate buffer, pH 7.1; 200 mM KCl; 20% *v/v* glycerol; 2 mM MgSO4; 4% *w/v* CHAPS) and 2% (*w/w*) ASB14 detergent (Calbiochem, UK). Samples were homogenised thoroughly before the addition of 0.025% (*v/w*) DNase and the resulting mix stirred at 4°C for 45 min. Subsequently, 23% *v/w* buffer B (700 mM 1,4-dithiothreitol (DTT), 7 M urea, 2 M thiourea) was added and the homogenate was stirred at room temperature for 45 min. The homogenate was then centrifuged for 60 min at 22,000 g at 17°C and the resulting supernatant containing membrane and nucleic acid-associated proteins (fractions II and III) was removed and stored at −80°C.

The proteins were prepared for 2-D DIGE using the 2D Clean-up kit (Amersham Biosciences), re-suspended in lysis buffer (30°mM Tris pH 8.5; 4% *w/v* 3-[(3-cholamidopropyl) dimethylammonio]-1-propane-sulphonate) (CHAPS); 7 M urea; 2 M thiourea) and quantified using the 2D Quant kit (Amersham Biosciences) according to the manufacturer's instructions. Protein samples were stored at −80°C until further analysis.

CyDye labelling

Samples were labelled using fluorescent cyanine dyes (Amersham Biosciences) according to the manufacturer's protocols. The cyanine dyes were reconstituted in fresh 99.8% anhydrous dimethyl formamide. Aliquots of 50 μg of protein were labelled with 400 pmol of amine reactive CyDye for 30 min on ice in the dark, then 1 μl of 10 μM lysine was added to the tube and incubated on ice in the dark to halt the reaction. The samples were made up to 100 μl with rehydration solution (8 M urea; 2% *w/v* CHAPS; 0.002% *w/v* bromophenol blue; 0.2% *w/v* DTT; 2% *w/v* immobilised pH gradient (IPG) buffer (pH 3–10, Amersham Biosciences)).

2-D DIGE

Samples were subjected to isoelectric focusing (IEF) using IPG strips (24 cm, Amersham Biosciences) in the pH 3–10 non-linear range, with rehydration loading to separate proteins in the first dimension according to isoelectric point. The IPG strips were rehydrated overnight at room temperature in the protein sample made up to 450 μl with rehydration solution and covered with mineral oil. The strips were transferred to an Ettan IPGphor II (Amersham Biosciences) and IEF was performed with a 50 μA limit/IPG strip. IEF voltage conditions were 300 V step and hold for 3 h, 1,000 V gradient for 6 h, 8,000 V gradient for 3 h and 8,000 V step for 4 h 40 min.

Sodium dodecyl sulfate (SDS) polyacrylamide gel electrophoresis was used to separate proteins in the second dimension according to molecular weight. Following focusing in the first dimension, each strip was removed from the IEF unit and equilibrated in 15 ml equilibration buffer (50 mM Tris pH 8.8; 6 M urea; 30% *v*/*v* glycerol; 2% *w*/*v* SDS; 0.002% *w*/*v* bromophenol blue) amended with 150 mg DTT with gentle shaking for 15 min at room temperature. The strips were equilibrated further in 15 ml equilibration buffer amended with 375 mg iodoacetamide with gentle shaking for 15 min at room temperature in the dark. Finally, the strips were equilibrated in 10 ml equilibration buffer alone for 5 min at room temperature. The strips were loaded onto a 12.5% acrylamide gel (dimensions 24 cm×20 cm×1 mm) and overlaid with 1% agarose in SDS running buffer (25 mM Tris pH 8.3; 192 mM glycine; 0.1% SDS) amended with 0.002% (*w*/*v*) bromophenol blue. The gels were electrophoresed in SDS buffer at 2.5 W per gel for 30 min, followed by 100 W until the bromophenol blue dye front had run off the bottom of the gels. A minimum of three biological samples was used in these experiments, with two replicate gels produced for fraction I and the combined fractions II plus III of the extracted proteins from each sample. A minimum of three additional, non-CyDye-labelled gels were run for each sample for protein spot picking.

Image analysis

Gels were scanned on a Typhoon 9400 imager (Amersham Biosciences) to visualise CyDye-labelled proteins. Cy3 scans were obtained using a 532 nm laser and emission filter of 580 nm BP30. Cy5 scans were obtained using a 633 nm laser and a 670 nm BP30 emission filter. Scans were performed at 100 μm resolution with the photomultiplier tube voltage set for a maximum pixel intensity of 60 to 80,000 pixels. All images were cropped using ImageQuant V5.2 software prior to analysis to remove areas outside the gel. Analysis of each of the gels was performed with DeCyder Differential In-gel Analysis module software (V5.0; Amersham Biosciences) using the double detection setting and an estimated protein spot number of 2,500. Parameters for an exclusion filter were determined and applied according to the manufacturer's instructions, with resulting spots confirmed individually by visual inspection. Protein spots altering in abundance by at least two-fold consistently on all gels, and one protein that remained unchanged, were selected for analysis.

MALDI-TOF MS and ESI Q-TOF MS/MS analyses

Protein spots were excised from the gels using an Ettan Spot Picker (Amersham Biosciences) and subsequently digested using an Ettan Digester (Amersham Biosciences) with 10 μl trypsin (20 ng μl^{-1}; Promega Sequencing Grade Porcine Modified) in 20 mM ammonium bicarbonate (Sigma) overnight at room temperature. Following tryptic digestion, the peptides were extracted in 50% acetonitrile/0.1% trifluoroacetic acid to a clean microtitre plate and transferred to an Ettan Spotter (Amersham Biosciences). The peptides were mixed with matrix (10 mg ml^{-1} α-cyano-4-hydroxycinnamic acid in 50:50 *v*/*v* methanol/acetonitrile) for spotting onto Micromass target plates for analysis in a MALDI-TOF mass spectrometer (Waters-Micromass, UK). Peptide mixtures were analysed using a nitrogen UV laser (337 nm). MS data were acquired in the MALDI reflector positive ion mode in the mass range 800–3,500 Da. Identification of proteins from the mass fingerprints generated was performed using Proteinlynx Global Server software (V2.0.5, Waters-Micromass, UK) for searching against the SwissProt and National Centre for Biotechnology Information (NCBI, Bethesda, USA) databases. Search parameters included a peptide mass tolerance of 100 ppm, estimated calibration error of +0.025 Da, one missed cleavage per peptide, fixed carbamidomethylation of cysteine, and variable oxidation of methionine.

Nanoelectrospray ionization tandem mass spectra were acquired using a Q-TOF Micro mass spectrometer (Waters-Micromass, UK) coupled to a LC Packings capillary liquid chromatography system. Aliquots (15 μl) of peptide solutions prepared as before were injected using an auxiliary solvent flow of 30 μl min^{-1} and desalted on a C_{18} PepMap Nano-Precolumn (5× 0.3 mm internal diam (i.d.), 5 μm particle size; Dionex, Amsterdam, The Netherlands) for 4 min. Peptides were eluted and separated using a C_{18} PepMap100 nano column (15 cm×75 μm i.d., 3 μm particle size) with a gradient flow of 200 nl min^{-1} and solvent system of: auxiliary solvent, 0.1% HCOOH; solvent A, 5% *v*/*v* CH_3CN/95% *v*/*v* 0.1% *v*/*v* aqueous HCOOH; solvent B, 80% *v*/*v* CH_3CN/20% *v*/*v* 0.1% *v*/*v* aqueous HCOOH. The solvent gradient was 4 min at 5% aqueous solvent B, 5% to 55% B over 40 min, 55% to 80% B over 1 min, maintained at 80% B for 5 min, then reduced to 5% B in 0.1 min and the column washed with solvent A for 9.9 min before the next sample injection. The column was connected to the nanosprayer of the Z-spray ion source using a short length of 75 μm i.d. capillary. Voltages used were 3,500 V for the capillary, 45 V for the sample cone and 2.5 V for the extraction cone. MS spectra were acquired throughout the chromatographic run, while MS/MS spectra were acquired in data-dependent mode on the most abundant ions having charge states of 2+, 3+ and 4+ between *m/z* 400–2,000. The collision cell was pressurised with 1.38 bar ultra-pure argon (99.999%, BOC) and collision voltages depended on the m/z and charge states of the parent ions. The mass spectrometer was calibrated daily using MS/MS fragment ions from [Glu^1]-fibrinopeptide B (Sigma). Processed data were submitted to ProteinLynx Global Server (V2.0.5) and also to MASCOT (Matrix Science) for searching against SwissProt and NCBI databases. Search criteria were: peptide tolerance of 100 ppm; fragment tolerance of 0.1 Da; two missed cleavages per peptide; fixed carbamidomethylation of cysteine and variable oxidation of methionine modifications.

Results

The proteins from pea plants inoculated with *P. viciae* and sterile distilled water (SDW) controls were visualised using 2D-DIGE, with each gel comprising two experimental samples labelled with Cy3 (control) and Cy5 (*P. viciae*-infected). DeCyder software detected between 977 and 1337 protein spots on two representative gel images (Table 1). The total number of spots detected on the gels varied by 19.7% between the two replicates of fraction I (enriched cytosolic soluble proteins), and by 26% between the replicates of combined fractions II plus III (enriched membrane-associated and nucleic acid-associated proteins). The proportion of proteins with decreased abundance following *P. viciae* infection was 1.7% and 3.5% for the replicates of fraction I, and 0.38% and 0.92% for the replicates of fractions II plus III. The proportion of proteins with increased abundance was 5.2% and 7.2% for fraction I, and 2.3% and 4.5% for fraction II plus III. Therefore the proteins for MALDI-TOF MS and ESI Q-TOF MS/MS were selected on the basis that their abundance altered significantly and reproducibly on all gel replicates of the different biological samples (two CyDye-labelled replicates plus a minimum of three additional unlabelled replicate gels) thus removing potential biological and gel artefacts from the analysis. This resulted in 12 proteins that increased in abundance during *P. viciae* infection as indicated in Fig. 1, a representative gel from this experiment. In contrast, no proteins decreased in abundance significantly (greater than two-fold) and consistently following infection. An additional protein (spot number 6) is indicated on the gel as an example of a protein that does not alter in abundance upon infection by *P. viciae*. The relative fold abundance of protein spots from infected versus SDW-inoculated leaves determined by DeCyder software, and selected individual spot images, are illustrated in Figs. 2 and 3. The molecular weight, pI, matched peptides, sequence coverage and score (either Proteinlynx Global Server or MASCOT) of each protein is indicated in Table 2.

Table 1 The relative abundance of proteins detected in 2D gels for fractions I (cytosolic proteins) and fractions II + III (enriched in membrane and nucleic acid-associated proteins) from pea leaves 4 dpi with conidia of *P. viciae*, compared to SDW, in two separate experiments (a and b)

Fraction	Relative abundance of proteins (no. of spots detected)			
	Decreased	Similar	Increased	Total
Ia	22	1,245	70	1,337
Ib	37	960	77	1,074
II + IIIa	5	1,285	30	1,320
II + IIIb	9	924	44	977

Protein 1 (Figs. 1 and 2) was identified as the disease resistance response protein PI176 from pea (accession number P13239). Compared to control plants, its abundance increased by 3.4 and 6.6-fold in fractions I and fractions II plus III respectively. The protein has similar predicted and observed molecular weights and iso-electric points. The three peptides matching published sequences of PI176 represented 16.4% of its amino acid sequence. Protein 2 was identified as abscisic acid responsive protein ABR17 from pea (accession number Q06931). The abundance of this protein resembled that of PI176, with increases of 2.9 and 5.8-fold for fractions I and II plus III respectively. Theoretical and observed values for molecular weight and pI are in accord and the seven matched peptides provided 51% coverage of the amino acid sequence. Protein 3 was matched following de novo sequencing by ESI Q-TOF MS/MS to a glycine-rich RNA binding protein from *Sinapis alba*

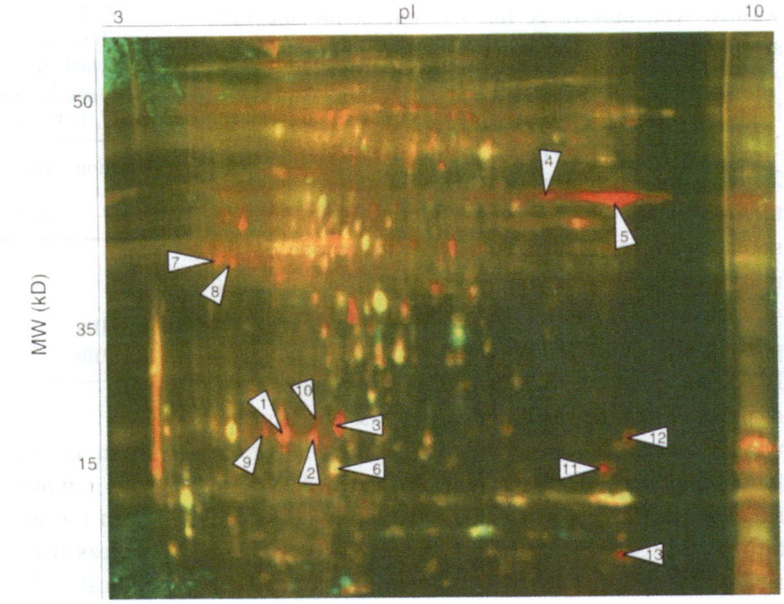

Fig. 1 A typical 2D-DIGE gel obtained from analysis of the pea leaf proteome during early stage infection (4 dpi) by *P. viciae*, compared to proteins from control leaves treated with SDW, for cytosolic fraction I. *Spot colour* indicates the effect of *P. viciae* on protein abundance: *red* = increased; *green* = decreased; *yellow* = no change. Proteins that increased in abundance reproducibly on all gels are indicated by the numbers *1–13*, except for *spot 6* which represents a protein showing no change in abundance during infection by *P. viciae* (see Fig. 2)

(accession number P49311) and, compared to controls, had increased abundances of 6.4 and 6.5-fold for fraction I and II plus III respectively. The theoretical and observed molecular weight and pI values agreed, and the single peptide covered 4.7% of the amino acid sequence. Protein 4 matched to cytosolic GAPDH from pea (accession number P34922). In comparison to control plants, the abundance of the protein increased by 6.1-fold in fraction I and by 6.3-fold in fractions II plus III. Predicted and observed molecular weight and pI values were similar, with the nine matched peptides covering 34.6% of the amino acid sequence. Protein 5 was identified as a chloroplastic precursor of GAPDH A from pea (accession number P12858). Compared to control samples, the protein increased in abundance by 7.7-fold for fraction I and 8.6-fold for fractions II plus III. Whilst 15 peptides were matched, covering 44.7% of the amino acid sequence, and the predicted and observed pI values were in accord, the observed molecular weight was significantly less than predicted. This suggests that processing of the precursor

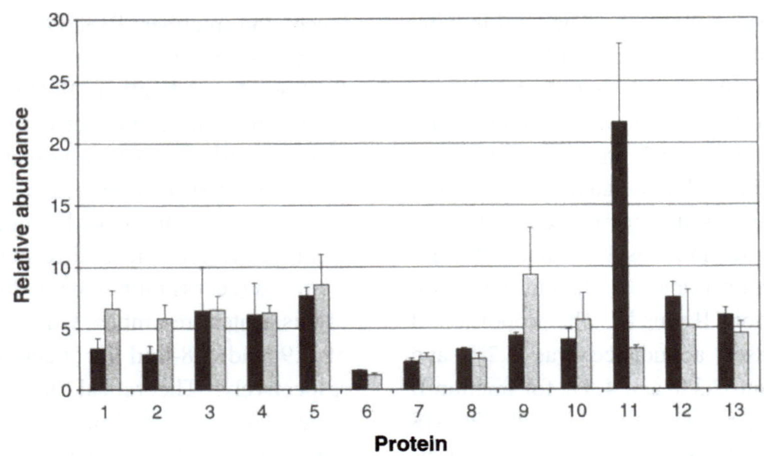

Fig. 2 Relative abundance of proteins *1–13* that increase in pea leaves at 4 dpi after inoculation with *P. viciae*, compared to proteins from control leaves treated with SDW. Protein abundances for fraction I (*black bar*, cytosolic proteins) and fractions II plus III (*gray bar*, enriched in membrane and nucleic acid-associated proteins) were calculated using DeCyder Software (Amersham Biosciences) from two replicate experiments. Standard errors are indicated

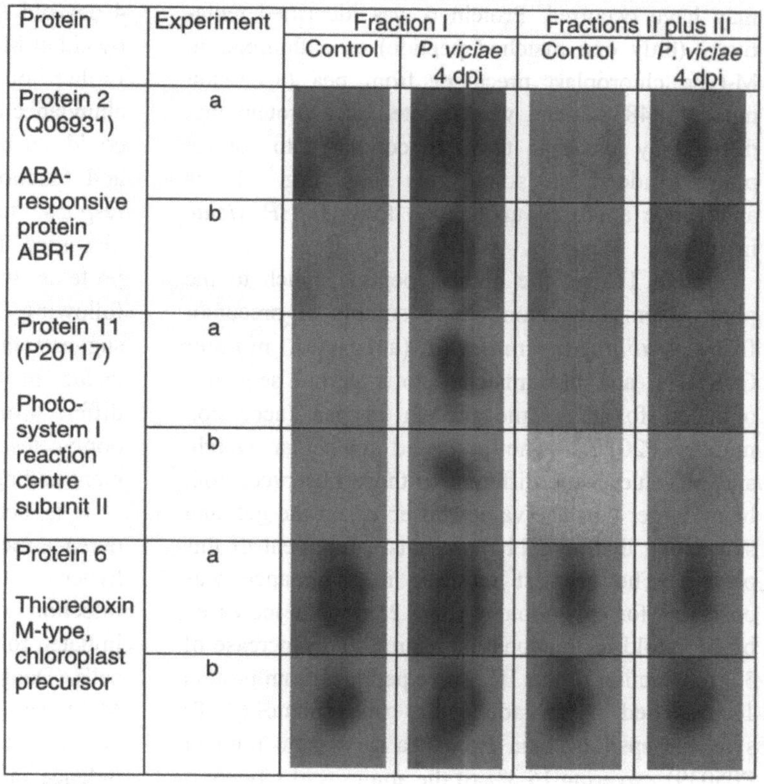

Fig. 3 Images of representative protein spots on 2D gels, from two replicate experiments (*a* and *b*). *Protein 2* (*ABR17*) increases in abundance during *P. viciae* infection of pea leaves in fractions I and fractions II plus III; *protein 11* (*photosystem I reaction centre subunit II*) increases to a greater extent in fraction I than fractions II plus III; *protein 6* (tentatively identified as *thioredoxin M-type, chloroplast precursor*) does not differ in abundance between control (SDW) and infected pea leaves

Table 2 Proteins identified using MALDI-TOF MS and ESI Q-TOF MS/MS that differ in abundance in response to infection by *P. viciae* at 4 dpi

Spot number	Matching protein	Protein accession no.	Observed/predicted mW (kDa)	Observed/ predicted pI	Matched peptides	Sequence coverage	Score
1	PI176	P13239	17.0/16.9	5.1/5.1	3	16.4	11.2[b]
2	ABR17	Q06931	16.9/16.6	5.2/5.1	7	51.0	9.1[b]
3	Glycine-rich RNA binding protein	P49311	17.0/16.4	5.5/5.5	1[a]	4.7[a]	NA
4	Cytosolic GAPDH	P34922	36.0/36.6	7.0/7.0	9	34.6	11.6[b]
5	Chloroplastic GAPDH	P12858	36.0/43.3	8.5/9.0	15	44.7	11.9[b]
6	Thioredoxin M-type precursor	P48384	14.0/12.5	5.4/5.4	1	5.2	11.2[b]
11	Photosystem I reaction centre subunit II	Q9S7H1	14.0/23.1	8.0/9.8	3	10.1	121[c]
12	ATP synthase epsilon chain	P05039	15.0/15.2	8.6/6.6	3	15.3	99[c]
13	Photosystem I iron sulphur centre	P10793	9.0/9.2	8.5/7.5	3	17.3	89[c]

All proteins increased in abundance by more than two-fold (see Fig. 2), except protein 6 which was essentially unchanged between treatments. Spot numbers relate to Fig. 1; all accession species were *P. sativum*, except *S. alba* for spot 3 and *A. thaliana* for spot 11

NA Not applicable

[a] Protein matched by de novo sequencing using ESI Q-TOF MS/MS

[b] Proteinlynx Global Server

[c] MASCOT

may have occurred. Protein 6 was identified tentatively (only one matched peptide) as a thioredoxin M-type chloroplast precursor from pea (accession number P48384) and was selected as a protein that differed by less than two-fold compared to control plants. Indeed, in some gels (e.g. Fig. 1), its abundance appeared unchanged following *P. viciae* infection.

Protein 11 had the highest peptide match to the photosystem I reaction centre subunit II precursor from *Arabidopsis thaliana* (accession number Q9S7H1), and also matched to a partial sequence obtained for the same protein in pea (accession number P20117). The predicted molecular weight and pI values were different to those observed, both being larger than the values observed on the gel, and suggesting that spot 11 represents a fragment of this protein. The greatest increase in abundance was observed for this protein (Fig. 2), which increased by 21.7-fold in fraction I compared to an increase of 3.3 for fraction II plus III. Three peptides from protein 12 matched to an adenosine triphosphate (ATP) synthase epsilon chain from pea (accession number P05039), covering 15.3% of the amino acid sequence. The molecular weight for the protein observed on the gel matched the predicted, but the pI differed. Protein 13 matched to the photosystem I iron sulphur centre from pea (accession number P10793). Three peptides were matched, covering 17.3% of the amino acid sequence, with predicted pI and molecular weight values matching those observed.

The abundances of the unidentified proteins 7 and 10 (Figs. 1 and 2) increased similarly in both fractions. The abundance of protein 8 was greater in the soluble fraction I than the membrane and nucleic acid-associated proteins of fraction II plus III, having a 3.3 and 2.5-fold increase in abundance compared to control samples, respectively. In fraction II plus III, protein 9 increased in abundance by more than twice that observed in fraction I (9.3 and 4.3-fold respectively). Thus differences between the relative abundance of proteins isolated from the two fractions were apparent for four of the 13 proteins (Fig. 2).

Discussion

Eight proteins whose abundance was observed to have increased consistently by at least two-fold in 4 day-old *P. viciae* infections of pea were identified by either MALDI-TOF MS or ESI Q-TOF MS/MS. A further four proteins were observed to increase in abundance consistently during *P. viciae* infection, yet could not be identified using MS. The possible roles and functions of the identified proteins during the response to infection by biotrophic pathogens are discussed. It is notable, however, that none of these proteins were reported to increase in abundance following *E. pisi* infection (Curto et al. 2006), which suggests that they reflect a specific response to *P. viciae* in this compatible interaction. This study differs from previous proteomic investigations of oomycetes in that it examines the proteome of the plant-pathogen interaction in leaves.

It is believed widely that the majority of plant defence mechanisms, such as basal resistance and the hypersensitive response, are induced early during infection via a complex network of signals that is initiated following perception of the pathogen by host cells (Dangl and Jones 2001; Kamoun et al. 1999b; McDowell and Dangl 2000). The identification of proteins with potential and proven roles in plant defence in plants harvested at 4 dpi would correlate with these concepts. Proteins were harvested at this stage as well-developed colonies are present, even though external symptoms are not apparent (El-Gariani and Spencer-Phillips 2004). Indeed, conidiophore initials are first observed at 4.5 dpi in this host-pathogen system (Clark and Spencer-Phillips 2004), with sporulation visible macroscopically by 7 dpi. Further proteomic analyses of these later stages of infection are underway to identify additional proteins involved in the *P. viciae*-pea interaction.

The difference in numbers of proteins identified as increased or decreased in abundance between the two fractions implies that a slightly larger number of cytosolic proteins increase in abundance than in a fraction enriched in membrane plus nuclear-associated proteins (6.1% and 3.2%, respectively). However, it should be noted that the same proteins were identified in both fraction types, indicating that the method of extraction may not be effective at specifically selecting cytosolic, membrane-associated and nucleic acid-associated proteins for pea samples. Whilst it may be more appropriate to pool all fractions in further studies to facilitate comparison of proteomes, some proteins (e.g. protein 11) showed significant differential increases between the fractions.

Of the 12 proteins observed to have abundances significantly and consistently increased during the *P. viciae* interaction with pea, eight were matched to pea or other plant proteins in the SWISS-PROT database. The observed and theoretical isoelectric points and molecular weights of the proteins were mostly in accord, except for proteins 5 and 11 where the predicted molecular weight values were much greater than those observed. For protein 11, this may reflect the match to the *A. thaliana* protein rather than the pea protein which has not been sequenced fully. Additionally, this protein was identified as photosystem I reaction centre subunit II (Table 2) and therefore should be membrane bound, but it was mostly present in fraction I which Giavalisco et al. (2003) suggests should contain soluble proteins. Together with the discrepancy in the predicted and observed molecular weights, this suggests that a protein fragment has been identified. The match of protein 5 to a precursor of chloroplastic glyceraldehyde 3-phosphate dehydrogenase (GAPDH) may indicate that the processed protein was present here.

Even in transcriptional profiling studies, it is often the case that not every gene that alters in expression during a specific interaction can be identified from databases, as shown in the yeast *Saccharomyces cerevisiae* (Gygi et al. 1999) and in *M. truncatula* roots following infection by the oomycete *A. euteiches* (Colditz et al. 2004; Nyamsuren et al. 2003). Indeed, further studies are needed to identify the four unidentified proteins, as they are likely to play significant roles in this plant-pathogen interaction. Identification would provide information about probable function and thereby help elucidate the molecular mechanisms of pathogenicity and host response in downy mildew infections.

Three of the proteins with increased abundance are likely to be involved in initiating and maintaining a defence response by pea during *P. viciae* infection. Disease resistance response protein PI176 was identified originally by Fristensky et al. (1988), and is a member of the pathogenesis-related class 10 (PR10) family of proteins which are found exclusively in plants. This set of proteins generally is induced by abiotic and biotic factors such as wounding (Liu et al. 2003; Warner et al. 1992, 1993), salt stress (Moons et al. 1997), pathogenic infections (Fristensky et al. 1988; Liu et al. 2003; Matton and Brisson 1989; McGee et al. 2001; Pinto and Ricardo 1995; Schmelzer et al. 1989; Somssich et al. 1988), drought (Dubos and Plomion 2001), chemicals such as copper (Utriainen et al. 1998) and plant hormones (Moons et al. 1997; Wang et al. 1999a). PI176 is a variant of the pea PI49 protein, differing in four amino acid substitutions and one amino acid deletion (Fristensky et al. 1988). There are conflicting views on the potential biological action of the PR10 proteins. Some studies suggest they have RNase activity (Bantignies et al. 2000; Moiseyev et al. 1994; Park et al. 2004; Swoboda et al. 1996), with others (Biesiadka et al. 2002) suggesting that this is not the case, as the crystal structure of some PR10 proteins such as LIPR10.1A and LIPR10.1B from *Lupinus luteus* does not support RNase activity. Additionally, Biesiadka et al. (2002) noted that the *L. luteus* PR10 proteins studied have little or no RNase activity. This discrepancy in the literature may reflect the large number of PR10 protein homologues within plants. For example, 13 different PR10 cDNA clones have been identified in *Pinus monticola* (Liu et al. 2003) and, although some redundancy is likely to exist in the activity of these proteins, it is also possible that some of the proteins have significantly different functions within plants. Similar proteins in bean (*Phaseolus vulgaris*) are thought to act intracellularly as no signal peptides have been identified for the proteins (von Heijne 1985; Walter et al. 1990), whilst transcripts of the parsley (*Petroselinum crispum*) *PcPR1* gene accumulate rapidly and to a large extent in cells adjacent to the site of pathogen infection (Schmelzer et al. 1989; Somssich et al. 1988). From the literature and the protein abundance pattern of PI176 observed in this study, it appears that PI176 is important in the intracellular response of pea to infection by *P. viciae* in this compatible interaction, and further analysis of this family of proteins is merited.

Abscisic acid response protein 17 (ABR17) is an additional member of the pathogenesis-related class 10 (PR10) family and has a similar pattern of abundance to PI176. This has also been observed for *PR10* genes in other pea-microbe interactions, such as infection by the arbuscular-mycorrhizal fungus *G. mosseae* where transcript levels of the *PI49* and *PI176* genes increase in pea roots up to 15 dpi (Ruiz-Lozano et al. 1999), and for their proteins in *A. euteiches* infection of *M. truncatula* (Colditz et al. 2004). The gene for ABR17 is closely related to

ABR18 having significant similarities at the DNA and amino acid levels, a trait shared by abscisic acid-inducible proteins from other plant species such as alfalfa (*Medicago sativa*; Luo et al. 1992) and the barrel medic (*M. truncatula*; Colditz et al. 2004). Abscisic acid is thought to play a key role in mediating adaptive plant responses to environmental stress, plant development, seed dormancy and germination, as well as plant defence (Luo et al. 1992; Moons et al. 1995). It is thought that the ABR proteins in pea increase in abundance in response to both abscisic acid and environmental cues, with the present study indicating a role also in response to pathogens.

The increased abundance of both ABR and PI proteins would indicate that ABA-mediated signalling is important in the *P. viciae*–pea interaction. This notion concurs with Nyamsuren et al. (2003) who observed increased transcripts of these genes in *A. euteiches* infections of *M. truncatula*, a close relative of pea. Although ABA has been shown to play a role in signalling, Colditz et al. (2004) showed that of six PR10 proteins identified in *M. truncatula*, only three increased in abundance in response to ABA, and none altered in abundance in response to drought stress. Therefore, three of the PR10 proteins increased in abundance in response to *A. euteiches* alone, for the limited range of stresses assessed. It would seem that some proteins may play a specific role in response to biotic stress such as pathogen invasion, perhaps providing opportunities for the development of pathogen-tolerant crop species. Further studies should compare both plant-microbe interactions, especially as *A. euteiches* is also a serious pathogen of pea crops (Pfender 1989). This may determine whether the induction of the *ABA*-responsive genes is a general plant response to oomycete pathogens, resulting from increased ABA content as a side-effect of infection due to senescence, reduced water availability and cell death, or perhaps a combination of several factors.

Glycine-rich proteins (GRPs) have been implicated in numerous roles in plants. In pea, GRPs have been associated with dormancy and have similarity to proteins that are stimulated by auxin and numerous abiotic stresses (Luo et al. 1991; Laberge et al. 1993; Stafstrom et al. 1998). Structurally related GRPs are often components of the cell walls of higher plants (Showalter 1993), and accumulate in vascular tissues as part of the defence mechanism against pathogens and wounding (Mousavi and Hotta 2005). A second class of GRPs, the glycine-rich RNA-binding proteins (GR-RBP) such as protein 3 identified tentatively in the present study, are thought to play an important role in post-transcriptional regulation of gene expression. Some evidence exists for altered transcript and/or protein abundance of GR-RBPs in response to virus infection (Geri et al. 1999; Naqvi et al. 1998), acute hypersensitive response and salicylic acid treatment (Naqvi et al. 1998), abscisic acid treatment (Aneeta Sanan-Mishra et al. 2002; Baudo et al. 1999; Bergeron et al. 1993; Carpenter et al. 1994; Gomez et al. 1988; Kim et al. 2005) and methyl-jasmonate treatment (Richard et al. 1999), thus providing evidence of a role in the response of plants to pathogens. There are no previous reports regarding the response of these proteins to attack by oomycete pathogens, and the tentative identification in the present study indicates that further investigation is needed.

Cytosolic GAPDH (protein 4 in the present study) is one of three forms of GAPDH in plants, and is involved in the second phase of glycolysis, catalysing the conversion of D-glyceraldehyde 3-phosphate into 3-phospho-D-glyceroyl phosphate. Evidence for GAPDH having a role in defence is provided by Laxalt et al. (1996) who noted that cytosolic GAPDH transcripts accumulated in potato plants infected by *P. infestans*, and treated with both the *P. infestans* elicitor eicosapentaenoic acid and salicylic acid. Interestingly, although GAPDH transcripts increased to their highest levels at 2 dpi in potato, it was not until 3 dpi that a corresponding increase in enzyme activity was observed. This would indicate that there is a substantial time-lag between induction of genes involved in the pathogen response and the production of their corresponding proteins. It has been suggested that these proteins be given the term 'stress-induced metabolic response' proteins (Laxalt et al. 1996). The time-lag also suggests that the selection of 4 dpi as the time point for extracting proteins in the present study is suitable for investigating host-pathogen interactions in plants capable of adjusting their metabolism to survive various stresses in this short time period.

The possibility that the increase in GAPDH abundance is due simply to an increase in metabolic turnover of the protein, and not a specific stress response, should not be overlooked. The abundance of this protein and the other proteins, both identified and unidentified, in other plant parts such as roots, stems and tendrils of pea is also worth investigating. Elevated levels of plant defence proteins in roots, for example, would help prevent downy mildew infection via soil-borne oospores. Investigations are in progress to establish whether the increase in GAPDH abundance is also evident in pea in response to abiotic stresses such as wounding and water-deficit.

The discovery that levels of a chloroplastic GAPDH A were also elevated in response to *P. viciae* infection (protein 5) is surprising, as to our knowledge there is no previous evidence for this in other biotrophic infections. It is likely that chloroplastic GAPDH plays a similar role to cytoplasmic GAPDH in that an overall increased metabolic state is induced during infection by *P. viciae*. Alternatively, GAPDH may have a role in plant signalling pathways, with increasing evidence indicating that plant enzymes may be multi-functional (Moore 2004). For example, GAPDH has been demonstrated to be an inhibitory target of hydrogen peroxide in *A. thaliana* with potential roles in mediating reactive oxygen species signalling (Hancock et al. 2005). Interestingly, mRNA and enzyme levels of GAPDH are also induced by abscisic acid treatment (Velasco et al. 1994), indicating that ABA signalling is important in the plant response to pathogens, especially as ABR17 protein levels also increased following *P. viciae* infection. GAPDH is also associated with the cell wall in *A. thaliana* (Chivasa et al. 2002), and found in the peribacteroid membrane of root nodules in *Lotus japonicus* (Wienkoop and Saalbach 2003). The specific function of GAPDH remains to be elucidated, and it seems that proteins involved in making adjustments to plant metabolism during stress are under-studied.

Three proteins increasing in abundance in response to *P. viciae* infection are involved in photosynthesis. Two are integral to photosystem I, the reaction centre subunit and the iron sulphur centre, whilst the third is involved in the synthesis of ATP (ATP synthase epsilon chain). Photosystem I is a chloroplastic, multimolecular complex that uses ferredoxin-like iron-sulphur cluster proteins as terminal electron acceptors. The ATP synthase epsilon chain is a small sub-unit of the chloroplast ATP synthase, which produces ATP from ADP in the presence of a proton gradient across the chloroplast thylakoid membrane (Hopkins and Hüner 2004). Whilst most studies indicate that infections by plant pathogens result in a reduced photosynthetic rate, transient increases in the rate of photosynthesis during infection by biotrophic pathogens have been observed (reviewed in Scholes 1992), which may reflect the increase in proteins of the photosynthetic apparatus observed in this study. In the biotrophic pathogen-plant interactions where photosynthetic rate initially increased during infection, an overall reduction in photosynthesis was eventually observed, when compared to control plants. An analysis of the proteins observed to alter in abundance at later stages of infection by *P. viciae*, along with experimental data on photosynthetic rates, would reveal if a similar pattern occurs. The hypothesis that a fragment of the photosystem I reaction centre subunit II has been identified in the present study, however, suggests that the intact protein has been degraded as a result of infection. This would be expected to cause a decrease, rather than an increase, in photosynthetic activity. The possibility that photosynthetic rates in adjacent, uninfected tissues and cells may increase to compensate for a reduction in colonised areas also merits investigation. These effects have been observed in *Botrytis fabae* infections of *Vicia faba* leaves, uninfected leaves on *P. vulgaris* plants infected by the rust fungus *Uromyces phaseoli*, and uninfected leaves of pea plants inoculated with powdery mildew spores (Lucas 1998).

Thus it would appear that pea leaves are capable of implementing a response to *P. viciae* infection through the up-regulation of various pathogen-related and metabolic pathway proteins. Whether these proteins act in concert rather than individually is not known, although previous studies have shown that plants synthesise a number of proteins in response to stresses that may work individually and/or in conjunction with other proteins (Luo et al. 1992).

It is noticeable that no *P. viciae* proteins were identified in this study, which may be because there is very little database coverage of DNA and proteins from this pathogen. This was also the case with the proteomic study of *A. euteiches* infection of *M. truncatula* (Colditz et al. 2004), where no *A. euteiches* proteins were found. Reasons given included the restricted amount of pathogen-infected root cells and the restricted growth of *A. euteiches* in host cells. The difficulty in identifying oomycete proteins was illustrated by Shepherd et al. (2003), who identified just three *P. palmivora* proteins from 2D gels loaded with proteins extracted from hyphae grown in vitro, all of which corresponded to actin. Indeed it can be difficult to identify genes specifically expressed during infection by oomycetes (Beyer et al. 2002 and references therein). Despite improved techniques, up to 70% of gene sequences cannot be identified using genomic studies (Beyer et al. 2002), and house-keeping genes are more easily identified than those that may be involved in infection. Proteomic approaches will always tend to be more successful if supported by extensive and complete genome databases. *P. infestans, Phytophthora sojae* and *Hyaloperonospora parasitica* genome projects are currently underway or are near completion (Beynon, personal communication; http://www.pfgd.org/; Kamoun et al. 1999a; Qutob et al. 2000) which may enhance the success of proteomic analyses of oomycetes. Indeed, the peptide mass fingerprints obtained for the unidentified proteins in this study may yield matches when used to search these new data sets, and also any proteins identified from analysis of the *P. viciae* conidial proteome (Chuisseu Wandji et al. 2007).

To our knowledge, the data presented here represent the first application of 2-D DIGE to investigate a plant–oomycete pathogen interaction, and indicate the sensitivity and accuracy of this method for identifying proteins that are important to the plant's response to infection. An increased understanding of these mechanisms may accelerate progress in developing novel control strategies for downy mildew diseases. Indeed, the DRR49 protein from pea has been used to increase potato tuber production in the presence of pathogen-infested soil (Chang et al. 1993), and to increase levels of resistance to *Leptosphaeria maculans* in canola (Wang et al. 1999b). It is therefore possible that the proteins identified in the current study may be of use in the development of novel plant lines with enhanced resistance to biotic and abiotic stresses.

Acknowledgments This work was funded by DEFRA grant HH3216SFV.

References

Amey, R. C., & Spencer-Phillips, P. T. N. (2006). Towards developing diagnostics for downy mildew diseases. *Outlooks on Pest Management, 17*, 4–8.

Aneeta Sanan-Mishra, N., Tuteja, N., & Sopory, S. K. (2002). Salinity- and ABA-induced up-regulation and light-mediated modulation of mRNA encoding glycine-rich RNA-binding protein from *Sorghum bicolor. Biochemical and Biophysical Research Communications, 296*, 1063–1068.

Bantignies, B., Seguin, J., Muzac, I., Dedaldechamp, F., Gulick, P., & Ibrahim, R. (2000). Direct evidence of ribonucleolytic activity of a PR-10-like protein from white lupin roots. *Plant Molecular Biology, 42*, 871–881.

Baudo, M. M., Meza-Zepeda, L. A., Palva, E. T., & Heino, P. (1999). Isolation of a cDNA corresponding to a low temperature- and ABA-responsive gene encoding a putative glycine-rich RNA-binding protein from *Solanum commersonii. Journal of Experimental Botany, 50*, 1867–1868.

Beranova-Giorgianni, S. (2003). Proteome analysis by two-dimensional gel electrophoresis and mass spectrometry: strengths and limitations. *Trends in Analytical Chemistry, 22*, 273–281.

Bergeron, D., Beauseigle, D., & Bellemare, G. (1993). Sequence and expression of a gene encoding a protein with RNA-binding and glycine-rich domains in *Brassica napus. Biochimica et Biophysica Acta, 1216*, 123–125.

Bestel-Corre, G., Dumas-Gaudot, E., Poinsot, V., Dieu, M., Dierick, J. F., van Tuinen, D., et al. (2002). Proteome analysis and identification of symbiosis-related proteins from *Medicago truncatula* Gaertn. by two-dimensional electrophoresis and mass spectrometry. *Electrophoresis, 23*, 122–137.

Beyer, K., Jiménez, S., Randall, T. A., Lam, S., Binder, A., Boller, T., et al. (2002). Characterization of *Phytophthora infestans* genes regulated during the interaction with potato. *Molecular Plant Pathology, 3*, 473–485.

Biesiadka, J., Bujacz, G., Sikorski, M. M., & Jaskolski, M. (2002). Crystal structures of two homologous pathogenesis-related proteins from yellow lupine. *Journal of Molecular Biology, 319*, 1223–1234.

Carpenter, C. D., Kreps, J. A., & Simon, A. E. (1994). Genes encoding glycine-rich *Arabidopsis thaliana* proteins with RNA-binding motifs are influenced by cold treatment and an endogenous circadian rhythm. *Plant Physiology, 104*, 1015–1025.

Castillejo, M. Á., Amiour, N., Dumas-Gaudot, E., Rubiales, D., & Jorrín, J. V. (2004). A proteomic approach to studying plant response to crenate broomrape (*Orobanche crenata*) in pea (*Pisum sativum*). *Phytochemistry, 65*, 1817–1828.

Chang, M. M., Chiang, C. C., Martin, M. W., & Hadwiger, L. A. (1993). Expression of a pea disease resistance response gene in the potato cultivar Shepody. *American Potato Journal, 70*, 635–647.

Chivasa, S., Ndimba, B. K., Simon, W. J., Robertson, D., Yu, X. L., Knox, J. P., et al. (2002). Proteomic analysis of the *Arabidopsis thaliana* cell wall. *Electrophoresis, 23*, 1754–1765.

Chuisseu Wandji, J. L., Amey, R. C., Butt, E., Harrison, J., Macdonald, H., & Spencer-Phillips, P. T. N. (2007). Towards proteomic analysis of *Peronospora viciae* conidiospores. In A. Lebeda, & P. T. N. Spencer-Phillips (Eds.), *Advances in Downy Mildew Research* (vol. 3, pp. 95–100). Kostelec na Hane (Czech Republic): Palacky University in Olomouc and JOLA.

Clark, J. S. C., & Spencer-Phillips, P. T. N. (2000). Downy Mildews. In J. Lederberg, M. Alexander, B. R. Bloom, D. Hopwood, R. Hull, B. H. Inglearski, A. I. Laskia, S. G. Oliver, M. Schaechter, & W. C. Summers (Eds.), *Ency-*

clopaedia of microbiology (vol. 2, pp. 117–129). San Diego: Academic.

Clark, J. S. C., & Spencer-Phillips, P. T. N. (2004). The compatible interaction in downy mildew infections. In P. T. N. Spencer-Phillips, & M. J. Jeger (Eds.), *Advances in downy mildew research* (vol. 2, pp. 1–34). Dordrecht: Kluwer Academic.

Colditz, F., Nyamsuren, O., Niehaus, K., Eubel, H., Braun, H-P., & Krajinski, F. (2004). Proteomic approach: identification of *Medicago truncatula* proteins induced in roots after infection with the pathogenic oomycete *Aphanomyces euteiches*. *Plant Molecular Biology, 55*, 109–120.

Corbett, M., Virtue, S., Bell, K., Birch, P., Burr, T., Hyman, L., et al. (2005). Identification of a new quorum-sensing-control led virulence factor in *Erwinia carotovora* subsp. *atroseptica* secreted via the type II targeting pathway. *Molecular Plant-Microbe Interactions, 18*, 334–342.

Coulthurst, S. J., Lilley, K. S., & Salmond, G. P. C. (2006). Genetic and proteomic analysis of the role of *luxS* in the enteric phytopathogen, *Erwinia carotovora*. *Molecular Plant Pathology, 7*, 31–45.

Curto, M., Camafeita, E., Lopez, J. A., Maldonado, A. M., Rubiales, D., & Jorrín, J. V. (2006). A proteomic approach to study pea (*Pisum sativum*) responses to powdery mildew (*Erysiphe pisi*). *Proteomics, 6*, S163–S174.

Dangl, J. L., & Jones, J. D. G. (2001). Plant pathogens and integrated defence responses to infection. *Nature, 411*, 826–833.

Djordjevic, M. A., Chen, H. C., Netera, S., Van Noorden, G., Menzel, C., Taylor, S., et al. (2003). A global analysis of protein expression profiles in *Sinorhizobium meliloti*: discovery of new genes for nodule occupancy and stress adaptation. *Molecular Plant-Microbe Interactions, 16*, 508–524.

Dubos, C., & Plomion, C. (2001). Drought differentially affects expression of a PR-10 protein in needles of maritime pine (*Pinus pinaster* Ait.) seedlings. *Journal of Experimental Botany, 52*, 1143–1144.

Ebstrup, T., Saalbach, G., & Egsgaard, H. (2005). A proteomics study of *in vitro* cyst germination and appressoria formation in *Phytophthora infestans*. *Proteomics, 5*, 2839–2848.

El-Gariani, N. K., & Spencer-Phillips, P. T. N. (2004). Isolation of viable *Peronospora viciae* hyphae from infected *Pisum sativum* leaves and accumulation of nutrients *in vitro*. In P. T. N. Spencer-Phillips, & M. J. Jeger (Eds.), *Advances in downy mildew research: Volume 2* (pp. 249–264). Dordrecht: Kluwer Academic.

Fristensky, B., Horovitz, D., & Hadwiger, L. A. (1988). cDNA sequences for pea disease resistance response genes. *Plant Molecular Biology, 11*, 713–715.

Geri, C., Cecchini, E., Giannakou, M. E., Covey, S. N., & Milner, J. J. (1999). Altered patterns of gene expression in *Arabidopsis* elicited by *Cauliflower mosaic virus* (CaMV) infection and by a CaMV gene VI transgene. *Molecular Plant–Microbe Interactions, 12*, 377–384.

Giavalisco, P., Nordhoff, E., Lehrach, H., Gobom, J., & Klose, J. (2003). Extraction of proteins from plant tissues for two-dimensional electrophoresis analysis. *Electrophoresis, 24*, 207–216.

Gomez, J., Sanchez-Martinez, D., Stiefel, V., Rigau, J., Puigdomènech, P., & Pagès, M. (1988). A gene induced by the plant hormone abscisic acid in response to water stress encodes a glycine-rich protein. *Nature, 344*, 262–264.

Grenville-Briggs, L. J., Avrova, A. O., Bruce, C. R., Williams, A., Whisson, S. C., Birch, P. R. J., et al. (2005). Elevated amino acid biosynthesis in *Phytophthora infestans* during appressorium formation and potato infection. *Fungal Genetics and Biology, 42*, 244–256.

Gygi, S. P., Rochon, Y., Franza, B. R., & Abersold, R. (1999). Correlation between protein and mRNA abundance in yeast. *Molecular Cell Biology, 19*, 1720–1730.

Hancock, J. T., Henson, D., Nyirenda, M., Desikan, R., Harrison, J., Lewis, M., et al. (2005). Proteomic identification of glyceraldehyde 3-phosphate dehydrogenase as an inhibitory target of hydrogen peroxide in *Arabidopsis*. *Plant Physiology and Biochemistry, 43*, 828–835.

Hopkins, W. G., & Hüner, N. P. A. (2004). *Introduction to plant physiology*. New York: Wiley.

Kamoun, S., Hraber, P., Sobral, B., Nuss, D., & Govers, F. (1999a). Initial assessment of gene diversity for the oomycete pathogen *Phytophthora infestans* based on expressed sequences. *Fungal Genetics and Biology, 28*, 94–106.

Kamoun, S., Huitema, E., & Vleeshouwers, V. G. A. A. (1999b). Resistance to oomycetes: a general role for the hypersensitive response. *Trends in Plant Science, 4*, 196–200.

Kav, N. N. V., Srivastava, S., Goonewardene, L., & Blade, S. F. (2004). Proteome-level changes in the roots of *Pisum sativum* in response to salinity. *Annals of Applied Biology, 145*, 217–230.

Kim, Y-O., Kim, J. S., & Kang, H. (2005). Cold-inducible zinc finger-containing glycine-rich RNA-binding protein contributes to the enhancement of freezing tolerance in *Arabidopsis thaliana*. *Plant Journal, 42*, 890–900.

Laberge, S., Castonguay, Y., & Vezina, L. P. (1993). New cold- and drought-regulated gene from *Medicago sativa*. *Plant Physiology, 101*, 1411–1412.

Laxalt, A. M., Cassia, R. O., Sanllorenti, P. M., Madrid, E. A., Andreu, A. B., Daleo, G. R., et al. (1996). Accumulation of cytosolic glyceraldehyde-3-phosphate dehydrogenase RNA under biological stress conditions and elicitor treatments in potato. *Plant Molecular Biology, 30*, 961–972.

Liu, J. J., Ekramoddoullah, A. K. M., & Yu, X. S. (2003). Differential expression of multiple PR10 proteins in western white pine following wounding, fungal infection and cold-hardening. *Physiologia Plantarum, 119*, 544–553.

Lucas, J. A. (1998). *Plant pathology and plant pathogens*. Oxford: Blackwell.

Luo, M., Lin, L., Hill, R. D., & Mohapatra, S. S. (1991). Primary structure of an environmental stress and abscisic acid-inducible alfalfa protein. *Plant Molecular Biology, 17*, 1267–1269.

Luo, M., Liu, J-H., Mohapatra, S., Hill, R. D., & Mohapatra, S. S. (1992). Characterization of a gene family encoding abscisic acid- and environmental stress-inducible proteins of alfalfa. *Journal of Biological Chemistry, 267*, 15367–15374.

Matton, D. P., & Brisson, N. (1989). Cloning, expression, and sequence conservation of pathogenesis-related gene transcripts of potato. *Molecular Plant–Microbe Interactions, 2*, 325–331.

McDowell, J. M., & Dangl, J. L. (2000). Signal transduction in the plant immune response. *Trends in Biochemical Science, 25*, 79–82.

McGee, J. D., Hamer, J. E., & Hodges, T. K. (2001). Characterization of a PR-10 pathogenesis-related gene family induced in rice during infection with *Magnaporthe grisea. Molecular Plant–Microbe Interactions, 14*, 877–886.

Mence, M. J., & Pegg, G. F. (1971). The biology of *Peronospora viciae* on pea: factors affecting the susceptibility of plants to local infection and systemic colonisation. *Annals of Applied Biology, 67*, 297–308.

Mitchell, H. J., Kovac, K. A., & Hardham, A. R. (2002). Characterisation of *Phytophthora nicotianae* zoospore and cyst membrane proteins. *Mycological Research, 106*, 1211–1223.

Moiseyev, G. P., Beintema, J. J., Fedoreyeva, L. I., & Yakovlev, G. I. (1994). High sequence similarity between a ribonuclease from ginseng calluses and fungus-elicited proteins from parsley indicates that intracellular pathogenesis-related proteins are ribonucleases. *Planta, 193*, 470–472.

Moons, A., Bauw, G., Prinsen, E., van Montagu, M., & van der Straeten, D. (1995). Molecular and physiological responses to abscisic acid and salts in roots of salt-sensitive and salt-tolerant Indica rice varieties. *Plant Physiology, 107*, 177–186.

Moons, A., Prinsen, E., Bauw, G., & Van Montagu, M. (1997). Antagonistic effects of abscisic acid and jasmonates on salt-stress inducible transcripts in rice roots. *Plant Cell, 9*, 2243–2259.

Moore, B. D. (2004). Bifunctional and moonlighting enzymes: lighting the way to regulatory control. *Trends in Plant Science, 9*, 221–228.

Mousavi, A., & Hotta, Y. (2005). Glycine-rich proteins – a class of novel proteins. *Applied Biochemistry and Biotechnology, 120*, 169–174.

Naqvi, S. M. S., Park, K.-S., Yi, S.-Y., Lee, H.-W., Bok, S. H., & Choi, D. (1998). A glycine-rich RNA-binding protein gene is differentially expressed during acute hypersensitive response following *Tobacco mosaic virus* infection in tobacco. *Plant Molecular Biology, 37*, 571–576.

Nyamsuren, O., Colditz, F., Rosendahl, S., Tamasloukht, M., Bekel, T., Meyer, F., et al. (2003). Transcriptional profiling of *Medicago truncatula* roots after infection with *Aphanomyces euteiches* (oomycota) identifies novel genes upregulated during this pathogenic interaction. *Physiological and Molecular Plant Pathology, 63*, 17–26.

Park, C.-J., Kim, K.-J., Shin, R., Park, J. M., Shin, Y.-C., & Paek, K.-H. (2004). Pathogenesis-related protein 10 isolated from hot pepper functions as a ribonuclease in an antiviral pathway. *Plant Journal, 37*, 186–198.

Pinto, P. M., & Ricardo, C. P. P. (1995). *Lupinus albus* L. pathogenesis-related proteins that show similarity to PR10 proteins. *Plant Physiology, 109*, 1345–1351.

Pfender, W. F. (1989). *Aphanomyces* root rot. In D. J. Hagedorn (Ed.), *Compendium of pea diseases* (pp. 25–28). St. Paul: American Phytopathological Society.

Qutob, D., Hraber, P. T., Sobral, B. W. S., & Gijzen, M. (2000). Comparative analysis of expressed sequences in *Phytophthora sojae. Plant Physiology, 123*, 243–253.

Repetto, O., Bestel-Corre, G., Dumus-Gaudot, B. G., Gianinazzi-Pearson, V., & Gianinazzi, S. (2003). Targeted proteomics to identify calcium-induced protein modifications in *Glomus mossae*-inoculated pea roots. *New Phytologist, 157*, 555–567.

Richard, S., Drevet, C., Jouanin, L., & Séguin, A. (1999). Isolation and characterization of a cDNA clone encoding a putative white spruce glycine-rich RNA binding protein. *Gene, 240*, 379–388.

Ruiz-Lozano, J. M., Roussel, H., Gianinazzi, S., & Gianinazzi-Pearson, V. (1999). Defense genes are differentially induced by a mycorrhizal fungus and *Rhizobium* sp. in wild-type and symbiosis-defective pea genotypes. *Molecular Plant–Microbe Interactions, 12*, 976–984.

Schmelzer, E., Kruger-Lebus, S., & Hahlbrok, K. (1989). Temporal and spatial patterns of gene expression around sites of attempted fungal infection in parsley leaves. *Plant Cell, 1*, 993–1001.

Scholes, J. D. (1992). Photosynthesis: cellular and tissue aspects in diseased leaves. In P. G. Ayres (Ed.), *Pests and pathogens: plant responses to foliar attack* (pp. 85–106). Oxford: BIOS Scientific.

Shepherd, S. J., van West, P., & Gow, N. A. R. (2003). Proteomic analysis of asexual development of *Phytophthora palmivora. Mycological Research, 107*, 395–400.

Showalter, A. M. (1993). Structure and function of plant cell wall proteins. *Plant Cell, 5*, 9–23.

Somssich, I. E., Schmelzer, E., Kawalleck, P., & Hahlbrook, K. (1988). Gene structure and in situ transcript localization of pathogenesis-related protein 1 in parsley. *Molecular and General Genetics, 213*, 93–98.

Stafstrom, J. P., Ripley, B. D., Devitt, M. L., & Drake, B. (1998). Dormancy-associated gene expression in pea axillary buds. *Planta, 205*, 547–552.

Swoboda, I., Hoffmann-Sommergruber, K., O'Ríordáin, G., Scheiner, O., Heberle-Bors, E., & Vicente, O. (1996). Bet v1 proteins, the major birch pollen allergens and members of a family of conserved pathogenesis-related proteins, show ribonuclease activity in vitro. *Physiologia Plantarum, 96*, 433–438.

Ünlü, M., Morgan, M. E., & Minden, J. S. (1997). Difference gel electrophoresis: a single gel method for detecting changes in protein extracts. *Electrophoresis, 18*, 2071–2077.

Utriainen, M., Kokko, H., Auriola, S., Sarrazin, O., & Karenlampi, S. (1998). PR-10 protein is induced by copper stress in roots and leaves of a Cu/Zn tolerant clone of birch, *Betula pendula. Plant Cell and Environment, 21*, 821–828.

Velasco, R., Salamini, F., & Bartels, D. (1994). Dehydration and ABA increase mRNA levels and enzyme activity of cytosolic GAPDH in the resurrection plant *Craterostigma plantineum. Plant Molecular Biology, 26*, 541–546.

von Heijne, G. (1985). Signal sequences. The limits of variation. *Journal of Molecular Biology, 184*, 99–105.

Walter, M. H., Liu, J.-W., Grand, C., Lamb, C. J., & Hess, D. (1990). Bean pathogenesis-related (PR) proteins deduced from elicitor-induced transcripts are members of a ubiquitous new class of conserved PR proteins including pollen allergens. *Molecular and General Genetics*, *222*, 353–360.

Wan, J., Torres, M., Ganapathy, A., Thelen, J., DaGue, B. B., Mooney, B., et al. (2005). Proteomic analysis of soybean root hairs after infection by *Bradyrhizobium japonicum*. *Molecular Plant–Microbe Interactions*, *18*, 458–467.

Wang, C. S., Huang, J. C., & Hu, J. H. (1999a). Characterization of two subclasses of PR-10 transcripts in lily anthers and induction of their genes through separate signal transduction pathways. *Plant Molecular Biology*, *40*, 807–814.

Wang, Y. P., Nowak, G., Culley, D., Hadwiger, L. A., & Fristensky, B. (1999b). Constitutive expression of pea defense gene DRR206 confers resistance to blackleg (*Leptosphaeria maculans*) disease in transgenic canola (*Brassica napus*). *Molecular Plant–Microbe Interactions*, *12*, 410–418.

Warner, S. A. J., Scott, R., & Draper, J. (1992). Characterisation of a wound-induced transcript from the monocot asparagus that shares similarity with a class of intracellular pathogenesis-related PR10 proteins. *Plant Molecular Biology*, *19*, 555–561.

Warner, S. A. J., Scott, R., & Draper, J. (1993). Isolation of an asparagus intracellular PR gene (AoPR1) wound-responsive promoter by the inverse polymerase chain reaction and its characterization in transgenic tobacco. *Plant Journal*, *3*, 191–201.

Wienkoop, S., & Saalbach, G. (2003). Proteome analysis. Novel protein identified at the peribacteroid membrane from *Lotus japonicus* root nodules. *Plant Physiology*, *131*, 1080–1090.

Eur J Plant Pathol (2008) 122:57–69
DOI 10.1007/s10658-008-9288-z

Tapping into molecular conversation between oomycete plant pathogens and their hosts

Mahmut Tör

Received: 20 November 2007 / Accepted: 31 January 2008

Abstract Several plant pathogenic oomycetes have been under investigation using modern molecular approaches. Genome sequencing and annotations are underway or near to completion for some of the species. Pathogen-associated molecular pattern molecules (PAMPs) and effector molecules perform inter- and intracellular tasks as adaptation factors and manipulators of the defence network. Hundreds of secreted putative effectors have been discovered and conserved molecular patterns such as RXLR and EER motifs have been identified and used for classifications. PAMPs and effectors are recognized directly or indirectly by the pattern recognition receptors at the cell surface including receptor-like kinases and receptor-like proteins, and/or by nucleotide binding site–leucine rich repeat proteins within the cytoplasm. The current knowledge of effectors, immune receptors and the defence network, will help us understand the 'intricate genetic dance' between the oomycete pathogens and their hosts. This review concentrates on the recent findings in oomycete-plant interactions.

Keywords PAMPs · Effector · Oomycete · Receptor · Biotrophy

M. Tör (✉)
Warwick HRI, University of Warwick,
Wellesbourne,
Warwick CV35 9EF, UK
e-mail: Mahmut.tor@warwick.ac.uk

Introduction

The oomycetes include a unique group of biotrophic and hemibiotrophic plant pathogens including *Plasmopara viticola* (grapevine downy mildew), *Albugo candida* (white rust), *Bremia lactucae* (lettuce downy mildew), *Hyaloperonospora arabidopsis* (downy mildew on *Arabidopsis*, formerly *Hyaloperonospora parasitica*; Göker et al. 2004) and *Phytophthora infestans* (potato and tomato late blight; Kamoun 2003; Hardham 2007). These pathogens establish intimate relations with their hosts by forming haustoria during the infection, which are well known structures used for obtaining nutrients from the plant, redirecting host metabolism and suppressing host defence in biotrophy (Hahn and Mendgen 2001; Voegele and Mendgen 2003; O'Connell and Panstruga 2006). *Bremia lactucae*, *P. viticola* or *P. infestans* have a significant economic importance in agriculture (Agrios 1997). *Hyaloperonospora arabidopsis* in *Arabidopsis thaliana* has been developed as an important model system to study plant–microbe interactions (Holub 2008). In addition, *A. candida* provides an alternative model on *Arabidopsis*; however, it has not been fully explored despite interesting characteristics such as the suppression of *R*-gene mediated and non-host resistance mechanisms to allow the growth of a second parasite like *H. arabidopsis* and causing a hormonal imbalance, which induces 'green island' formation (Cooper et al. 2002; Holub 2006, 2008).

Understanding the mechanisms of microbial pathogenesis and plant–microbe interactions has motivated plant pathologists for a long time. In nature, plants are generally resistant to most pathogens due to their innate ability to recognize pathogen-derived molecules and to mount a series of carefully orchestrated and highly evolved defence responses. Much of the progress in the field of molecular plant–pathogen interactions has been led by the research on prokaryotic bacterial plant pathogens. This may have been due to the fact that bacterial pathogens have the ability to secrete effectors including avirulence proteins with their type III secretion system into the cytoplasm of their host plant cell (Van den Ackerveken et al. 1996; Mudgett and Staskawicz 1998; Chisholm et al 2005). However, in the last few years, significant progress has also been made in the understanding of interactions between eukaryotic pathogens, including oomycetes, ascomycetes and basidiomycetes, and their host plants (Allen et al. 2004; Catanzariti et al. 2006; Shen et al. 2007). Results from these studies led to the establishment of a general consensus on plant-microbe interactions, which is that; (a) pathogens have pathogen associated molecular pattern molecules (PAMPs) and effector molecules that modulate the immune system (Kamoun 2006; Lotze et al. 2007); (b) the plant innate immune system is a collection of subsystems that carry out distinct functions in the host's defence; (c) the cell surface receptors or pattern recognition receptors (PRRs) and cytoplasmic receptors or nucleotide binding site-leucine rich repeat (NB-LRR) proteins play a significant role in the detection of these PAMPs and effectors (Chisholm et al. 2006), and (d) effector molecules are virulence factors and have the ability to suppress the immune system of the plant (Bos et al. 2006; Jones and Dangl 2006; Fig 1).

Recent genomic studies on the oomycete pathogens including *Phytophthora sojae, Phytophthora ramorum*, *P. infestans* and *H. arabidopsis* (Win et al. 2007; Whisson et al. 2007) have revealed hundreds of hypothetical effectors that are secreted into the apoplast or the cytoplasm of the host plants. This review focuses on the contributions of recent findings to the understanding of oomycete pathogenesis with emphasis on PAMPs, effectors and receptors rather than repeating the existing reviews on gene-for-gene interactions.

PAMPs play a significant role in pathogenesis and trigger the innate immune system

These molecules were originally described as microbial elicitors and could be present in pathogenic and non-pathogenic microorganisms. They are unique to microbes, invariant among the given class of microorganisms and seen as foreign molecules by plants. They are important for microbial fitness and are able to elicit innate immune responses in a non-cultivar specific manner. Their conserved nature makes it difficult for the pathogen to avoid recognition through adaptive evolution of these molecules (Hahn 1996; Medzhitov and Janeway 2002; Ingle et al. 2006; Medzhitov 2007).

Several bacterial PAMPs including flagellin and Ef-Tu have been identified and studied in detail (Felix et al. 1999; Zipfel et al. 2006). The majority of PAMP studies on oomycete pathogens have been carried out with *Phytophthora* species. For example, studies on *P. parasitica* var. *nicotianae* have identified the cell wall elicitor protein cellulose binding elicitor lectin (CBEL), which enables the pathogen to attach to the host cell and contains two cellulose-binding domains 1 and 2. When recombinant CBEL is expressed in *Escherichia coli* and the protein injected into tobacco (*Nicotiana tabacum*) leaves, activation of the defense gene expression and formation of necrotic lesions have been observed. In addition, CBEL production *in planta* induced necrosis and synthetic peptides derived from CBEL activated the defence response in tobacco and *A. thaliana* leaves, indicating that these molecules have the necessary molecular patterns to be recognized by the innate immune system of plants (Gaulin et al. 2002, 2006).

Another molecule that has the characteristics of a PAMP is the β-glucan obtained from the cell wall of the soybean oomycete pathogen *P. sojae* (formerly known as *Phytophthora megasperma* f. sp. *glycinea*, Sharp et al. 1984). Treatment of the soybean cell suspension cultures with β-glucan has been shown to induce defence reactions including an increase in the cytosolic calcium concentration, the production of reactive oxygen species (ROS), and the activation of genes such as two mitogen-activated protein kinases (MAPKs) and one MAPK kinase, which play a role in signal transduction (Mithofer et al. 2001; Yamamizo et al. 2006; Daxberger et al. 2007).

Pep-13 is also a cell wall product from the soybean pathogen *P. sojae* and considered to be a PAMP. Initial

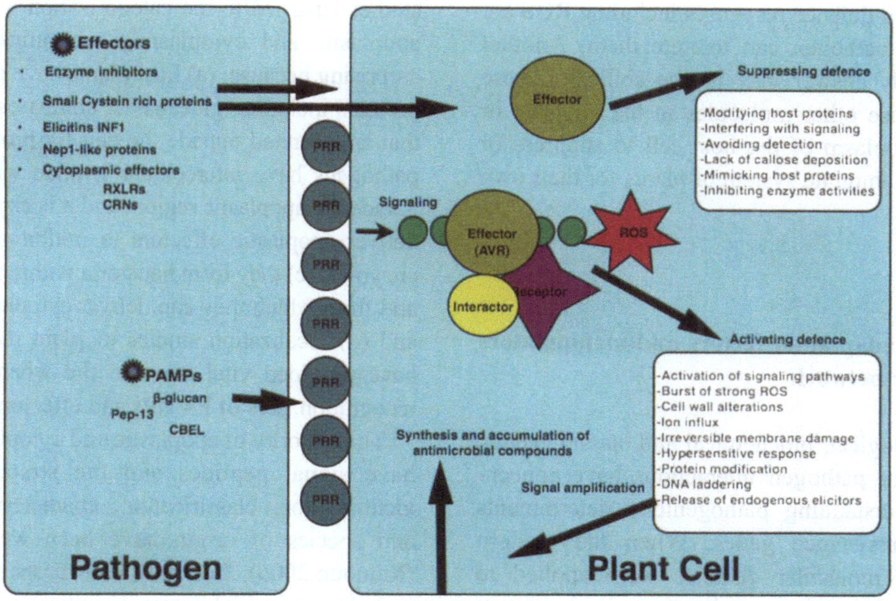

Fig. 1 Pathogen-associated molecular pattern molecules (*PAMPs*) and effector molecules help pathogen adapt to its niche. Oomycete pathogens including *P. infestans*, *B. lactucae*, *A. candida* and *H. arabidopsis* signal their presence with PAMPs such as β-glucan or Pep13. Recognition of these PAMPs by yet unidentified cell surface receptors (or pattern recognition receptors) activates a signalling cascade leading to innate immune responses including a small burst of reactive oxygen species (*ROS*). These pathogens have effector molecules, such as enzyme inhibitors, small cysteine rich proteins and RXLR type cytoplasmic effectors, which are encoded by the pathogen genome and are delivered into the apoplast or cytoplasm of the plant cell. These effectors are usually virulence factors and have the capability of suppressing the plant's immune response by modifying host proteins, interfering with signalling and inhibiting enzyme activities. However, similar to PAMPs, these effectors (which are then termed as AVR proteins) could also be recognized directly or indirectly by RLK/RLP type receptors at the cell surface or NB-LRR type receptors within the cytoplasm. Recognition of AVRs triggers a defence response including activation of signalling pathways, generation of a strong ROS, cell wall alterations, irreversible membrane damage, protein modifications, DNA laddering, hypersensitive response and the release of endogenous elicitors. These local defence responses are further amplified by secondary signalling. Antimicrobial compounds are synthesised within and at the neighbouring cells and accumulate at the infection sites. RLK, receptor-like kinase; RLP, receptor-like protein; NB-LRR, nucleotide binding site-leucine rich repeat proteins

studies with this molecule have been carried out with parsley cells, which are not normally a host for this pathogen. Activation of complex defence responses including ion influxes and effluxes, generation of oxidative burst, elevated expression of defense-related genes, and phytoalexin formation have been reported (Nurnberger et al. 1994). Later, Brunner et al. (2002) showed that the Pep-13 pattern is conserved in all *Phytophthora* species including *P. infestans* and forms part of the cell wall calcium-dependent transglutaminase (TGase) enzyme. However, it is still not clear what type of role TGase plays in the fitness of the pathogen. When cells of a susceptible host plant such as potato have been used in experiments with Pep-13, responses similar to those reported in parsley cells have been observed. Although, Brunner et al. (2002) reported the absence of TGase-related transcripts from other oomycete pathogens such as *H. arabidopsis* and *Pythium*, bioinformatic investigation into the recently available genome sequences of *H. arabidopsis* showed at least five copies of TGase elicitor precursor (M. Tör, unpublished).

One of the main reasons why PAMPs have been reported only from *Phytophthora* species could be the life-style of these pathogens, being easy to grow in axenic culture without contaminants from the host plants. As new and refined techniques are developed for the identification of new PAMPs, we would expect to see more from other oomycetes particularly the obligate biotrophs.

If the PAMPs are triggering a defence response in both a susceptible host and resistant non-host plants, the important question then arises as to why these defence responses are not sufficient to stop pathogen invasion. One possible explanation might be that the

PAMP-activated defence responses including ROS are weak and the pathogen can tolerate them. Another and more likely answer may be the ability of these pathogens to use other molecules in the apoplast or within the cytoplasm of the host cell to suppress or manipulate the immune system of plants for their own purpose.

Effectors are adaptation factors and manipulators of the defence network

Earlier physiological, biochemical and classic genetic studies in plant–pathogen interactions have concentrated on understanding pathogenicity determinants and disease resistance genes. When the modern techniques of molecular genetics were applied to analyse the pathogen, especially bacteria, important pathogenicity factors including strong attachment of bacteria to the host cell and hydrolytic enzymes, such as pectinases and cellulases, that facilitate pathogen invasion into host tissues, were identified. Studies on avirulence proteins in bacteria led to the discovery of the trafficking of effectors from the pathogen into host cells via the Type III secretion system. These molecules were found to bind to a protein and thereby alter the activity of that protein (Mudgett and Staskawicz 1998). This finding helped the establishment of a common link in the mechanisms of pathogenicity between plant and animal pathogens. It has also brought a change in our thinking. Rather than killing the host cell from outside, pathogens inject effector proteins as virulence factors into the host cell to adapt to a particular niche (Medzhitov 2007) and manipulate it for its own purpose (Xiao et al. 2007). When these effectors are somehow recognized by the cytoplasmic receptors, they are termed avirulence (AVR) proteins (Jones and Dangl 2006). Furthermore, it also promoted the question whether effector trafficking could also be observed in eukaryotes, such as oomycete or fungal pathogens (Birch et al. 2006; Ellis et al. 2006). The identification of the effector proteins, their function and their corresponding molecular targets in the host has been a challenge for the scientists working on oomycete pathogens.

In the past few years, genome sequencing and annotations, genome mapping and associated genetic studies have led to the categorization of effectors into apoplastic and cytoplasmic groupings. This is not surprising because: (a) host plants have defence-related proteins including glucanases, chitinases and proteases that are secreted outside the cell for protection; (b) the pathogens have intercellular hyphae with which they invade the apoplastic region, and it is expected that they deliver apoplastic effectors to inhibit or escape plant enzymes, (c) they form haustoria within individual cells and from which they can deliver cytoplasmic effectors, and (d) localization studies of plant receptor proteins have provided vital clues to the whereabouts of the recognition sites of PAMPs and effectors.

The majority of apoplastic and cytoplasmic effectors have signal peptides and the strategies for their identification, classification, characteristic properties and species of origin have been well documented (Kamoun 2006). Currently known apoplastic effectors include enzyme inhibitors, members of the NEP1-like protein family and small cysteine-rich proteins (Qutob et al. 2006; Kamoun 2006). We should also expect to find that some effectors are delivered to the apoplast but function within the cytoplasm after being brought into the cell through endocytosis or membrane trafficking.

Recently, cloning of four *Avr* genes, *Avr1b-1*, *ATR13* and *ATR1*^{NdWs} and *Avr3a* from three oomycetes, *P. sojae, H. arabidopsis* and *P. infestans*, respectively (Shan et al. 2004; Allen et al. 2004; Rehmany et al. 2005; Armstrong et al. 2005) has enabled the identification of common conserved regions including the N-terminal RXLR (for arginine (Arg), any amino acid, leucine (Leu), Arg) and EER (for glutamine (Glu), Glu, Arg.) motifs (Fig. 2). These motifs along with the signal peptide have been used in bioinformatic studies to analyse the available sequence data and identify a class of RXLR cytoplasmic effectors from the oomycete plant pathogens (Kamoun 2007; Morgan and Kamoun 2007). Bhattacharjee et al. (2006) were quick to explore the resemblances of these two motifs to that (RXLXE/D/Q) used for translocation of the malaria parasite (*Plasmodium*) into host erythrocytes (Hiller et al. 2004) and they demonstrated the function of the *P. infestans* RXLR motif in the *Plasmodium* system. In addition, Bos et al (2006) used the *P. infestans* Avr3a effector and demonstrated that the C-terminal-half activated the *R*-gene mediated resistance and suppressed INF1-induced cell death in tobacco (*Nicotiana benthamiana*) plants. These findings

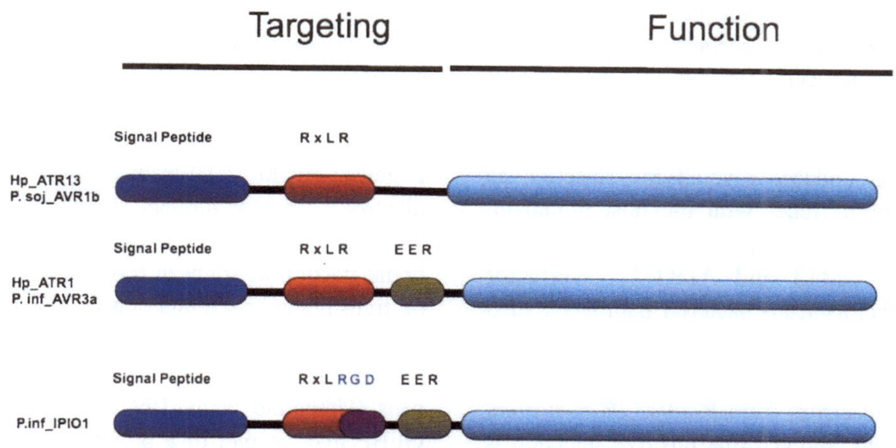

Fig. 2 Predicted structures of RXLR type oomycete effectors. RXLR family effectors from *P. infestans*, *P. sojae* and *H. arabidopsis* have targeting and functional domains. These proteins have a signal peptide and an RXLR motif (for Arg, any amino acid, Leu, Arg). Detailed analyses of the genome sequences of *P. sojae*, *P. ramorum*, and *H. arabidopsis* showed that the majority of these RXLR proteins have also an EER motif (Win et al. 2007). Some of the effectors with and without EER motif are shown. *Hyaloperonospora arabidopsis* ATR1NdWsB (311aa) and *P. infestans* AVR3a (147aa) have the EER motif, whereas *H. arabidopsis* ATR13 (187aa) and *P. sojae* AVR1b (138aa) do not. *Phytophthora infestans* IPI-O1 protein has the EER motif but also has the RGD motif that overlaps with the RXLR motif. In these effectors, the targeting domains are involved in secretion of the effectors out of the pathogens and translocation into the plant cell and are usually well conserved. The C-terminal is responsible for manipulating the host immune system

suggest that, similar to bacterial effectors, oomycete effectors also play a role in altering signalling in the defence network during their adaptation into the given niche. In addition, the N-terminal region of these effectors that contains the signal peptide, RXLR and EER motifs has been shown to be responsible for the delivery of the effectors in the elegant studies by Whisson et al. (2007). Using *Avr3a* from *P. infestans*, Whisson et al. (2007) fused the N-terminal region containing the RXLR and EER motifs, to the GUS protein and showed that the GUS protein could be delivered from the pathogen into the host cell. In addition, *P. infestans* failed to deliver the Avr3a or Avr3a-GUS fusion into the host cell when the RXLR and EER motifs were modified. However, silencing of *Avr3a* in *P. infestans* has not been carried out, which would be a key experiment to reveal whether this effector has a role in pathogenicity.

Although a great deal of information has been accumulating on these effectors, the method of their transmission from extracellular space into the host cytoplasm is still not clear. Morgan and Kamoun (2007) have proposed that RXLR binding proteins, chaperons or translocons, originating either from the pathogen or plant, may be required for the delivery of these effectors into the host cytoplasm.

Phytophthora infestans IPI-B and *IPI-O* genes are expressed at an early stage of the infection (Pieterse et al. 1994) and IPI-O1 belongs to the RXLR family of effector proteins. In addition to the signal peptide, RXLR and EER motif, IPI-O1 has an RGD tripeptide, which overlaps with the RXLR motif (Fig. 2). The RGD motif has been described as a cell adhesion motif found in several mammalian extracellular matrix proteins and has been proposed to reduce plant defence responses by disrupting adhesions between the cell wall and plasma membrane (Senchou et al. 2004). Furthermore, detailed studies showed that RGD-containing proteins could be ligands for some of the receptor-like kinase type cell surface receptors described below (Gouget et al. 2006).

Bacterial flagellin is a ligand for the receptor-like kinase (RLK)-type pattern recognition receptor flagellin-sensitive 2 (FLS2) and the ligand-stimulated receptor endocytosis is a kind of trafficking at the plasma membrane (Robatzek et al. 2006; Robatzek 2007). It can be proposed that at least some of the RXLR family effectors may be translocated into the extrahaustorial matrix with the help of RXLR and EER motifs and physically interact with cell surface receptors through the RGD motif as was shown with IPI-O1 (Gouget et al. 2006) and could be internalized by these receptors as was observed with FLS2.

Effectors often undergo diversifying selection as a result of an 'arms race' with the host organism. This has been well documented with studies on *H. arabidopsis* ATR13 and ATR1^{NdWsB} effector proteins and effectors from bacterial pathogens such as *Pseudomonas syringae* (Allen et al. 2004, Rehmany et al. 2005; Guttman et al. 2006). Recent work on RXLR effectors from *P. sojae, P. ramorum*, and *H. arabidopsis*, showed that positive selection is mainly on the C-terminal region, which is responsible for the function (Win et al. 2007).

Recognition and beyond

Cell biological research of oomycete–plant interactions has entered a new phase with the identification of hundreds of putative effector molecules. Since both PAMPs and effectors are foreign molecules to the host plant, we need to address several questions including; (a) whether the effectors from oomycete pathogens mimic the host plant protein as seen with the bacterial effector AvrPtoB (Abramovitch et al. 2006), (b) whether all the molecules with the same motif, such as RXLR, function as effectors, (c) which of these effector molecules are suppressors or activators of the immune response and which microbial patterns are recognized by the plant's sensor mechanism.

Although the oomycete effector delivery system is different from that of bacteria, nematodes and aphids, the end result, recognition of these effectors and activation of the defence response, would probably use similar mechanisms. It has been well established that plants have sensors at the cell surface and within the cytoplasm (Fig. 3). PRRs localised at the cell surface play a significant role in connecting the cell wall, plasma membrane and cytoskeleton. They are also major players in the perception and transmission of external signals. They include several classes such as polygalacturonase inhibitor-like proteins, receptor-like proteins (RLPs) and, RLKs (Shiu et al. 2004; Fritz-Laylin et al. 2005). Some of these PRRs have been shown to recognize PAMPs such as Ef-Tu (Zipfel and Felix 2005) and effector molecules such as AvrXa21 (Lee et al. 2006) from bacterial pathogens. Until now, no PRR that recognizes an oomycete PAMP has been identified. However, reports are emerging that Pep-13 could be recognized in parsley by an RLK-type receptor (Altenbach and Robatzek 2007). In addition, β-glucan elicitor (GE) from *P. sojae* has been used to identify a receptor protein. However, although a GE-binding protein (GEBP) was purified from the membrane fraction of soybean root cells, no signal peptide or transmembrane domain was identified. Nevertheless, immunolocalization assays indicated that the GEBPs are localized in the plasma membrane of root cells (Umemoto et al. 1997). This suggests that the GEBP may be part of a protein complex localized to the membrane, which somehow interacts with other membrane bound proteins including PRRs.

An *Arabidopsis* RLK with a lectin domain recognizes IPI-O1, an RXLR type effector protein with RGD motif (Fig. 2) from *P. infestans* (Gouget et al. 2006). This is an interesting finding since there is no report of an *Arabidopsis* lectin RLK orthologue from tomato or potato to suggest that IPI-O1 is recognized in tomato or potato, and secondly, the effector is from *P. infestans* for which *Arabidopsis* is a non-host. Thus, the immunity triggered by this effector may be that of non-host resistance, which would be a fascinating piece of *data*.

The expression level of *Arabidopsis* wall-associated kinase 1, an RLK-type PRR, increases when plants are challenged with the Hiks1 isolate of *H. arabidopsis* (Eulgem et al. 2007), which is a further confirmation that some of these plasma membrane–cell wall interacting PRRs are involved in signal transduction. Alterations of expression levels of the PRRs in *Arabidopsis* have also been reported in studies with Nep1-like protein (Qutob et al. 2006). Similarly, we have investigated the publicly available microarray databases on *H. parasitica–Arabidopsis* interactions and observed increased and decreased levels of expression in some of these PRRs (N. Holton and M. Tör, unpublished data). *Arabidopsis* Brassinosteroid insensitive 1-associated receptor kinase 1 (BAK1) is involved in the regulation of the containment of microbial infection-induced cell death. When *bak1* mutants were challenged with several compatible and incompatible isolates of *H. arabidopsis*, reduced sporulation of the pathogen was observed (Kemmerling et al. 2007) indicating that cell surface receptors may also play a role in compatibility.

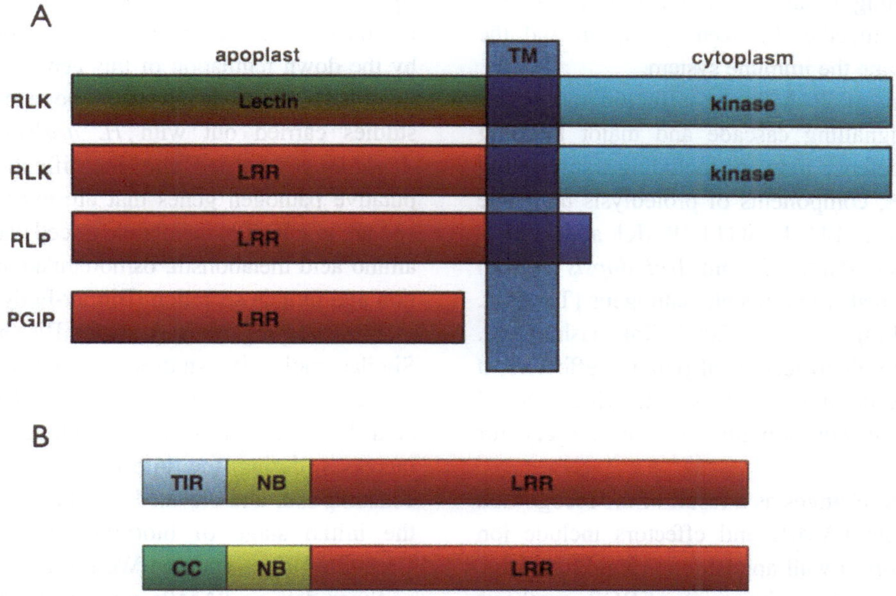

Fig. 3 Canonical domain structures of cell surface and cytoplasmic receptors. *A* Cell surface receptors show variation in their domains. Receptor-like kinases (*RLKs*) have varying types of extracellular domains such as leucine-rich repeat (*LRR*) or lectin-type, followed by a transmembrane spanning region (*TM*) and a cytoplasmic kinase domain. Receptor-like proteins (*RLPs*) are similar to RLKs but do not have the cytoplasmic kinase domains. Instead, they have a short cytoplasmic tail. Polygalacturonase inhibitor proteins have an LRR domain and are totally extra cellular. These cell surface receptors are also known as Pattern Recognition Receptors as they have been implicated in the recognition of PAMPs. To date, only one Lectin RLK type receptor from *Arabidopsis* (Gouget et al. 2006), has been implicated to play a role in oomycete–plant interactions. *B* Cytoplasmic receptors show variation at their N-terminal. These proteins have a central nucleotide binding (*NB*) region and an LRR domain at their C-terminal. The N-terminal region shows variations and either has a *TIR* domain, resembling the cytoplasmic signalling domain of the *T*oll and *I*nterleukin 1 transmembrane *r*eceptors (referred to henceforth *TIR-NB-LRR* genes) or has a *c*oiled-*c*oil domain, (referred to as *CC-NB-LRR* genes). Most of the receptors that are involved in the recognition of oomycete pathogens are cytoplasmic and include *Arabidopsis RAC1* (Borhan et al. 2004) for recognition of *A. candida*, *Arabidopsis RPP1*, *RPP4*, *RPP5* and *RPP13* for *H. arabidopsis* (Tör et al. 2003), potato *R3a* (Huang et al. 2005) for *P. infestans* and lettuce *RGC2B* (Shen et al. 2002) for *B. lactucae*

Another mode of direct and indirect effector detection and recognition takes place within the cytoplasm by NB-LRR proteins (Fig. 3). Traditionally, the genes encoding these proteins have been known as disease resistance genes or *R*-genes and form one of the largest gene families within the plant kingdom. Several members of the R-proteins that provide resistance to oomycete pathogens have been identified or cloned. For example, *RPP* and *RAC* genes from *Arabidopsis* confer resistance to isolates of *H. arabidopsis* and *A. candida* (Tör et al. 1994, 2003; Holub 2001; Borhan et al. 2004), *DM3* and *RGC2* gene clusters in lettuce confers resistance to *B. lactucae* (Shen et al. 2002; Wroblewski et al. 2007), *R1* and *R3a* in potato provide resistance to *P. infestans* (Ballvora et al. 2002; Huang et al. 2005). Domain structures of these proteins (Fig. 3) are known (Tör et al. 2003) and the ways in which they activate the immune response are beginning to emerge. Recent findings suggest that although these NB-LRR proteins are residents of the cytoplasm, the majority of them have a nuclear localization signal (Meyers et al. 2003). Some of them including barley MLA, tobacco N and *Arabidopsis* RPS4, have been shown to move into the nucleus and it has been proposed they activate defence expression by de-repressing basal defence through association with a WRKY transcription factor (Dangl 2007; Shen and Schulze-Lefert 2007; Shen et al. 2007). Two *Arabidopsis* NB-LRR proteins RPP2a and RPP2b are required for the recognition of Cala2 isolate of *H. arabidopsis* (Sinapidou et al. 2004). In this case, it will be fascinating to elucidate whether both NB-LRR proteins travel together between cytoplasm and the nucleus to activate the immune system.

Recognition of PAMPs or effector molecules activate the signalling cascade and major building blocks of the defence network including transcription factors, kinases, components of proteolysis or innate immunity such as EDS1, SGT1, RAR1 and NDR1, which have been identified from *Arabidopsis* or from plants that are hosts to oomycete pathogens (Tör et al. 2002, 2003; Eulgem et al. 2007, Takahashi et al. 2007). With the identification of putative effectors, it should now be possible to investigate which one of these signalling components are the targets for suppression.

Physiological changes as a result of the recognition of the oomycete PAMPs and effectors include ion influx, formation of wall apposition around haustoria, hypersensitive response, formation of ROS, synthesis of phytoalexins and PR proteins and production of salicylic acid. These have been well documented elsewhere (Hardham 2007).

Role of PAMPs and effectors in biotrophy

A common denominator for the important oomycete pathogens is the biotrophic phase in their life cycles. While *H. arabidopsis, B. lactucae* and *A. candida* are obligate biotrophs and cause minimum injury to their hosts, *P. infestans* and *P. sojae* are hemi-biotrophs being biotrophic for the initial stage of up to 36 h after inoculation and subsequently becoming necrotrophic killing the host tissue to consume the cell content (Grenville-Briggs and van West 2005). One of the most distinguishing features of the biotrophic phase in these pathogens, as well as some of the fungal pathogens including powdery mildews and rusts, is the formation of haustoria, which are used in nutrient acquisition (Catanzariti et al. 2007, Voegele et al. 2001). Recent studies on the flux–rust interaction identified effector molecules such as AvrL567, AvrM, AvrP4, and AvrP123 within the haustorium, indicating that haustoria act as reservoirs for effector molecules during the infection process (Catanzariti et al. 2006).

Using a viral-based expression system, Qutob et al. (2002) identified a necrosis-inducing protein (PsojNIP) from *P. sojae* and proposed that this protein plays a significant role in the transition from biotrophy to necrotrophy. However, complementation of this study by the down regulation of this gene to show that it is involved in biotrophy has yet to be reported. Molecular studies carried out with *H. arabidopsis* infecting *Arabidopsis* helped the identification of several putative pathogen genes that are expressed *in planta* and are involved in membrane or cell wall biosynthesis, amino acid metabolism, osmoregulation, phosphorylation and protein secretion (Bittner-Eddy et al. 2003) or in housekeeping roles (van der Biezen et al. 2000). Similar molecular studies coupled with proteomics carried out with *P. infestans* showed that the amino acid biosynthesis in both pathogen and the host increases during the infection. In addition, energy consumption, and elevated metabolism are required at the initial stage of biotrophy (Guo et al. 2006; Grenville-Briggs and van West 2005).

Since data on PAMPs and effector molecules from oomycetes have been accumulating, their role in the establishment of biotrophy rather than as the activator of immunity can be re-evaluated. Although flagella on zoospores of *A. candida* and *P. infestans* provide motility for the establishment of biotrophy and are an important part of the structure, no PAMP associated with these flagella has yet been identified. Attachment of these pathogens to the host cell wall is important in the early stage of infection for initiation of appressoria and haustoria. In this regard, the role of PAMPs such as cell wall binding proteins cannot be ignored.

The major players for the establishment of biotrophy will undoubtedly be the effector molecules. Working on the expression of RXLR and EER motif-containing effectors from *P. infestans*, Whisson et al. (2007) divided these effectors into three groups according to the stage of infection at which they are induced; during pre-infection, throughout infection and during biotrophy only. Silencing of those effectors induced pre-infection and during the biotrophic phase would help to understand the contribution of effectors towards biotrophy.

Although expression of some effectors such as ATR13 from *H. arabidopsis* has been found to be present in spores (Allen et al. 2004), it is not yet known whether the effector protein is localised in the spore (Rebecca Allen, personal communication). When working on an incompatible *H. arabidopsis*–*Arabidopsis* interaction, we observed that in most cases the resistance response is triggered after the formation of haustoria (Tör et al. 2002) indicating that the pathogen is able to develop the necessary structures such as appressoria and intercellular hyphae and establish a limited biotrophy before recognition. These findings, along with those from *P. infestans* infections indicate that (a) some effectors are delivered from hyphae and appressoria into the apoplast (see above) and are used by the pathogen as pioneering molecules to suppress the initial innate immune response and adapt the pathogen to the surrounding niche; and (b) other effectors that are delivered through haustoria into the cytoplasm may be used for diverting nutrients towards the pathogen.

The question as to whether the pathogen is solely responsible for initiating biotrophy is one of the central problems in the interactions between obligate pathogens and their host plant. If the pathogen has PAMPs and effectors to establish a compatible interaction, what is the contribution of the host plant in the compatibility? A great deal of information on defence responses and disease resistance is available. However, the knowledge on 'susceptibility' is very limited. Until now, a few host genes required for susceptibility have been isolated through mutant screens and subsequent genetic analysis. Some examples of these include *POWDERY MILDEW RESISTANT 4* genes (Vogel and Somerville 2000; Vogel et al. 2002, Nishimura et al. 2003) and oomycete *DOWNY MILDEW RESISTANT* genes in *Arabidopsis* (van Damme et al. 2005).

Concluding remarks and future prospects

In the last few years, genome sequencing of several oomycete plant pathogens, including *H. arabidopsis*, *Phytophthora capsici*, *P. infestans*, *P. sojae*, and *P. ramorum* has been carried out and annotations are underway (Tyler et al. 2006). Molecular genomic studies, including large-scale expressed sequence tag sequencing or generating genomic libraries, are also being carried out with other oomycete pathogens including *Bremia*. Arrival of new technologies such as use of Solexa machines should be a great help in these studies. For those species where sequence information is available, bioinformatics studies are being carried out to identify putative effector molecules and classify them according to their functional locations (apoplastic or cytoplasmic), mode of actions, (e.g. enzymatic or transcription factor), their motifs or domains (RXLR, RGD). These studies should also consider whether these effectors are constitutive or are induced *in planta*. Although we are in the middle of stock counting and cataloguing these effectors, we have seen some excellent studies towards functional analysis with a few of the known effector molecules such as the RXLR family members from different oomycete species (Allen et al. 2004, Shan et al. 2004, Rehmany et al. 2005; Sohn et al. 2007). In general, it is assumed that the effector response depends on the pathogen type. Therefore, the initial studies on the known oomycete effectors can be used as a starting point to launch large-scale, high throughput effector analyses to uncover whether there are common lines of communication between oomycete pathogens and their host plants.

Several laboratories around the world are adopting the bacterial type III secretion system to study these oomycete effectors on a large scale (Sohn et al. 2007). This technique may be very suitable for the study of cytoplasmic effectors, but, other techniques should also be adopted to investigate the effectors that are delivered to the apoplast.

Results obtained from high throughput studies that concentrate on individual effectors should be compared and contrasted with those obtained from native, pathogen-delivered effectors. A given pathogen would deliver multiple effectors some of which would act as suppressors of the others. Therefore, pathogen delivery should not be ignored, and if necessary the same effector should be put back into the same pathogen with a known tag and investigated further to obtain a clear picture.

Secretion and translocation of the RXLR type effectors have been attributed to the N-terminal of these effectors (Win et al. 2007). However, it would be interesting to find out which plant proteins, if any, at the cell surface are involved in the endocytosis or transmission of these effector proteins into the cytoplasm of the plant cell.

Micro-array studies have been employed to investigate the plant's defence network and to understand the modulation of signalling in plants by pathogens. However, in the next few years, we should also expect to see microarray studies on these oomycete pathogens. A great deal of information should then be obtained about how the host plant can modulate gene expression in the pathogen genome. A systems biology approach can then be used to look at the interaction from both the pathogen and the host's side.

Although effectors are receiving much attention in current investigations, the next few years should also bring more publications on oomycete PAMPs. These would be particularly useful in the elucidation of non-host resistance.

Another area of great importance is the genetic manipulation of these oomycete pathogens. Although *Phytophthora* species can be transformed and subjected to genetic manipulations, routine genetic transformation methods have not been established for the obligate species. In the next few years, we expect to see the development of different stable transformation methods for these pathogens including *Bremia*, *Hyaloperonospora* and *Albugo*. Development of the RNA interference method to silence these effectors within the pathogen may be an alternative way to stable transformation. It would then be possible to study pathogen genetics.

Although development of new technologies is vital to investigate the interactions between oomycete pathogens and their hosts, the ultimate aim of these studies, in the longer term, should be the development of intelligent systems to control economically important crop pathogens.

Acknowledgements I am grateful to colleagues in the oomycete field for invigorating discussions. I would like to thank Prof Eric Holub and Dr. Alison-Woods Tör for critically reading the manuscript and three anonymous referees for their constructive suggestions. Related research in my laboratory has been supported by grants BB/D000750/1, BB/C509490/1 and BB/E02484X/1 from the UK Biotechnology and Biological Sciences Research Council.

References

Abramovitch, R. B., Janjusevic, R., Stebbins, C. E., & Martin, G. B. (2006). Type III effector AvrPtoB requires intrinsic E3 ubiquitin ligase activity to suppress plant cell death and immunity. *Proceedings of the National Academy of Sciences of the United States of America, 103*, 2851–2856.

Agrios, G. N. (1997). *Plant pathology.* San Diego: Academic.

Allen, R. L., Bittner-Eddy, P. D., Grenville-Briggs, L. J., Meitz, J. C., Rehmany, A. P., Rose, L. E., et al. (2004). Host-parasite coevolutionary conflict between *Arabidopsis* and downy mildew. *Science, 306*, 1957–1960.

Altenbach, D., & Robatzek, S. (2007). Pattern recognition receptors: From the cell surface to intracellular dynamics. *Molecular Plant–Microbe Interactions, 20*, 1031–1039.

Armstrong, M. R., Whisson, S. C., Pritchard, L., Bos, J. I., Venter, E., Avrova, A. O., et al. (2005). An ancestral oomycete locus contains late blight avirulence gene *Avr3a*, encoding a protein that is recognized in the host cytoplasm. *Proceedings of the National Academy of Sciences of the United States of America, 102*, 7766–7771.

Ballvora, A., Ercolano, M. R., Weiss, J., Meksem, K., Bormann, C. A., Oberhagemann, P., et al. (2002). The *R1* gene for potato resistance to late blight (*Phytophthora infestans*) belongs to the leucine zipper/NBS/LRR class of plant resistance genes. *Plant Journal, 30*, 361–371.

Bhattacharjee, S., Hiller, N. L., Liolios, K., Win, J., Kanneganti, T. D., Young, C., et al. (2006). The malarial host-targeting signal is conserved in the Irish potato famine pathogen. *PLoS Pathogens, 2*, e50.

Birch, P. R., Rehmany, A. P., Pritchard, L., Kamoun, S., & Beynon, J. L. (2006). Trafficking arms: Oomycete effectors enter host plant cells. *Trends in Microbiology, 14*, 8–11.

Bittner-Eddy, P. D., Allen, R. L., Rehmany, A. P., Birch, P., & Beynon, J. L. (2003). Use of suppression subtractive hybridization to identify downy mildew gene expressed during infection of *Arabidopsis thaliana. Molecular Plant Pathology, 4*, 501–507.

Borhan, M. H., Holub, E. B., Beynon, J. L., Rozwadowski, K., & Rimmer, S. R. (2004). The *Arabidopsis* TIR-NB-LRR gene *RAC1* confers resistance to *Albugo candida* (white rust) and is dependent on *EDS1* but not *PAD4*. *Molecular Plant-microbe Interactions, 17*, 711–719.

Bos, J. I., Kanneganti, T. D., Young, C., Cakir, C., Huitema, E., Win, J., et al. (2006). The C-terminal half of *Phytophthora infestans* RXLR effector AVR3a is sufficient to trigger R3a-mediated hypersensitivity and suppress INF1-induced cell death in *Nicotiana benthamiana*. *Plant Journal, 48*, 165–176.

Brunner, F., Rosahl, S., Lee, J., Rudd, J. J., Geiler, C., Kauppinen, S., et al. (2002). Pep-13, a plant defense-inducing pathogen-associated pattern from *Phytophthora* transglutaminases. *EMBO Journal, 21*, 6681–6688.

Catanzariti, A. M., Dodds, P. N., Lawrence, G. J., Ayliffe, M. A., & Ellis, J. G. (2006). Haustorially expressed secreted proteins from flax rust are highly enriched for avirulence elicitors. *Plant Cell, 18*, 243–256.

Catanzariti, A. M., Dodds, P. N., & Ellis, J. G. (2007). Avirulence proteins from haustoria-forming pathogens. *FEMS Microbiology Letters, 269*, 181–188.

Chisholm, S. T., Dahlbeck, D., Krishnamurthy, N., Day, B., Sjolander, K., & Staskawicz, B. J. (2005). Molecular characterization of proteolytic cleavage sites of the *Pseudomonas syringae* effector AvrRpt2. *Proceedings of the National Academy of Sciences of the United States of America, 102*, 2087–2089.

Chisholm, S. T., Coaker, G., Day, B., & Staskawicz, B. J. (2006). Host-microbe interactions: Shaping the evolution of the plant immune response. *Cell, 124*, 803–814.

Cooper, A. J., Woods-Tör, A., & Holub, E. B. (2002). *Albugo candida* (White rust) suppresses resistance to downy mildew pathogens in *Arabidopsis thaliana*. Proceedings 6th Conference of EFPP, *Plant Protection Science, 38* (Special issue), 474–476.

Dangl, J. L. (2007). Plant science. Nibbling at the plant cell nucleus. *Science, 315*, 1088–1089.

Daxberger, A., Nemak, A., Mithofer, A., Fliegmann, J., Ligterink, W., Hirt, H., et al. (2007). Activation of members of a MAPK module in beta-glucan elicitor-mediated non-host resistance of soybean. *Planta, 225*, 1559–1571.

Ellis, J., Catanzariti, A. M., & Dodds, P. (2006). The problem of how fungal and oomycete avirulence proteins enter plant cells. *Trends in Plant Science, 11*, 61–63.

Eulgem, T., Tsuchiya, T., Wang, X. J., Beasley, B., Cuzick, A., Tör, M., et al. (2007). EDM2 is required for RPP7-dependent disease resistance in *Arabidopsis* and affects RPP7 transcript levels. *Plant Journal, 49*, 829–839.

Fritz-Laylin, L. K., Krishnamurthy, N., Tör, M., Sjolander, K. V., & Jones, J. D. (2005). Phylogenomic analysis of the receptor-like proteins of rice and *Arabidopsis*. *Plant Physiology, 138*, 611–623.

Felix, G., Duran, J. D., Volko, S., & Boller, T. (1999). Plants have a sensitive perception system for the most conserved domain of bacterial flagellin. *Plant Journal, 18*, 265–276.

Gaulin, E., Jauneau, A., Villalba, F., Rickauer, M., Esquerre-Tugaye, M. T., & Bottin, A. (2002). The CBEL glycoprotein of *Phytophthora parasitica* var *nicotianae* is involved in cell wall deposition and adhesion to cellulosic substrates. *Journal of Cell Science, 115*, 4565–4575.

Gaulin, E., Drame, N., Lafitte, C., Torto-Alalibo, T., Martinez, Y., Ameline-Torregrosa, C., et al. (2006). Cellulose binding domains of a *Phytophthora* cell wall protein are novel pathogen-associated molecular patterns. *Plant Cell, 18*, 1766–1777.

Gouget, A., Senchou, V., Govers, F., Sanson, A., Barre, A., Rouge, P., et al. (2006). Lectin receptor kinases participate in protein–protein interactions to mediate plasma membrane–cell wall adhesions in *Arabidopsis*. *Plant Physiology, 40*, 81–90.

Göker, M., Reithmuller, A., Voglmayr, H., Weiss, M., & Oberwinkler, F. (2004). Phylogeny of *Hyaloperonospora* based on nuclear ribosomal internal transcribed spacer sequences. *Mycological Progress, 3*, 83–94.

Grenville-Briggs, L. J., & van West, P. (2005). The biotrophic stages of oomycete-plant interactions. *Advances in Applied Microbiology, 57*, 217–243.

Guo, J., Jiang, R. H., Kamphuis, L. G., & Govers, F. (2006). A cDNA-AFLP based strategy to identify transcripts associated with avirulence in *Phytophthora infestans*. *Fungal Genetics and Biology, 43*, 111–123.

Guttman, D. S., Gropp, S. J., Morgan, R. L., & Wang, P. W. (2006). Diversifying selection drives the evolution of the type III secretion system pilus of *Pseudomonas syringae*. *Molecular Biology and Evolution, 23*, 2342–2354.

Hahn, M. G. (1996). Microbial elicitors and their receptors in plants. *Annual Review of Phytopathology, 34*, 387–412.

Hahn, M., & Mendgen, K. (2001). Signal and nutrient exchange at biotrophic plant fungus interfaces. *Current Opinion in Plant Biology, 4*, 322–327.

Hardham, A. R. (2007). Cell biology of plant–oomycete interactions. *Cellular Microbiology, 9*, 31–39.

Hiller, N. L., Bhattacharjee, S., van Ooij, C., Liolios, K., Harrison, T., Lopez-Estrano, C., et al. (2004). A host-targeting signal in virulence proteins reveals a secretome in malarial infection. *Science, 306*, 1934–1937.

Holub, E. B. (2001). The arms race is ancient history in *Arabidopsis*, the wildflower. *Nature Reviews Genetics, 2*, 516–527.

Holub, E. B. (2006). Evolution of parasitic symbioses between plants and filamentous microorganisms. *Current Opinion in Plant Biology, 9*, 397–405.

Holub, E. B. (2008). Natural history of *Arabidopsis thaliana* and oomycete symbioses. *European Journal of Plant Pathology*, (in press).

Huang, S., van der Vossen, E. A., Kuang, H., Vleeshouwers, V. G., Zhang, N., Borm, T. J., et al. (2005). Comparative genomics enabled the isolation of the *R3a* late blight resistance gene in potato. *Plant Journal, 42*, 251–261.

Ingle, R. A., Carstens, M., & Denby, K. J. (2006). PAMP recognition and the plant–pathogen arms race. *Bioessays, 28*, 880–889.

Jones, J. D., & Dangl, J. L. (2006). The plant immune system. *Nature, 444*, 323–329.

Kamoun, S. (2003). Molecular genetics of pathogenic oomycetes. *Eukaryotic Cell, 2*, 191–199.

Kamoun, S. (2006). A catalogue of the effector secretome of plant pathogenic oomycetes. *Annual Review of Phytopathology, 44*, 41–60.

Kamoun, S. (2007). Groovy times: Filamentous pathogen effectors revealed. *Current Opinion in Plant Biology, 10*, 358–365.

Kemmerling, B., Schwedt, A., Rodriguez, P., Mazzotta, S., Frank, M., Qamar, S. A., et al. (2007). The BRI1-associated kinase 1, BAK1, has a brassinolide-independent role in plant cell-death control. *Current Biology, 17*, 1116–1122.

Lee, S. W., Han, S. W., Bartley, L. E., & Ronald, P. C. (2006). Colloquium review. Unique characteristics of *Xanthomonas oryzae* pv. *oryzae* AvrXa21 and implications for plant innate immunity. *Proceedings of the National Academy of Sciences of the United States of America, 103*, 18395–18400.

Lotze, M. T., Zeh, H. J., Rubartelli, A., Sparvero, L. J., Amoscato, A. A., Washburn, N. R., et al. (2007). The grateful dead: Damage associated molecular pattern molecules and reduction/oxidation regulate immunity. *Immunological Reviews, 220*, 60–82.

Medzhitov, R. (2007). Recognition of microorganisms and activation of the immune response. *Nature, 449*, 819–826.

Medzhitov, R., & Janeway, C. A. (2002). Decoding the patterns of self and non-self by the innate immune system. *Science, 296*, 298–300.

Meyers, B. C., Kozik, A., Griego, A., Kuang, H., & Michelmore, R. W. (2003). Genome-wide analysis of NBS-LRR-encoding genes in Arabidopsis. *Plant Cell, 15*, 809–834.

Mithofer, A., Fliegmann, J., Daxberger, A., Ebel, C., Neuhaus-Url, G., Bhagwat, A. A., et al. (2001). Induction of H(2)O (2) synthesis by beta-glucan elicitors in soybean is independent of cytosolic calcium transients. *FEBS Letters, 508*, 191–195.

Morgan, W., & Kamoun, S. (2007). RXLR effectors of plant pathogenic oomycetes. *Current Opinion in Microbiology, 10*, 332–338.

Mudgett, M. B., & Staskawicz, B. J. (1998). Protein signaling via type III secretion pathways in phytopathogenic bacteria. *Current Opinion in Microbiology, 1*, 109–114.

Nishimura, M. T., Stein, M., Hou, B. H., Vogel, J. P., Edwards, H., & Somerville, S. C. (2003). Loss of a callose synthase results in salicylic acid-dependent disease resistance. *Science, 301*, 969–972.

Nurnberger, T., Nennstiel, D., Jabs, T., Sacks, W. R., Hahlbrock, K., & Scheel, D. (1994). High affinity binding of a fungal oligopeptide elicitor to parsley plasma membranes triggers multiple defense responses. *Cell, 78*, 449–460.

O'Connell, R. J., & Panstruga, R. (2006). Tete a tete inside a plant cell. Establishing compatibility between plants and biotrophic fungi and oomycetes. *New Phytologist, 171*, 699–718.

Pieterse, C. M., van West, P., Verbakel, H. M., Brasse, P. W., van den Berg-Velthuis, G. C., & Govers, F. (1994). Structure and genomic organization of the *ipiB* and *ipiO* gene clusters of *Phytophthora infestans*. *Gene, 138*, 67–77.

Qutob, D., Kamoun, S., & Gijzen, M. (2002). Expression of a *Phytophthora sojae* necrosis-inducing protein occurs during transition from biotrophy to necrotrophy. *Plant Journal, 32*, 361–373.

Qutob, D., Kemmerling, B., Brunner, F., Kufner, I., Engelhardt, S., Gust, A. A., et al. (2006). Phytotoxicity and innate immune responses induced by Nep1-like proteins. *Plant Cell, 18*, 3721–3744.

Rehmany, A. P., Gordon, A., Rose, L. E., Allen, R. L., Armstrong, M. R., Whisson, S. C., et al. (2005). Differential recognition of highly divergent downy mildew avirulence gene alleles by *RPP1* resistance genes from two *Arabidopsis* lines. *Plant Cell, 17*, 1839–1850.

Robatzek, S. (2007). Vesicle trafficking in plant immune responses. *Cellular Microbiology, 9*, 1–8.

Robatzek, S., Chinchilla, D., & Boller, T. (2006). Ligand-induced endocytosis of the pattern recognition receptor FLS2 in *Arabidopsis*. *Genes and Development, 20*, 537–542.

Senchou, V., Weide, R., Carrasco, A., Bouyssou, H., Pont-Lezica, R., Govers, F., et al. (2004). High affinity recognition of a *Phytophthora* protein by Arabidopsis via an RGD motif. *Cellular and Molecular Life Sciences, 61*, 502–509.

Shan, W., Cao, M., Leung, D., & Tyler, B. M. (2004). The *Avr1b* locus of *Phytophthora sojae* encodes an elicitor and a regulator for avirulence on soybean plants carrying resistance gene *Rps1b*. *Molecular Plant–Microbe Interactions, 17*, 394–403.

Sharp, J. K., Valent, B., & Albersheim, P. (1984). Purification and partial characterization of a beta-glucan fragment that elicits phytoalexin accumulation in soybean. *Journal of Biological Chemistry, 259*, 11312–11320.

Shen, Q. H., & Schulze-Lefert, P. (2007). Rumble in the nuclear jungle: Compartmentalization, trafficking, and nuclear action of plant immune receptors. *EMBO Journal, 26*, 4293–4301.

Shen, K. A., Chin, D. B., Arroyo-Garcia, R., Ochoa, O. E., Lavelle, D. O., Wroblewski, T., et al. (2002). *Dm3* is one member of a large constitutively expressed family of nucleotide binding site-leucine-rich repeat encoding genes. *Molecular Plant–Microbe Interactions, 15*, 251–61.

Shen, Q. H., Saijo, Y., Mauch, S., Biskup, C., Bieri, S., Keller, B., et al. (2007). Nuclear activity of MLA immune receptors links isolate-specific and basal disease-resistance responses. *Science, 315*, 1098–1101.

Shiu, S. H., Karlowski, W. M., Pan, R., Tzeng, Y. H., Mayer, K. F., & Li, W. H. (2004). Comparative analysis of the receptor-like kinase family in Arabidopsis and rice. *Plant Cell, 16*, 1220–1234.

Sinapidou, E., Williams, K., Nott, L., Bahkt, S., Tör, M., Bittner-Eddy, P., et al. (2004). Two TIR:NB:LRR genes are required to specify resistance to *Peronospora parasitica* isolate Cala2 in *Arabidopsis*. *Plant Journal, 38*, 898–909.

Sohn, K. H., Lei, R., Nemri, A., & Jones, J. D. (2007). The downy mildew effector proteins ATR1 and ATR13 promote disease susceptibility in *Arabidopsis thaliana*. *Plant Cell, 19*, 4077–4090. DOI 10.1105/tpc.107.054262.

Takahashi, Y., Nasir, K. H., Ito, A., Kanzaki, H., Matsumura, H., Saitoh, H., et al. (2007). A high-throughput screen of cell-death-inducing factors in *Nicotiana benthamiana* identifies a novel MAPKK that mediates INF1-induced cell death signaling and non-host resistance to *Pseudomonas cichorii*. *Plant Journal, 49*, 1030–1040.

Tör, M., Holub, E. B., Brose, E., Musker, R., Gunn, N., Can, C., et al. (1994). Map positions of three loci in *Arabidopsis thaliana* associated with isolate-specific recognition of *Peronospora parasitica* (downy mildew). *Molecular Plant-microbe Interactions, 7*, 214–222.

Tör, M., Gordon, P., Cuzick, A., Eulgem, T., Sinapidou, E., Mert, F., et al. (2002). *Arabidopsis* SGT1b is required for defence signaling conferred by several downy mildew (*Peronospora parasitica*) resistance genes. *Plant Cell, 14*, 993–1003.

Tör, M., Yemm, A., & Holub, E. (2003). Role of proteolysis in *R*-gene mediated defence in plants. *Molecular Plant Pathology, 4*, 287–296.

Tyler, B. M., Tripathy, S., Zhang, X., Dehal, P., Jiang, R. H., Aerts, A., et al. (2006). *Phytophthora* genome sequences uncover evolutionary origins and mechanisms of pathogenesis. *Science, 313*, 1261–1266.

Umemoto, N., Kakitani, M., Iwamatsu, A., Yoshikawa, M., Yamaoka, N., & Ishida, I. (1997). The structure and function of a soybean beta-glucan-elicitor-binding protein. *Proceedings of the National Academy of Sciences of the United States of America, 94*, 1029–1034.

Van Damme, M., Andel, A., Huibers, R. P., Panstruga, R., Weisbeek, P. J., & Van den Ackerveken, G. (2005). Identification of *Arabidopsis* loci required for susceptibility to the downy mildew pathogen *Hyaloperonospora parasitica*. *Molecular Plant-microbe Interactions, 18*, 583–592.

Van den Ackerveken, G., Marois, E., & Bonas, U. (1996). Recognition of the bacterial avirulence protein AvrBs3 occurs inside the host plant cell. *Cell, 87*, 1307–1316.

van der Biezen, E. A., Juwana, H., Parker, J. E., & Jones, J. D. G. (2000). cDNA-AFLP display for the isolation of *Peronospora parasitica* genes expressed during infection in *Arabidopsis thaliana*. *Molecular Plant–Microbe Interactions, 13*, 895–898.

Voegele, R. T., & Mendgen, K. (2003). Rust haustoria: Nutrient uptake and beyond. *New Phytologist, 159*, 93–100.

Voegele, R. T, Struck, C., Hahn, M., & Mendgen K. (2001). The role of haustoria in sugar supply during infection of broad bean by the rust fungus *Uromyces fabae*. *Proceedings of the National Academy of Sciences of the United States of America, 98*, 8133–8138.

Vogel, J., & Somerville, S. (2000). Isolation and characterization of powdery mildew-resistant *Arabidopsis* mutants. *Proceedings of the National Academy of Sciences of the United States of America, 97*, 1897–1902.

Vogel, J. P., Raab, T. K., Schiff, C., & Somerville, S. C. (2002). PMR6, a pectate lyase-like gene required for powdery mildew susceptibility in *Arabidopsis*. *Plant Cell, 14*, 2095–2106.

Whisson, S. C., Boevink, P. C., Moleleki, L., Avrova, A. O., Morales, J. G., Gilroy, E. M., et al. (2007). A translocation signal for delivery of oomycete effector proteins into host plant cells. *Nature, 450*, 115–118.

Win, J., Morgan, W., Bos, J., Krasileva, K. V., Cano, L. M., Chaparro-Garcia, A., et al. (2007). Adaptive evolution has targeted the C-Terminal domain of the RXLR effectors of plant pathogenic oomycetes. *Plant Cell, 19*, 2349–2369.

Wroblewski, T., Piskurewicz, U., Tomczak, A., Ochoa, O., & Michelmore, R. W. (2007). Silencing of the major family of NBS-LRR-encoding genes in lettuce results in the loss of multiple resistance specificities. *Plant Journal, 51*, 803–818.

Xiao, F., Giavalisco, P., & Martin, G. B. (2007). *Pseudomonas syringae* type III effector AvrPtoB is phosphorylated in plant cells on serine 258, promoting its virulence activity. *Journal of Biological Chemistry, 282*, 30737–30744.

Yamamizo, C., Kuchimura, K., Kobayashi, A., Katou, S., Kawakita, K., Jones, J. D., et al. (2006). Rewiring mitogen-activated protein kinase cascade by positive feedback confers potato blight resistance. *Plant Physiology, 140*, 681–692.

Zipfel, C., & Felix, G. (2005). Plants and animals: A different taste for microbes. *Current Opinion in Plant Biology, 8*, 353–360.

Zipfel, C., Kunze, G., Chinchilla, D., Caniard, A., Jones, J. D., Boller, T., et al. (2006). Perception of the bacterial PAMP EF-Tu by the receptor EFR restricts *Agrobacterium*-mediated transformation. *Cell, 125*, 749–760.

Eur J Plant Pathol (2008) 122:71–89
DOI 10.1007/s10658-008-9292-3

Diversity of defence mechanisms in plant–oomycete interactions: a case study of *Lactuca* spp. and *Bremia lactucae*

Aleš Lebeda · Michaela Sedlářová · Marek Petřivalský · Jitka Prokopová

Received: 8 September 2007 / Accepted: 18 February 2008

Abstract Plant pathogenic oomycetes, including biotrophic downy mildews and hemibiotrophs/necrotrophs such as *Phytophthora* and *Pythium*, cause enormous economic losses on cultivated crops. Lettuce breeders and growers face the threat of *Bremia lactucae*, the causal agent of lettuce downy mildew. This pathogen damages leaf tissues and lettuce heads and is also frequent on wild Asteraceae plants. The interactions of *Lactuca* spp. with *B. lactucae* (abbr. lettuce–*Bremia*) display extreme variability, due to a long co-evolutionary history. For this reason, during the last 30 years, the lettuce–*Bremia* pathosystem has been used as a model for many studies at the population, individual, organ, tissue, cellular, physiological and molecular levels, as well as on genetic variability and the genetics of host–parasite interactions. The first part of this review summarizes recent data on host–parasite specificity, host variability, resistance mechanisms and genetics of lettuce–*Bremia* interactions. The second part focuses on the development infection structures. Phenotypic expression of infection, behaviour of *B. lactucae* on leaf surfaces, the process of penetration, development of primary infection structures, hyphae and haustoria are discussed in relation to different resistance mechanisms. In the third part, the components of host resistance and the variability of defence responses are analysed. The role of reactive oxygen species (ROS), antioxidant enzymes, nitric oxide (NO), phenolic compounds, reorganization of cytoskeleton, electrolyte leakage, membrane damage, cell wall disruption, hypersensitive reaction and plant energetics are discussed in relation to defence responses. In general, the extreme variability of interactions between lettuce and *Bremia*, and their phenotypic expression, results from diversity of the genetic background. Different mechanisms of resistance are conditioned by an orchestra of defence responses at the tissue, cell, and molecular levels. The various events responsible for defence involve a complex interaction of the processes and reactions mentioned above. This review also provides an overview on the timing of pathogen development, host pathological anatomy, cytology and physiology

A. Lebeda (✉) · M. Sedlářová
Faculty of Science, Department of Botany,
Palacký University,
Šlechtitelů 11,
783 71 Olomouc-Holice, Czech Republic
e-mail: ales.lebeda@upol.cz

M. Petřivalský
Faculty of Science, Department of Biochemistry,
Palacký University,
Šlechtitelů 11,
783 71 Olomouc-Holice, Czech Republic

J. Prokopová
Faculty of Science, Department of Experimental Physics,
Palacký University,
Tř. Svobody 26,
771 46, Olomouc, Czech Republic

of lettuce–*Bremia* associations. The significance of these factors on the expression of different resistance mechanisms (non-host and host resistance, race-specific and race non-specific resistance, field resistance) is discussed.

Keywords Cytoskeleton · Genetics · Host-and non-host resistance · Hypersensitive reaction · Infection structures · Lettuce · Lettuce downy mildew · Nitric oxide · Phenolic compounds · Photosynthesis · Plant energetics · Reactive oxygen species · Specificity

Abbreviations

ATP	adenosine triphosphate
BAP	benzylaminopurine
CKs	cytokinins
dai	days after inoculation
EHM	extrahaustorial membrane
ER	endoplasmatic reticulum
H	intercellular hypha
HA	haustorium
hai	hours after inoculation
HR	hypersensitive reaction
IH	intracellular hypha
IMD	irreversible membrane damage
MTs	microtubules
NADPH	reduced nicotinamide adenine dinucleotide phophate
NO	nitric oxide
NOS	nitric oxide synthase
PAL	phenylalanine ammonium lyase
PAs	phenolic acids
POX	peroxidase
PSII	photosystem II
PTIO	2-phenyl-4,4,5,5-tetramethylimidazoline-1-oxyl-3-oxide
QTL	quantitative trait loci
ROS	reactive oxygen species
TEM	transmission electron microscopy
3-MeOBAPR	6-(3-methoxy-benzylamino)purine-9-β-ribofuranoside

Introduction

Aspects of plant defence against pathogens have been studied in different model pathosystems, including interactions between plants and oomycetes (Glazebrook 2005; Göker et al. 2007; Hardham 2007; Hulbert et al. 2001; Kamoun 2006). One of the most important pathosystems in terms of economic loss is the *Lactuca* spp.–*Bremia lactucae* (abbr. lettuce–*Bremia*) pathosystem (Crute 1992a; Lebeda et al. 2007, 2008). This paper highlights several aspects of defence mechanisms in this pathosystem from the viewpoint of phenotypic expression, cytology, physiology, biochemistry and biophysics.

Lettuce is the common name used for about 100 species of the genus *Lactuca*; they are prevalently distributed in Asia and Africa, but occur also in Europe, North and Central America (Lebeda et al. 2004). Only one species, cultivated lettuce (*Lactuca sativa*), is grown as a crop worldwide. Lettuce ranks as one of the earliest domesticated vegetables (8,000 years ago; Lebeda et al. 2007). The centre of *Lactuca* spp. biodiversity is in southwestern Asia, in the Tigris–Euphrates region, and lettuce probably originates as a food plant from this region. Systematic breeding of cultivated lettuce started in the 19th century and nowadays it is an extremely variable species, both morphologically and genetically (Lebeda et al. 2007).

Many diseases of lettuce have been described (Davis et al. 1997), but only a few are important enough to be considered in crop protection and resistance breeding programmes (Lebeda et al. 2007). One such disease is lettuce downy mildew, caused by the biotrophic oomycete *B. lactucae* (Lebeda et al. 2002). The breeding of lettuce for resistance to *B. lactucae* started in 1920s and now is considered as a priority among the vegetable features. Several different mechanisms of resistance to *B. lactucae* have been identified within cultivated and wild lettuce (Lebeda et al. 2001b). Because of limited durability of race-specific resistance (Lebeda et al. 2002, 2007; Lebeda and Zinkernagel 2003a), the search has focused on field resistance (Grube and Ochoa 2005) and new sources of resistance in wild *Lactuca* species (Lebeda et al. 2002, 2007).

This paper deals with the complexity of pathological processes induced in host *Lactuca* spp. plants following infection by *B. lactucae*. The main aim was to analyse critically the recent knowledge in this area, within a wider context of plant–oomycete interactions and with the main focus on the lettuce–*Bremia* pathosystem.

Host–parasite specificity

General aspects of specificity in plant–oomycete interactions have been at the centre of mycological and phytopathological research during the 1990s and early 2000s (Clark and Spencer-Phillips 2004; Glazebrook 2005; Göker et al. 2007; Grenville-Briggs and van West 2005; Hardham 2007; Holub and Cooper 2004; Latijnhouwers et al. 2003; Lebeda and Schwinn 1994; Lipka et al. 2005; Lucas et al. 1995; Mauch-Mani 2002; O'Connell and Panstruga 2006). For the lettuce–*Bremia* pathosystem, a detailed survey of host–parasite specificity can be found in Lebeda et al. (2001b, 2002). The enormous variability in the specificity of interactions can be explained by the lengthy coevolution of the host–parasite association between *Lactuca* spp. and *B. lactucae* (Lebeda 2002; Lebeda et al. 2002).

Taxonomy, host range and specialization of *B. lactucae*

It has become apparent that taxa in the Peronosporaceae are polyphyletic or paraphyletic (Thines et al. 2006; Voglmayr 2008; Voglmayr et al. 2004). As a result of molecular phylogenetic investigations, six new genera have been described in the Peronosporaceae, with features revealed by scanning electron microscopy (Thines 2007). The genera *Bremia, Protobremia, Paraperonospora, Plasmoverna, Basidiophora, Benua* and *Plasmopara* form a dense cluster due to the uniting aspects of similarities in the morphology of their haustoria (vesicular to pyriform) and ultimate branches (Göker et al. 2007; Thines 2007; Voglmayr 2008).

Bremia lactucae, a pathogen of cultivated and wild lettuce (Lebeda et al. 2002), has been reported to infect plants of more than two hundred species from about 40 genera of the Asteraceae (Crute and Dixon 1981; Lebeda et al. 2002). New host species continue to be reported (e.g. Koike and Ochoa 2007). Based on cross-inoculation experiments and morphological observations, the specialization of *B. lactucae* into 11 formae speciales was accepted (Lebeda et al. 2002; Skidmore and Ingram 1985). These experiments showed a high specificity of the formae speciales, with each almost exclusively limited to an individual host genus (Lebeda et al. 2002). However, several previous (Lebeda and Syrovátko 1988) and recent (Vieira and Barreto 2006) experiments have suggested the possibility that infection of lettuce (*L. sativa*) by *B. lactucae* originates from *Sonchus* spp. and *vice versa*. *Sonchus oleraceus* was the most common weed hosting *B. lactucae* outside of *Lactuca* spp. (Lebeda et al. 2008). From about 100 wild species described within the genus *Lactuca* (Lebeda et al. 2004), only 14 are reported as natural hosts of *B. lactucae* (Lebeda et al. 2002). *Lactuca serriola* is considered to be the most common weed host in central Europe (Lebeda et al. 2008; Petrželová and Lebeda 2004). Specificity of the interactions between wild *Lactuca* spp. and *B. lactucae* is still not completely understood. Comprehensive data are available only for *L. serriola* (primary lettuce gene pool), *L. saligna* and *L. virosa* (secondary and tertiary gene pools, respectively; Lebeda et al. 2002, 2007).

Molecular phylogenetic studies, using nuclear large subunit rDNA sequences with D1/D2 regions (Voglmayr et al. 2004) and internal transcribed spacer rDNA (Choi et al. 2007), revealed the presence of several highly supported clades within the *B. lactucae* complex. These lineages match partially to formae speciales (Skidmore and Ingram 1985). Most importantly, the genetic distance of isolates originating from *Lactuca* spp. and those from other hosts was clearly demonstrated, suggesting a lack of interbreeding. Therefore, infected wild Asteraceae plants other than *Lactuca* are unlikely to serve as a source of inoculum for infections in *Lactuca* spp. populations (Lebeda et al. 2002, 2008; Voglmayr et al. 2004). Previous individual (Lebeda 1986) and recent population (Lebeda 2002; Lebeda et al. 2008) studies, however, have shown that isolates of *B. lactucae* from *L. serriola* are significantly more pathogenic to *L. serriola* than to *L. sativa*. Data from these phenotypic studies demonstrate the possibility of very close genetic affinity between *Lactuca* host species and *B. lactucae* and *vice versa*, and this must be considered in the planning of experiments focused on research of resistance mechanisms.

Resistance mechanisms in *Lactuca* spp.

In most interactions, the resistance of *Lactuca* spp. to *B. lactucae* is considered as host-resistance, according to the phenotypic, tissue and cellular expression. Only *L. saligna* appears to differ in several features, thus raising the possible existence of non-host resistance (basic incompatibility; Lebeda et al. 2002). Recent studies with individual plants (Beharav et al. 2006; Lebeda and Zinkernagel 2003b) and populations (Petrželová et al. 2007) of *L. saligna* showed a high degree of resistance to all *B. lactucae* races originating from lettuce, and also those from *L. serriola* (Lebeda 1986; Lebeda and Boukema 1991). Moreover, studies at the tissue, cellular and physiological levels (Lebeda et al. 2001b, 2002, 2006; Lebeda and Pink 1998; Lebeda and Reinink 1994; Sedlářová and Lebeda 2001; Sedlářová et al. 2001b, 2007a, b) confirmed that the mechanism of resistance in *L. saligna* differs significantly from the mechanisms known in *L. sativa, L. serriola* and *L. virosa* (Lebeda et al. 2002).

Host resistance (basic compatibility) is a better known phenomenon in this pathosystem as it has been studied since the beginning of the 20th century from many perspectives. The most common three categories of host resistance are reviewed below, i.e. race-specific resistance, race non-specific resistance and field resistance (Lebeda et al. 2001b, 2002).

Race-specific resistance with its characteristic phenotypic expression and intensively studied genetics can be found in cultivars of *L. sativa* (e.g. Lebeda et al. 2007). The specificity is determined by dominant resistance genes and/or factors in the host (*Dm* genes and/or R-factors) which are matched by pathogen dominant factors of avirulence (Crute 1992b; for more details see Genetics of *Lactuca* spp.–*Bremia lactucae* interactions). Race-specificity is well documented also in wild *Lactuca* spp. (Table 1) and a few closely related genera (Lebeda et al. 2002). Recently, it was found as a common phenomenon in wild populations of *L. serriola* where enormous diversity of this type of resistance was described (Lebeda et al. 2008; Lebeda and Petrželová 2004).

Race non-specific (non-differential) resistance is conferred by several genes and characterized by effectiveness against a spectrum of *B. lactucae* races. *Lactuca* spp. genotypes with this type of resistance posses a certain level of non-specific resistance according to phenotypic expression (Lebeda et al. 2002). The presence of race non-specific resistance is not well-documented for *L. sativa* (Lebeda et al. 2001b). It has only been reported in some accessions of *L. serriola* (PI 281876 and PI 281877) for which the genetic background is not well known, and the presence of some major genes and modifiers is predicted (Lebeda et al. 2002).

Field resistance is a complex epidemiological phenomenon (Lebeda et al. 2002), expressed by reduced susceptibility of mature plants grown in the field with natural infections of *B. lactucae* (Grube and Ochoa 2005). A search for sources of field resistance in *L. sativa* located a high level of this resistance in cvs Iceberg and Grand Rapids (Crute and Norwood 1981). Recent studies suggested simple inheritance of this trait, but the single gene models did not fit the data obtained (Grube and Ochoa 2005). Field resistance also is expected in wild *Lactuca* spp., with direct evidence existing for some *L. serriola* accessions (e.g. PI 281876; Lebeda et al. 2002).

Genetics of *Lactuca* spp.–*Bremia lactucae* interactions

Nowadays, more than 45 host race-specific resistance genes/factors (*Dm*/R) and complementary pathogen virulence (*v*) genes (factors) are predicted in the lettuce-*Bremia* pathosystem (Lebeda et al. 2006). Many of these are used for phenotypic screening of *B. lactucae* isolates and characterization of their virulence (Lebeda and Petrželová 2008). The number of host resistance genes is expected to increase further with continuation of extensive phenotypic characterization of *Lactuca* germplasm (Beharav et al. 2006; Lebeda and Petrželová 2004; Lebeda and Zinkernagel 2003b) and molecular investigations (Kuang et al. 2004, 2006; Michelmore, pers. comm.). At least 15 *Dm* genes have been characterized in lettuce spp. (*Dm*1, *Dm*2, *Dm*3, *Dm*4, *Dm*5/8, *Dm*6, *Dm*7, *Dm*10, *Dm*11, *Dm*12, *Dm*13, *Dm*14, *Dm*15, *Dm*16 and R18). These genes occur in distinct clusters within the lettuce genome, at least five clusters being recognized (Witsenboer et al. 1997). Some of these *Dm* genes originate from *L. serriola* (Lebeda et al. 2002), others were described and subjected to preliminary genetic characterisation (Bonnier et al. 1994), and numerous others are to be expected (Lebeda and Petrželová 2004). Some recently released R-factors (R36 and R37), which are located in *L. sativa* originate from *L. saligna* (Michelmore et al. 2005). Other resistance gene candidates (RGCs) are proposed and used for evolutionary studies to explore the diversity of *Lactuca* spp. germplasm (Kuang et al. 2004, 2006). However, the specificities of the genes need to be established and their effectiveness against given pathogen races must be demonstrated. This poses a

Table 1 Generalized overview on variability in formation of *B. lactucae* infection structures and reactions of *Lactuca* spp. tissues at 48 h after inoculation in various categories of resistance (compiled according to Lebeda et al. 2001b, 2002, 2006; Lebeda and Pink 1998; the data were obtained on leaf discs derived from adult plants)

Category of resistance	*Lactuca* spp. genotype	Genetical background/ resistance gene (factor)[a]	Response to *Bremia lactucae*[b]	Relative degree of infection structure development and tissue response[c]					
				Primary vesicle	Secondary vesicle	Hyphae	Haustoria	Hypersensitive reaction	Subepidermal necrosis
Non-host?	*L. saligna* (LSA/6)	?	–	3	2	0	0	1	1
Race-specific	*L. sativa* (Cobham Green)	R 0 (?)	+	4	4	4	4	1	0
	L. sativa (Dandie)	*Dm* 3	+	4	4	2	1	1	1
	L. sativa (Valmaine)	*Dm* 5/8	–	3/4	1	1	0	2	0
	L. sativa (Mariska)	R 18	–	2	2	2	2	4	3
	L. serriola (PIVT 1168)	R ?	–	2	1	0/1	0	4	2
	L. saligna (CGN 5147)	R ?	–	3	1	1	0	1	0
	L. virosa (LVIR/57/1)	R ?	(–)	3/4	3	1/2	1/2	4	2/3
Race-non-specific	*L. serriola* (PI 281876)	R ? (+modif.?)	(–)	4	4	2	2	4	2
Field	*L. sativa* (Iceberg)	nR?	+	3	3	3	3	1	0

? This category is still questionable for *L. saligna* (see discussion in: Jeuken and Lindhout 2002; Lebeda et al. 2001b, 2002)

[a] ? Not known or unspecified, *R* race-specific resistance factor, *Dm* race-specific resistance gene; *modif.* modifier gene(s); *n* more R-factors

[b] Categories of phenotypic expression of *Lactuca* spp. response to *B. lactucae*: – incompatible (no sporulation); *(–)* incompletely incompatible (very limited sporulation occurring mostly at the cutting edges of leaf discs); + compatible (profuse sporulation); a field resistance cannot be distinguished by screening either on cotyledons or leaf discs

[c] Relative degree of occurrence of pathogen infection structures and plant tissue response compared to susceptible control (details are given in Lebeda et al. 2002): *0* none recorded, *1* very low frequency, *2* low frequency, *3* medium frequency, *4* high frequency. Significant differences in frequency and timing are specific for given genotype-race interaction, usually *L. sativa* genotypes vary from other wild *Lactuca* spp.

barrier to the rapid engineering of durable resistance. Currently, much effort in lettuce resistance breeding is focused on deployment of non-durable R-genes (Lebeda et al. 2007; Pink 2002). Thus, breeders have to look for new sources in wild *Lactuca* spp. (Beharav et al. 2006; Lebeda et al. 2002; Lebeda and Zinkernagel 2003b), but there is still a lack of genetic and molecular data on variation and resistance in other wild *Lactuca* spp. (e.g. *L. virosa, L. saligna*; Kitner et al. 2008; Lebeda et al. 2002, 2007).

The first detailed genetic studies dealing with *L. saligna* resistance against *B. lactucae* were performed by Jeuken and Lindhout (2002) as a QTL analysis on plants of a *L. saligna* (resistant) × *L. sativa* (susceptible) cross. The phenotype of the F2 population showed a continuous range of resistance categories from completely resistant to completely susceptible, providing evidence that both qualitative and quantitative resistance were involved. Subsequent QTL mapping revealed a qualitative gene (R39) and three QTL (RBQ1, RBQ2 and RBQ3) accountable for the quantitative resistance. Some additional studies implied that resistance in *L. saligna* was quantitatively expressed and might be race non-specific. The current general view on *L. saligna* non-host resistance is that it is not explained by accumulation of race-specific resistance genes (*Dm* genes) but instead by resistance mechanisms based on QTL (Jeuken and Lindhout 2002).

Pyramiding of resistance genes in lettuce cultivars (Crute 1992b) forms a selection pressure that alters the structure of pathogen populations (Lebeda and Zinkernagel 2003a) and initiates the boom and bust cycle. On the other hand, gene-flow from cultivated to natural *Lactuca* spp.–*B. lactucae* populations and *vice versa* must also be considered (Lebeda 2002; Lebeda et al. 2008). An hypothesis of Hooftman et al. (2007) attributed the expanding distribution of prickly lettuce (*L. serriola*) in Europe to enhanced plant fitness by hybridisation with lettuce (*L. sativa*). However, this ecological study revealed that introgression of an important crop trait, downy mildew resistance, from lettuce into *L. serriola* hybrids was insignificant for plant reproductive fitness. In contrast, effectiveness of some resistance traits introduced to lettuce from wild *Lactuca* spp. (esp. *L. serriola*) might be broken by pathogen populations present in wild plant pathosystems (Lebeda 2002; Lebeda et al. 2008).

Development of *Bremia lactucae*

Symptoms of lettuce downy mildew

Description of downy mildew symptoms on lettuce (*L. sativa*) can be found elsewhere (e.g. Crute and Dixon 1981; Davis et al. 1997). Symptoms typically appear as areas of chlorotic tissue, mostly delimited by the main veins, which is accompanied by profuse sporulation on the abaxial side of leaves in compatible interactions. In field conditions, the air-borne asexual conidia are the most important means of disease spread throughout the growing season. Intensity of sporulation as well as viability of conidia is influenced substantially by environmental factors (Judelson and Michelmore 1992; Nordskog et al. 2007) and by concentration of primary inoculum (Crute and Dickinson 1976).

Broad variation of phenotypic expression of *B. lactucae* infection was reported in both susceptible and resistant *Lactuca* spp. genotypes (Table 1; for review see Lebeda et al. 2001b, 2002, 2008). The phenotype of non-host resistance (e.g. in some *L. saligna* accessions and most Asteraceae species) is characterized by a lack of symptoms (Crute and Dickinson 1976; Lebeda and Reinink 1994; Lebeda and Syrovátko 1988; Sedlářová et al. 2001b). Nevertheless, expression of macroscopic chlorosis (Crute and Dickinson 1976), necrosis (Lebeda and Reinink 1994; Norwood et al. 1981) or sub-epidermal necrosis (Lebeda and Reinink 1994; Lebeda et al. 2006) was also recorded. Expression of host resistance symptoms varies according to the ontogenetic stage of the host, as it was found to differ between the cotyledons and adult plants within the same interaction (Crute and Dickinson 1976; Lebeda and Reinink 1991; Lebeda et al. 2006). In lettuce–*Bremia* interactions a wide array of symptoms occurs, ranging from no visible symptoms to an extensive necrotization (incompatibility), and from limited sporulation (incomplete resistance) to profuse sporulation without any other visible symptoms (full compatibility; Lebeda et al. 2002). These categories of symptoms are highly specific and conditioned by race-specific resistance *Dm* genes in many cases. For example, *Dm7* conditions reduced sporulation and necrosis in some genotype–race interactions (Crute and Johnson 1976). Wild *Lactuca* spp., e.g. *L. serriola* and *L. virosa*, have also been reported to express a broad spectrum of symptoms (Lebeda and Pink 1998; Norwood et al. 1981). There are many intermediate phenotypes between the extremes with a substantial influence of experimental and environmental conditions (for overview see Lebeda et al. 2001b, 2002).

Characteristics of leaf surface: influence on conidial germination and appressorium formation

The characteristics of plant leaf surfaces, referred to here as the indumentum and including the number and character of trichomes, thickness and composition of waxes, number and position of stomata, determine success or failure of pathogen spore deposition and subsequent ingress by infection structures. Spores deposited on the leaf surface face several obstacles to gaining host nutrients: cuticle, cell wall and plasma membrane (Lebeda et al. 2001b).

The cuticle is known as a 'two-step' barrier comprising an internal and external layer. The internal cuticle on the inner periclinal walls of epidermal cells functions primarily in water exchange regulation (Pesacreta and Hasenstein 1999). More important from the pathogen perspective is the external cuticle. Its structure and function have been documented in many plants but experimental data are lacking for *Lactuca* species. Study of indumentum characteristics revealed substantial differences among *Lactuca* species (Lebeda et al. 1999), but there are no data relating the indumentum pattern to *B. lactucae* germination. However, evidence does exist for a relationship between leaf epidermal characteristics in cultivars of potato (*Solanum tuberosum*) and expression of resistance or susceptibility to *Phytophthora infestans* (Mahajan and Dhillon 2003).

There are several crucial steps required prior to the start of oomycete pathogenesis similar to fungal pathogens, i.e. adhesion of spores to the plant surface, and the formation of germ tubes, appressoria and penetration pegs (reviewed in Latijnhouwers et al. 2003). Attachment of germinating spores is mediated through secretion of an extra-conidial matrix. As soon as a germ tube emerges from the *Hyaloperonospora arabidopsidis* (*H. parasitica*) conidium, an 'adhesive coctail' composed of proteins, glycoproteins and β-1,3-glucans is released (Carzaniga et al. 2001). Recently, Hardham (2007) brought together microscopic and molecular data relating to early stages of the infection process in *Phytophthora*, *Pythium* and *Hyaloperonospora* spp.

The mechanisms of retention and adhesion of *B. lactucae* conidia to host leaf surfaces have not been elucidated in detail. The pre-penetration phase was studied by Andrews (1975) who reported the possibility that *B. lactucae* absorbs nutrients (e.g. glucose) from the leaf surface. Many papers deal with *B. lactucae* spore germination, penetration and the further development of infection structures (reviewed in Lebeda et al. 2001b). Conidia start germination mostly at 1–3 h after inoculation on both non-host and host plants (summarized in Lebeda et al. 2002). Germ-tube length is a highly variable parameter (compare Fig. 1a and g; Lebeda and Pink 1998) and does not relate directly to the host resistance, i.e. to *Dm* gene expression, but seems to be specific for each host genotype–parasite race interaction (Lebeda et al. 2001b, 2006). In general, significantly shorter germ tubes develop on wild *Lactuca* spp. than on lettuce (*L. sativa*) genotypes (Lebeda and Pink 1998; Lebeda et al. 2006).

The germination of *B. lactucae* conidia (Fig. 1a) is affected by environmental conditions, such as temperature (Sargent 1976; Sargent and Payne 1974). The peripheral cytoplasm of conidia in the 'dormant' stage contains lipid droplets which are dispersed during the phase of activation (preceding germination). This phase is followed by activation of dictyosomes and endoplasmatic reticulum. Mobilisation of reserves for development of the germ tube tip is accompanied by increased lipolytic activity in mitochondria and esterase activity in vacuoles (Duddridge and Sargent 1978).

Formation of oomycete appressoria, non-pigmented swellings of germ tube tips that differentiate penetration pegs, is not synchronised with germination and may be induced by topological features of the leaf surface (Latijnhouwers et al. 2003; Lebeda et al. 2001b). Significant differences in the frequency of appressorial formation were found between cotyledons (higher frequency) and leaf discs of adult plants (Lebeda and Reinink 1991). The influence of leaf surface character on appressorial formation was demonstrated in cv. Iceberg (genotype with high level of field resistance), where frequency of appressorial formation was significantly lower than on lettuce cultivars with ineffective race-specific resistance (Lebeda and Reinink 1991). A comparative study showed a higher incidence of *B. lactucae* appressorial development on lettuce (*L. sativa*) plants compared to wild relatives (*L. serriola,*

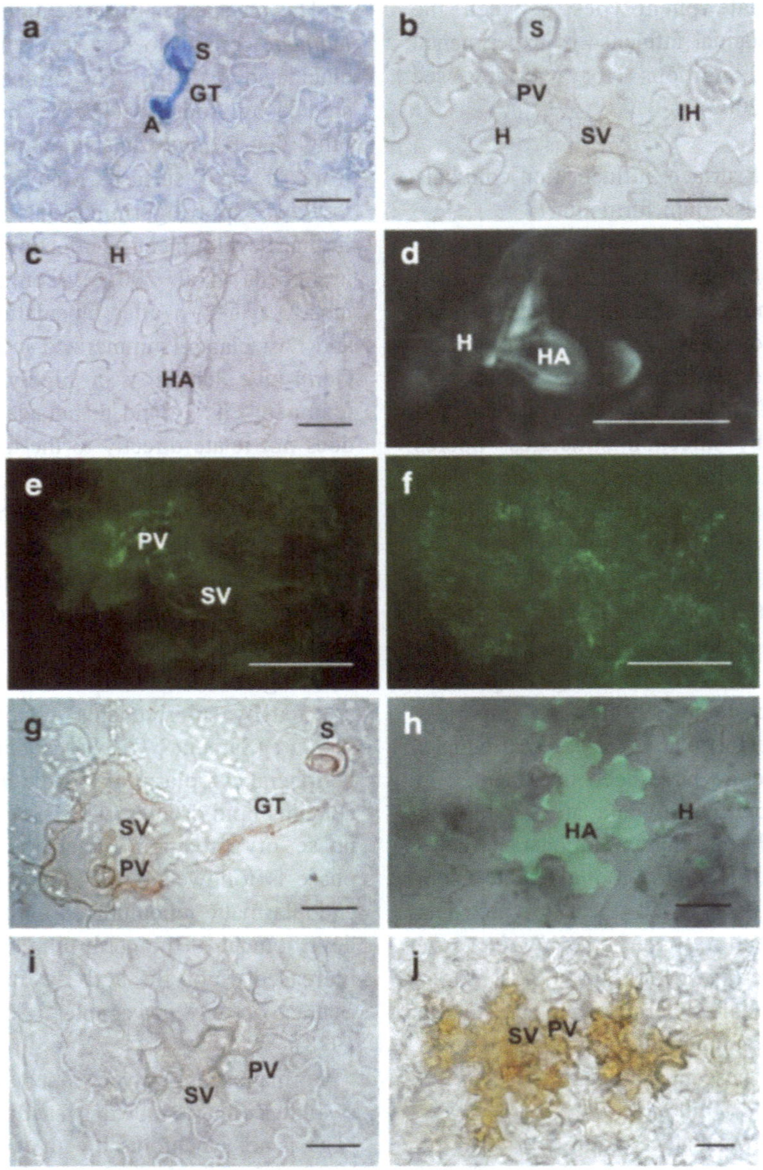

Fig. 1 Development of *Bremia lactucae* (race BL16) and the response of *Lactuca* spp. cells. **a–c**. Pathogen growth within tissues of susceptible *L. sativa* (Cobham Green); **a**, germination and appressorial formation (12 hai); **b**, colonisation of host tissues (48 hai); **c**, formation of numerous haustoria (120 hai). **d**. A detail of haustorium with callose deposited around its neck (168 hai), susceptible *L. sativa* (British Hilde). **e**, **f**. Realignment of host MTs due to infection (immunolocalisation of α-tubulin); **e**, microtubular 'basket' formed with host plasma membrane invagination (48 hai), susceptible *L. sativa* (UCDM2); **f**, depolymerization of cortical microtubules induced by initiation of HR (48 hai), resistant *L. serriola* (PIVT 1309). **g**. Peroxidase activity localised both in pathogen infection structures and in the cell of susceptible *L. sativa* (UCDM2; 24 hai). **h**. Signal for NO in epidermal cell of susceptible *L. sativa* (Cobham Green) penetrated by haustoria beneath the growing hypha (192 hai). **i**, **j**. Hypersensitive reaction (HR); **i**, initial stages of HR with granulation of cytoplasm (48 hai), *L. virosa* (NVRS 10.001602); **j**, necrosis, a visible outcome of HR is caused by oxidation of phenolics (336 hai), resistant *L. serriola* (PIVT 1309). Infection structures: spore (*S*), germ tube (*GT*), appressorium (*A*), primary (*PV*) and secondary vesicle (*SV*), intracellular hypha (*IH*), intercellular hypha (*H*), haustorium (*HA*). The bar corresponds to 20 μm. The micrographs were obtained by conventional light microscopy (**a–c**, **g**, **i**, **j**), fluorescence microscopy (**d–f**) and confocal laser scanning microscopy (**h**). Photo courtesy by M. Sedlářová

L. saligna, L. virosa), whereas no significant differences were found among the wild *Lactuca* spp. (Sedlářová et al. 2001b).

Penetration and development of primary and secondary vesicles

Penetration of plant surfaces by oomycetes is performed by a combination of mechanical force and secreted chemicals, as with other fungal pathogens (Latijnhouwers et al. 2003; Lebeda et al. 2001a). Appressoria of *B. lactucae* are a prerequisite for penetration, and exert high pressure on cell walls allowing penetration pegs to pierce the periclinal cell wall, and to colonize the underlying epidermal cell by the formation of primary and secondary vesicles (Sargent et al. 1973). To date, the turgor pressure exerted by appressoria has not been quantified in any oomycete species. The cuticular penetration predominates in *B. lactucae*, with penetration via stomata incident in about 1–5% of germ tubes (Lebeda and Reinink 1991).

Environmental factors are crucial for the *B. lactucae* penetration process. The effect of temperature is especially important, as seen from an optimum of 15–20°C for germination (Sargent 1976) but 12–15°C for penetration (MacLean and Tommerup 1979). Some details related to the timing of this process were summarized by Lebeda et al. (2001b). The penetration rate is a frequently studied parameter in screening studies (e.g. Lebeda and Pink 1998; Lebeda et al. 2006; Sedlářová et al. 2001b).

Chemical degradation of the cuticle and cell wall is the second necessity for successful penetration of plant cells. Secretion of a wide range of degradative enzymes has been described for fungi and oomycetes (Lebeda et al. 2001a), including hemibiotrophic *Phytophthora* and *Pythium* spp. (Hardham 2007). Extensive genomic studies have initiated the characterization of genes encoding these enzymes. However, direct demonstration of the action of cell wall-degrading enzymes is a perspective for future work (Hardham 2007).

Data on *B. lactucae*-derived enzymes are quite limited, and only polygalacturonase, esterase and protease activities have been reported (Van Pelt-Heerschap and Smit-Bakker 1993). Pathogen lipolytic enzymes, mentioned above in the context of germination (Characteristics of leaf surface: influence on conidial germination and appressorium formation), enable lipid degradation in the cuticle and enable subsequent pathogenesis (Sargent et al. 1973). Lipase activity is also elevated during penetration and the formation of primary vesicles (Duddridge and Sargent 1978). At this stage, the lipolytic activity is localized in lomasomes which occur mostly at the periphery of the expanding vesicle, as illustrated by transmission electron microscopy (Zinkernagel 1985; Zinkernagel and Bartscherer 1978).

Once the pathogen overcomes the barriers of the cuticle and cell wall by the activity of cell-wall degrading enzymes (Lebeda et al. 2001a; Sargent et al. 1973), it gains access to the cell lumen. *Bremia lactucae* forms primary infection structures within the host epidermal cell by invagination of the host plasma membrane. Formation of the primary vesicle (PV), secondary vesicle (SV), and intracellular hypha (IH; Fig. 1b) are not initially followed by destruction of the host cell plasma membrane (Ingram et al. 1976). Only subsequently is the plasma membrane perforated, with colonization of sub-epidermal tissues by intercellular hyphae (H) and haustoria (HA) ensuing (Fig. 1c). Various aspects of this process were reviewed by Lebeda et al. (2001b).

Early stages after inoculation are very important for recognition and initiation of defence responses. In lettuce–*Bremia* interactions, both pre-haustorial and post-haustorial recognition occurs, based on specific *Dm*/Avr gene combinations (Mansfield et al. 1997). Incidence of PVs and SVs, as well as the timing of their formation, may differ in interactions with an identical phenotype (Lebeda and Pink 1998; Lebeda et al. 2006; Sedlářová et al. 2001b). Formation of PVs and SVs also can be found in non-host plants, though at significantly lower frequencies than in host plants (summarized in Lebeda et al. 2002). In some *L. saligna* interactions with *B. lactucae*, the SV represents the final stage of oomycete development which is considered to distinguish non-host resistance (Lebeda and Reinink 1994; Lebeda et al. 2002, 2006; Sedlářová et al. 2001b). Quite intriguing is *B. lactucae* development in interspecific hybrids of lettuce (*L. sativa*) with wild *Lactuca* spp., where an 'heterosis' effect was recorded (increased rate of *B. lactucae* infection structures in F1 hybrids compared to both parents). Details can be found elsewhere (Lebeda and Reinink 1994; Lebeda et al. 2001b, 2006).

Development of hyphae and haustoria

The most important period for hyphal development of *B. lactucae* is 24–48 hai, but the extent and speed of formation of intra- (IH) and inter-cellular (H) hyphae and haustoria (HA) is extremely variable among non-host and host genotypes (Table 1; Lebeda et al. 2001b, 2002). The growth of intercellular hyphae of *B. lactucae* in mesophyll tissue even starts at 12 hai in some compatible interactions (Sedlářová et al. 2001b). Although delayed, the development of IH and H occur in most incompatible (non-host and host) interactions (Lebeda and Schwinn 1994).

An extreme variability of incompatible reactions in *Lactuca* spp. has been described (Lebeda et al. 2006). In *L. saligna* (LSA/6), which is considered as a non-host genotype (Lebeda et al. 2001b), *B. lactucae* did not form any intra- and intercellular hyphae (Lebeda et al. 2006). However, in many other incompatible interactions both types of hyphae were formed with significant differences in frequency (Lebeda and Pink 1998; Lebeda and Reinink 1994; Lebeda et al. 2006). Formation of IH and H was suggested as a crucial developmental stage for *B. lactucae*, and as a limiting factor in host–parasite communication as well as expression of various resistance mechanisms (Lebeda et al. 2001b, 2006). Quantitative comparative studies showed significant variation in size (length and width) of hyphae (Lebeda and Pink 1998; Lebeda and Reinink 1994), thus supporting the assumption that these features relate to differences in the physiology of resistance (Lebeda and Reinink 1991, 1994; Lebeda and Pink 1998; Lebeda et al. 2006; Sedlářová et al. 2001b).

Pyriform haustoria, characteristic for *B. lactucae* (Voglmayr et al. 2004), originate as hyphal side branches in penetrated cells to accomplish parasitic feeding. The haustorium remains outside the plant protoplast and an altered interface is developed that probably assists uptake of nutrients and the exchange of signals between both partners (Spencer-Phillips 1997). Composition of the extrahaustorial membrane (EHM), separating the haustorium from the host cytoplasm, differs from the semi-permeable plasma membrane as described in detail for *Hyaloperonospora arabidopsidis* (O'Connell and Panstruga 2006). In this respect, the EHM in both oomycete and fungal infections (e.g. in powdery mildews; Koh et al. 2005) are similar. Callose deposits may be formed around haustoria in lettuce–*Bremia* interactions (for detail, see Plasma membrane homeostasis and deposition of callose).

The frequency and timing of haustorial formation and the final size of haustoria are very specific features of host–parasite interactions (Lebeda et al. 2001b). In non-host resistance of *L. saligna*, haustoria form neither on IH nor directly on SV (Lebeda et al. 2006). Frequency of haustorial formation varies specifically among *Lactuca* spp. genotypes carrying different *Dm* genes and/or R-factors for host resistance. In compatible host–parasite interactions, the frequency and size of haustoria is significantly higher than in incompatible interactions (Lebeda et al. 2002, 2006; Sedlářová et al. 2001b). The 'heterosis' effect mentioned for PV and SV also was also reported for haustoria (Lebeda and Reinink 1994; Lebeda and Pink 1998).

Components of host resistance and variability of defence

An integrated approach is being adopted in plant science to understand intercellular signalling, i.e. how plants perceive and respond to external and internal stimuli. Combination of molecular, chemical and electrical components is essential (Birch et al. 2006; Mansfield 2005; O'Connell and Panstruga 2006; Robatzek 2007; Takemoto and Hardham 2004; Walters and McRoberts 2006). Several chemical and physical factors that condition shifts in plant metabolism and architecture induced by oomycete pathogenesis, as well as their importance for resistance of *Lactuca* spp. to *B. lactucae*, are considered below.

Reactive oxygen species (ROS), antioxidants and ROS-scavenging enzymes

Release of reactive oxygen (ROS), nitrogen (RNS) and sulphur (RSS) species intermediates, combined with transport of phytohormones, are amongst the early chemical signals during plant–pathogen interactions. ROS affect establishment of infection, enable redox signal transduction (e.g. hydrogen peroxide together with NO and SA amplifies resistance responses; Delledonne et al. 2003) and trigger programmed cell death (Kamoun et al. 1999). Therefore, generation of

ROS may serve as a marker of pathogenesis and/or plant defence initiation and progress.

Hydrogen peroxide (H_2O_2), a secondary messenger molecule, was accumulated in *Lactuca* spp. tissues challenged by *B. lactucae,* whereas superoxide (O_2^-) was not detected (Sedlářová et al. 2007a). Dramatic changes of H_2O_2 correlate with race-specific resistance, especially in *L. virosa* where it is characterized by early HR onset. In contrast, the supposed non-host resistance in *L. saligna* (CGN 05271) is accompanied by only minor changes in the level of H_2O_2, the content of which is generally lower compared to the other species (Sedlářová et al. 2007a).

High antioxidant status in plants was reported to hinder transportation of ROS across the cell (Neill et al. 2002); therefore our experiments have included the use of antioxidant enzymes and non-enzymatic ROS scavengers. Changes in peroxidase (POX), catalase and polyphenoloxidase activities in lettuce tissue, in relation to the infection process of *B. lactucae,* were demonstrated by Zinkernagel (1986). Our study focused on the dynamics and isozyme spectrum of three ROS-scavenging enzymes, catalase, peroxidase and superoxide dismutase, and unveiled the importance of peroxidase (POX). However, POX activity (Fig. 1g) was found only in the cytosolic fraction, with a higher basic level in wild *Lactuca* spp. compared to cultivated lettuce. Increase of POX activity was linked to expression of race-specific resistance in prickly lettuce (*L. serriola*) and great lettuce (*L. virosa*), with a two-peak timing (6–12 hai, the recognition phase, and from 24 hai at induction of HR; Sedlářová et al. 2007a). The relationship between the increase of pre-infection POX activity and level of field resistance to *B. lactucae* was demonstrated in lettuce (*L. sativa*) cultivars and accessions of *L. serriola* (Reuveni et al. 1991). Thus POX was proposed to serve as a marker in the selection for field resistance to different downy mildew pathogens (Lebeda and Schwinn 1994). From a large group of molecules with an antioxidative action, contents of quercetin and rutin were studied in the leaf extracts (for details see Flavonoids, phenolic acids and PAL).

Nevertheless, the complexity of leaf phytochemistry raises the possibility that many other antioxidants may be involved in the interplay between *Lactuca* spp. and *B. lactucae*. This merits investigation as it would provide a better understanding of host–parasite interactions.

Nitric oxide, NO synthase and NO modulators

Nitric oxide (NO) performs a variety of phytochemical roles during pathogenesis. NO and its metabolites mediate transcription of specific genes during pathogenesis (Neill et al. 2002); synchronized formation of NO and H_2O_2 co-regulates the cell death programme and the phenylpropanoid pathway (Dellendone et al. 2003).

In lettuce–*Bremia* interactions, the formation of NO was localized in cells penetrated by either primary infection structures or haustoria (Fig. 1h). NO synthase (NOS) activity was followed by the oxyhemoglobin method to detect NO production in lettuce and wild *Lactuca* spp. leaf extracts up to 216 hai. A significant increase of NOS activity was found in *L. virosa* early after inoculation (4–8 hai), with a second lower peak at 168 hai. Non-host resistance of *L. saligna* (CGN 05271) correlated with low amounts of NO production and relatively small-scale increase of NOS activity (Petřivalský, unpubl.).

Modulators of NO metabolism were applied to *L. sativa* tissues to follow their influence on *B. lactucae* development up to 48 hai (Petřivalský et al. 2007). Sodium nitroprusside, a model NO donor, decreased conidial germination rate at 4 hai and strongly inhibited further pathogen growth. On the contrary, PTIO as a specific NO scavenger, showed a strong stimulatory effect on pathogen development at 24–48 hai. However, no significant effect of either L-NAME (competitive inhibitor of animal nitric oxide synthases) or sodium tungstate (specific inhibitor of plant nitrate reductase) was found. This may be explained by either the possible contribution of another NO-generating system in *Lactuca* spp., or the lower bioavailability and chemical stability of these substances during leaf tissue treatment (for more details, see Petřivalský et al. 2007).

Flavonoids, phenolic acids and PAL

Phenolic compounds are abundant in plants of the Asteraceae family (Bohm and Stuessy 2001). Screening was conducted in the early 1980s in order to utilize flavoids and flavonols of the genus *Lactuca* in chemotaxonomy (Rees and Harborne 1984). Recently, the phenolic compounds in lettuce have been studied in relation to: leafy vegetable processing to avoid browning due to mechanical injury (Saltveita et al. 2005) and human medications for anti-inflammatory, anti-bacterial, anti-diabetic and anti-proliferative effects (Chen et al. 2007).

Quercetin, rutin, caffeic acid and chlorogenic acid and several other phenolic compounds are known as major components of lettuce extracts (e.g. Chen et al. 2007). Quercetin, one of the aglycones, was reported from *L. sativa, L. serriola* and *L. virosa*, whilst only traces were found in *L. saligna* and (Rees and Harborne 1984). Quercetin and rutin molecules operate as strong antioxidants. A study was conducted to measure their amount in all four of these *Lactuca* species during the course of *B. lactucae* pathogenesis, determined by the accumulation of autofluorescent phenolics near the plasma membrane of penetrated cells (Bennett et al. 1996; Sedlářová and Lebeda 2001). Quantitative analysis of leaf extracts led to the finding that *L. sativa* genotypes with non-effective race-specific resistance significantly differ in quercetin content (approx. 10 μmol g^{-1} FW) from *L. sativa* with effective race-specific resistance and three wild lettuce species (less than 1 μmol g^{-1} FW). This may help in the balancing of oxidative processes induced by *B. lactucae*. Content of rutin varied slightly from 0.28 μmol g^{-1} FW in *L. virosa* to 0.87 μmol g^{-1} FW in *L. sativa* (Petřivalský, unpubl.). Whilst no striking linkage between rutin level and genotype susceptibility/resistance was found, the external application of rutin solution to lettuce tissues delayed *B. lactucae* germination and penetration (Petřivalský et al. 2007).

The content of phenolic acids (PAs) changes during ontogenesis of *Lactuca* spp.; in cotyledons, chlorogenic acid prevails, whereas amounts of other PAs increase with plant development (Grúz, unpubl.). A time-course study of PAs in adult *L. sativa* (cv. Mariska) showed significant changes in caffeic acid and minor changes in chlorogenic acids after inoculation with incompatible *B. lactucae* race BL16. A two-peak (6–24 and 72 hai) decrease in their level (Grúz, unpubl.) corresponds with induction of oxidative processes (see also previously).

Preliminary studies of phenylalanine ammonium lyase (PAL), a key enzyme of the phenylpropanoid pathway, have not disclosed a relationship between PAL activity and *B. lactucae* colonization (Sedlářová, unpubl.). The phenylpropanoid pathway is also known to be connected with the formation of structural barriers to pathogen ingress by the deposition of lignin (Mauch-Mani and Slusarenko 1996). Although large pools of phenolic compounds that might serve as a source of precursors for incorporation in the lettuce cell wall were found due to *B. lactucae* challenge (Bennett et al. 1996), lignification has not been proved (Sedlářová and Lebeda 2001).

Reorganisation of the cytoskeleton

Host cells challenged by oomycetes undergo drastic changes similar to cells targeted by fungal pathogens (Latijnhouwers et al. 2003; Takemoto and Hardham 2004). Rapid rearrangements of cytoskeletal components (microtubules and F-actin) begin even prior to penetration of the cell wall (during maturation of the appressorium), and link to the relocation of cytoplasm, nuclei and other organelles within epidermal cells in contact with the pathogen (Koh et al. 2005; Takemoto et al. 2003). Pathogens that continue colonisation beyond the epidermis via intercellular hyphae induce similar alterations of architecture in mesophyll cells penetrated by haustoria (Spencer-Phillips 1997).

The multitude of binding proteins associated with the cytoskeleton and its extraordinary dynamics facilitate trafficking of many pathogen-derived signals. The vital role of the host cytoskeleton in non-pathogen and pathogen recognition (Takemoto et al. 2003), the binding of effector molecules (Binet et al. 2001), gene expression (Hamada 2007) and in relation to defects in cell wall microfibril orientation (Wasteneys 2004) are well documented. In lettuce–*Bremia* interactions, actin filaments were not detected in epidermal cells after contact with the pathogen, whereas cortical microtubules (MTs) supported invagination of the plasma membrane and formation of primary and secondary vesicles (Sedlářová et al. 2001a). In compatible interactions, such a unique layout resembles a 'microtubular basket' (Fig. 1e), and is characterised by a high density of MTs at the necks of vesicles (Sedlářová, unpubl.). This suggests a role in the deposition of callose at these locations (Sedlářová and Lebeda 2001). In resistant plants, the timing and extent of the destruction of MTs (Fig. 1f) is correlated with a hypersensitive reaction and typically affects one cell per infection site in *L. sativa*, and 2–3 cells in *L. virosa* (Lebeda and Pink 1998; Lebeda et al. 2006; Sedlářová et al. 2001b).

Construction of *Arabidopsis thaliana* mutants with GFP-tagged cytoskeleton or organelles made it possible to follow subcellular changes *in vivo* (Hardham 2007; Koh et al. 2005; Takemoto et al. 2003). Rapid and continuing intracellular realignment during *Hyaloperonospora arabidopsidis* challenge was shown elegantly by Takemoto et al. (2003), including 'focusing-to-pathogen' of F-actin below the penetration site and in neighbouring cells. Secretion of plant materials around the infection site, indicated by an aggregation of ER and Golgi bodies, did not stop penetration by an avirulent isolate of *H. arabidopsidis* and even the non-pathogen *Phytophthora sojae*.

A number of detailed experimental data raise the question: what facilitates the extreme plasticity of the plant cytoskeleton in reaction to oomycete and fungal pathogens? As well as the high degree of conservation of tubulin and actin throughout a variety of genomes, a wide array of associated molecules (proteins, RNA) was found in the cytoskeletal complexes. The rapid rearrangement of MTs in reaction to external/internal stimuli occurs because the nucleation sites of MTs are based on γ-tubulin anchors which can be relocated easily within the plant cell (Hamada 2007).

Plasma membrane homeostasis and deposition of callose

Similar adaptations, including haustorial development, have evolved in oomycetes and fungi to enhance a parasitic life strategy (Latijnhouwers et al. 2003). As the integrity of the plasma membrane is a crucial prerequisite for plant cell functionality, as well as the homestasis of cellular processes, the biotrophic pathogens deploy mechanisms to minimise disruption of the host cell (Glazebrook 2005; Grenville-Briggs and van West 2005; Koh et al. 2005; O'Connell and Panstruga 2006; Walters and McRoberts 2006). The plasma membrane regulates osmotic processes (Bennett et al. 1996). Host transmembrane proteins are engaged in the perception of pathogen-associated molecular patterns (O'Connell and Panstruga 2006; Robatzek 2007) and with the aid of the cytoskeleton, facilitate vesicle trafficking (Robatzek 2007; Takemoto and Hardham 2004).

After penetration of the host cell wall, primary and secondary vesicles of *B. lactucae* are formed in the first epidermal cell by invagination of the host plasma membrane (Fig. 1g), as described above. Initiation of intercellular growth is usually linked to membrane damage (Woods et al. 1988) and accumulation of autofluorescent phenolics (Bennett et al. 1996). Pathogen recognition in resistant cells results in irreversible loss of membrane integrity and initiation of the HR. In compatible interactions, the hyphae growing between mesophyll cells penetrate adjacent cell walls (Fig. 1c) to form haustoria that invaginate host plasma membranes. A new interface, the extra-haustorial membrane (EHM), arises from the secretion of proteinaceous and carbohydrate compounds by both partners (Koh et al. 2005; O'Connell and Panstruga 2006).

Formation of *Bremia* infection structures is associated with the deposition of callose, especially in the necks between PVs and SVs, and between SVs and haustoria, forming sheath-like structures around haustoria (Fig. 1d; Sargent et al. 1973). The callose is of host origin and the strongest deposition was reported in compatible interactions (Sedlářová and Lebeda 2001). The chemical composition of callose is very similar to components of extracellular matrices released by *H. arabidopsidis* spores upon germination, namely β-1,3-glucans which has a protective action (Carzaniga et al. 2001).

Hypersensitive response: extent and timing

The hypersensitive reaction (HR), a form of programmed cell death (Kamoun et al. 1999), is one of the most important features in race-specific resistance of lettuce to *B. lactucae* (Lebeda et al. 2001b). On a small scale, it has been reported to also occur in compatible or non-host interactions (Lebeda et al. 2002). Necrosis of affected plant cells and tissues (Fig. 1i,j) is used for phenotypic evaluation (Lebeda and Petrželová 2008). Contemporary methods are able to detect the onset of cell death before visible symptoms occur, either by measuring natural bioluminiscence as the emission of biophotons (Mansfield 2005) or by the application of osmotic stress to test the plasma membrane functionality (Bennett et al. 1996). It was concluded that lettuce cells undergoing the HR experience a prolonged oxidative stress (Bestwick et al. 2001).

In a wide range of lettuce–*Bremia* interactions, the substantial differences found in timing and rate of formation of infection structures correspond with detailed histological investigations of HR (Lebeda and Pink 1998; Lebeda and Reinink 1994; Lebeda et al. 2001b, 2002, 2006; Sedlářová et al. 2001b). Post-haustorial resistance in *Lactuca* spp. with race specificity includes an intensive HR. Although the extent of the HR is specific for a genotype–race interaction, the number of cells involved in the HR is generally higher than one in wild lettuce species (Fig. 1i,j), especially *L. virosa*, where cells of underlying mesophyll tissue often also show the HR (Lebeda et al. 2006; Sedlářová et al. 2001b). Conversely, the non-host resistance in *L. saligna* (CGN 05271) is expressed before haustorial formation (Sedlářová et al. 2001b), and is characterized by a lack of the HR which might relate to the previously mentioned adjustment of oxidative processes (see Reactive oxygen species (ROS), antioxidants and ROS-scavenging enzymes and nitric oxide, NO synthase and NO modulators; details in Sedlářová et al., 2007a).

Plant energetics

The life strategy of parasites is based on the need to derive nutrients from host tissues, thus affecting plant energetics. Economically damaging infections of crops by powdery mildews and rusts has led to intensive research in this area, with the aim of reducing losses in yield. Although the chlorotic symptoms typical of downy mildew infections are well known (Lebeda and Schwinn 1994), relatively little research has attempted to elucidate the consequent changes of host photosynthetic processes. Interactions with hemibiotrophic *Phytophthora* species, *P. capsici* (Aguirreolea et al. 1995), *P. citricola* and *P. cambivora* (Fleischmann et al. 2002, 2005), *P. infestans* (Restrepo et al. 2005; Schnabel et al. 1998) and *P. nicotianae* (Scharte et al. 2005), and the necrotroph *Pythium aphanidermatum* (Johnstone et al. 2005) have been investigated. As for biotrophs, the effect of white blister rust (*Albugo candida*) on the photosynthetic and carbohydrate metabolism of *Arabidopsis thaliana* was studied by Tang et al. (1996). They showed that infection caused a decrease in chlorophyll and Rubisco content, as well as an inhibition of photosynthetic rates which might result from accumulation of soluble carbohydrates and starch in infected leaves. Moriondo et al. (2005) indicated that *Plasmopara viticola* reduced functional green leaf area of grapevine (*Vitis vinifera*), with decreased chlorophyll content, and affected stomatal closure and transpiration in lesions and adjacent tissues. The reduced assimilation rate was not limited by changes in electron transport capacity and generation of ATP and NADPH.

Prior to Restrepo et al. (2005) reporting the suppression of a large group of photosynthesis-related genes in susceptible potato following *Phytophthora infestans* infection, photosynthesis was supposed to be affected indirectly. Loss of photosynthetic activity has been attributed to the reduction of photosynthetically-active leaf area and lower pigment content (Moriondo et al. 2005; Tang et al. 1996), and to changes in stomatal aperture and transpiration (Aguirreolea et al. 1995). Impairment of the photosynthetic apparatus also was reported by other authors (Koch et al. 1994; Fleischmann et al. 2005).

Recently, we investigated the impact of *B. lactucae* (race BL16) on photosynthetic parameters of *Lactuca* spp. plants. Analyses of chlorophyll fluorescence induction curves and content of photosynthetic pigments (chlorophylls and carotenoids) revealed a linkage between deterioration of the photosynthetic apparatus and compatibility by 13 dai. In susceptible genotypes of *L. sativa* (Cobham Green and UCMD2), an impairment of photosystem II (PSII) photochemistry and decreased content of photosynthetic pigments were noticed due to profuse growth of *B. lactucae*. In resistant *L. virosa* (NVRS 10.001602), characterised by rapid pathogen elimination via the HR (Sedlářová et al. 2001b), no significant influence of inoculation was observed (Prokopová, unpubl.). Our data are in agreement with results of other authors such as Schnabel et al. (1998), who showed a correlation between degree of resistance and changes in PSII photochemistry. Transmission electron microscopy (TEM) studies of susceptible *L. sativa* (Cobham Green) cells revealed a decrease of chloroplast area in sections of cells following challenge by *B. lactucae* (race BL16). On the other hand, the number and area of starch granules within host cells did not change within 13 dai (Novotný and Sedlářová, unpubl.).

Further experiments addressed the effect of treating host tissues with cytokinins (*meta*-topolin, BAP and 3-MeOBAPR) before inoculation. These compounds significantly reduced *B. lactucae* sporulation on susceptible lettuce tissues, and influenced optical parametres of the leaves (Prokopová et al. 2007; Sedlářová et al. 2007b). All cytokinins (CKs) applied retained the maximal quantum yield of PSII photochemistry and content of photosynthetic pigments in infected leaf discs, whereas they slightly reduced these parameters in non-infected controls (Prokopová, unpubl.). CKs also increased the area and number of chloroplasts and starch granules within infected tissues alone (Novotný and Sedlářová, unpubl.). These results relate to several physiological aspects. In plant–parasite interactions the parasitic partner operates as sink to which carbohydrates are relocated. CKs are also known to enhance sinks for the transport of solutes, e.g. from older to younger parts of a plant. Some biotrophs themselves produce CKs to modulate nutrient transport, tissue senescence and even host morphology (Walters and McRoberts 2006). Either infection or application of CKs alone disturbs the normal functioning of processes linked to photosynthesis. In the concurrent presence of both factors, CKs might suppress the effect of pathogenesis. Increasing concentration of CKs in leaf tissues by external application preceding inoculation thus represents a competitor with pathogen-driven relocation of photosynthates. Although the feedback mechanisms of assimilates on the enzymes of photosynthesis are quite complex and still not completely understood, the increase of invertase leading to carbohydrate accumulation seems to be a principal mechanism (Walters and McRoberts 2006).

Future perspectives

From the data summarized in this review it is evident that relationships between plants and oomycetes are heterogeneous and complex. During the last two decades, the understanding of the biology of these associations has advanced significantly. However, much basic information is still needed before the very complicated mosaic of components, processes, interactions and feedbacks can be assembled to obtain a more complete view about host–oomycete specificity. In particular, the *Lactuca* spp. and *B. lactucae* interactions have been recognized as reflecting a very diverse and complicated system (Lebeda et al. 2001b, 2002), and provide a suitable model for studies of host-parasite specificity and variability of plant defence mechanisms (Lebeda et al. 2001b).

Several investigations worldwide on the lettuce–*Bremia* pathosystem are expected to contribute substantially to complement our present fragmentary knowledge. These include: (1) molecular and ultrastructural studies on pathogen taxonomy and phylogeny (Choi et al. 2007; Göker et al. 2007; Voglmayr, 2008; Voglmayr et al. 2004); (2) characterization of *B. lactucae* population structure, virulence evolution (Lebeda and Zinkernagel 2003a) and gene-flow between natural and cultivated plant pathosystems (Lebeda et al. 2008); (3) collecting and characterization of *Lactuca* spp. germplasm variation (Lebeda et al. 2007), screening of lettuce germplasm for resistance to *B. lactucae* (Lebeda and Petrželová 2008) and detection of new sources of resistance (Beharav et al. 2006; Lebeda and Zinkernagel 2003b; Petrželová et al. 2007), their genetic characterization (Bonnier et al. 1994) and utilization in lettuce breeding (Lebeda et al. 2007); (4) detailed studies of resistance mechanisms within lettuce genotypes and *Lactuca* spp. (Bestwick et al. 2001; Grube and Ochoa 2005; Jeuken and Lindhout 2002; Lebeda et al. 2002, 2006; Sedlářová et al. 2007a); (5.) molecular mapping of genes responsible for resistance/susceptibility and virulence/avirulence both in host plants and the pathogen (Kuang et al. 2004, 2006; Michelmore and Wong 2008); (6.) development of new control methods based on screening of various 'natural' compounds for their activity to suppress lettuce downy mildew (Portz et al. 2008; Sedlářová et al. 2007b), utilization of new generations of fungicides (Cohen et al. 2008; Gisi and Sierotzki 2008) and induced resistance.

Acknowledgements The work was funded by grants from the Czech Ministry of Education (MSM 6198959215) and Grant Agency of the Czech Republic (GP 522/02/D011). The authors thank Dr. P.T.N. Spencer-Phillips (UWE, Bristol, UK) for critical reading of the first draft of the manuscript, participants at the 2nd International Downy Mildews Symposium (Olomouc, Czech Republic, 2007) for valuable discussions, and Olympus C&S (Prague) for supporting the arrangement and development of the Laboratory of Confocal Microscopy in the Department of Botany at Palacky University in Olomouc.

References

Aguirreolea, J., Irigoyen, J., Sanchez-Diaz, M., & Salaverri, J. (1995). Physiological alterations in pepper during wilt induced by *Phytophthora capsici* and soil water deficit. *Plant Pathology, 44*, 587–596.

Andrews, J. H. (1975). Distribution of label from ^{3}H-glucose and ^{3}H-leucine in lettuce cotyledons during the early stages of infection with *Bremia lactucae*. *Canadian Journal of Botany, 53*, 1103–1115.

Beharav, A., Lewinsohn, D., Lebeda, A., & Nevo, E. (2006). New wild *Lactuca* genetic resources with resistance against *Bremia lactucae*. *Genetic Resources and Crop Evolution, 53*, 467–474.

Bennett, M., Gallagher, M., Fagg, J., Bestwick, C., Paul, T., Beale, M., et al. (1996). The hypersensitive reaction, membrane damage and accumulation of autofluorescence phenolics in lettuce cells challenged by *Bremia lactucae*. *The Plant Journal*, *9*, 851–865.

Bestwick, C. S., Adam, A. L., Puri, N., & Mansfield, J. W. (2001). Characterisation of changes to pro- and anti-oxidant enzyme activities during the hypersensitive reaction in lettuce (*Lactuca sativa* L.). *Plant Science*, *161*, 497–506.

Binet, M. N., Humbert, C., Lecourieux, D., Vantard, M., & Pugin, A. (2001). Disruption of microtubular cytoskeleton induced by cryptogein, an elicitor of hypersensitive response in tobacco cells. *Plant Physiology*, *125*, 564–572.

Birch, P. R. J., Rehmany, A. P., Pritchard, L., Kamoun, S., & Beynon, J. L. (2006). Trafficking arms: Oomycete effectors enter host plant cells. *Trends in Microbiology*, *14*, 8–11.

Bohm, B. A., & Stuessy, T. F. (2001). *Flavonoids of the sunflower family (Asteraceae)*. Wien: Springer.

Bonnier, F. J. M., Reinink, K., & Groenwold, R. (1994). Genetic analysis of *Lactuca* accessions with new major gene resistance to lettuce downy mildew. *Phytopathology*, *84*, 462–468.

Carzaniga, R., Bowyer, P., & O'Connell, R. J. (2001). Production of extracellular matrices during development of infection structures by the downy mildew *Peronospora parasitica*. *New Phytologist*, *149*, 83–93.

Chen, Y. H., Chen, H. Y., Hsu, C. L., & Yen, G. C. (2007). Induction of apoptosis by *Lactuca indica* L. in human leukemia cell line and its active components. *Journal of Agricultural and Food Chemistry*, *55*, 1743–1749.

Choi, Y. J., Hong, S. B., & Shin, H. D. (2007). Extreme size and sequence variation in the ITS rDNA of *Bremia lactucae*. *Mycopathologia*, *163*, 91–95.

Clark, J. S. C., & Spencer-Phillips, P. T. N. (2004). The compatible interaction in downy mildew infections. In P. T. N. Spencer-Phillips, & M. Jeger (Eds.) *Advances in downy mildew research, vol. 2* (pp. 1–34). Dordrecht: Kluwer.

Cohen, Y., Rubin, E., & Gotlieb, D. (2008). Activity of carboxylic acid amide (CAA) fungicides against *Bremia lactucae*. *European Journal of Plant Pathology* (this issue).

Crute, I. R. (1992a). Downy mildew of lettuce. In H. S. Chaube, J. Kumar, A. N. Mukhopadhyay, & U. S. Singh (Eds.) *Plant diseases of international importance, vol. II. Diseases of vegetable and oil seed crops* (pp. 165–185). New Jersey: Prentice Hall.

Crute, I. R. (1992b). From breeding to cloning (and back again?): A case study with lettuce downy mildew. *Annual Review of Phytopathology*, *30*, 485–506.

Crute, I. R., & Dickinson, C. R. (1976). The behaviour of *Bremia lactucae* on cultivars of *Lactuca sativa* and on other Composites. *Annals of Applied Biology*, *82*, 433–450.

Crute, I. R., & Dixon, G. R. (1981). Downy mildew diseases caused by the genus *Bremia* Regel. In D. M. Spencer (Ed.) *The downy mildews* (pp. 421–460). London: Academic.

Crute, I. R., & Johnson, A. G. (1976). The genetic relationship between races of *Bremia lactucae* and cultivars of *Lactuca sativa*. *Annals of Applied Biology*, *83*, 125–137.

Crute, I. R., & Norwood, J. M. (1981). The identification and characteristics of field resistance to lettuce downy mildew (*Bremia lactucae* Regel). *Euphytica*, *30*, 707–717.

Davis, R. M., Subbarao, K. V., Raid, R. N., & Kurtz, E. A. (1997). *Compendium of lettuce diseases*. St. Paul, MN. USA: APS.

Delledonne, M., Polverari, A., & Murgia, I. (2003). The functions of nitric oxide-mediated signalling and changes in gene expression during the hypersensitive response. *Antioxidants and Redox Signaling*, *5*, 33–41.

Duddridge, J. A., & Sargent, J. A. (1978). A cytochemical study of lipolytic activity in *Bremia lactucae* Regel during germination of the conidium and penetration of the host. *Physiological Plant Pathology*, *12*, 289–296.

Fleischmann, F., Koehl, J., Portz, R., Beltrame, A. B., & Oßwald, W. F. (2005). Physiological changes of *Fagus silvatica* seedlings infected with *Phytophthora citricola* and the contribution of its elicitin "Citricolin" to pathogenesis. *Plant Biology*, *7*, 650–658.

Fleischmann, F., Schneider, D., Matyssek, R., & Oßwald, W. F. (2002). Investigation on net CO_2 assimilation, transpiration and root growth of *Fagus sylvatica* infected with four different *Phytophthora* species. *Plant Biology*, *4*, 144–152.

Gisi, U., & Sierotzki, H. (2008). Fungicides modes of action and resistance in downy mildews. *European Journal of Plant Pathology* DOI 10.1007/s10658-008-9290-5.

Glazebrook, J. (2005). Contrasting mechanisms of defense against biotrophic and necrotrophic pathogens. *Annual Review of Phytopathology*, *43*, 205–227.

Göker, M., Voglmayr, H., Riethmüller, A., & Oberwinkler, F. (2007). How do obligate parasites evolve? A multi-gene phylogenetic analysis of downy mildews. *Fungal Genetics and Biology*, *44*, 105–122.

Grenville-Briggs, L. J., & van West, P. (2005). The biotrophic stages of oomycete–plant interactions. *Advances in Applied Microbiology*, *57*, 217–243.

Grube, R. C., & Ochoa, O. E. (2005). Comparative genetic analysis of field resistance to downy mildew in the lettuce cultivars 'Grand Rapids' and 'Iceberg'. *Euphytica*, *142*, 205–215.

Hamada, T. (2007). Microtubule-associated proteins in higher plants. *Journal of Plant Research*, *120*, 79–98.

Hardham, A. R. (2007). Cell biology of plant–oomycete interactions. *Cellular Microbiology*, *9*, 31–39.

Holub, E. B., & Cooper, A. (2004). Matrix, reinvention in plants: How genetics is unveiling secrets of non-host disease resistance. *Trends in Plant Science*, *9*, 211–214.

Hooftman, D. A. P., Nieuwenhuis, B. P. S., Posthuma, K. I., Oostermeijer, J. G. B., & den Nijs, H. J. C. M. (2007). Introgression potential of downy mildew resistance from lettuce to *Lactuca serriola* and its relevance for plant fitness. *Basic and Applied Ecology*, *8*, 135–146.

Hulbert, S. H., Webb, C. A., Smith, S. M., & Sun, Q. (2001). Resistance gene complexes: Evolution and utilization. *Annual Review of Phytopathology*, *39*, 285–312.

Ingram, D. S., Sargent, J. A., & Tommerup, I. C. (1976). Structural aspects of infection by biotrophic fungi. In J. Friend & D. R. Threlfall (Eds.), *Biochemical aspects of plant–parasitic relationships* (pp. 43–78). *Annual Proceedings of Phytochemical Society, 13*.

Jeuken, M., & Lindhout, P. (2002). *Lactuca saligna*, a non-host for lettuce downy mildew (*Bremia lactucae*), harbors a new race-specific *Dm* gene and three QTLs for resistance. *Theoretical and Applied Genetics*, *105*, 384–391.

Johnstone, M., Chatterton, S., Sutton, J. C., & Grodzinski, B. (2005). Net carbon gain and growth of bell peppers, *Capsicum annuum* 'Cubico', following root infection by *Pythium aphanidermatum*. *Phytopathology*, *95*, 354–361.

Judelson, H. S., & Michelmore, R. W. (1992). Temperature and genotype interactions in the expression of host resistance in lettuce downy mildew. *Physiological and Molecular Plant Pathology, 40*, 233–245.

Kamoun, S. (2006). A catalogue of the effector secretome of plant pathogenic oomycetes. *Annual Review of Phytopathology, 44*, 41–60.

Kamoun, S., Huitema, E., & Vleeshouwers, V. G. A. A. (1999). Resistance to oomycetes: A general role for the hypersensitive response. *Trends in Plant Science, 4*, 196–200.

Kitner, M., Lebeda, A., Doležalová, I., Maras, M., Křístková, E., Nevo, E., Pavlícek, T., Meglic, V., & Beharav, A. (2008). AFLP analysis of *Lactuca saligna* gemplasm collections from four European and three Middle East Countries. *Israel Journal of Plant Sciences* (in press)

Koch, C., Noga, G., & Strittmatter, G. (1994). Photosynthetic electron transport is differentially affected during early stages of cultivar/race-specific interactions between potato and *Phytophthora infestans. Planta, 193*, 551–557.

Koh, S., André, A., Edwards, H., Ehrhardt, D., & Somerville, S. (2005). *Arabidopsis thaliana* subcellular responses to compatible *Erysiphe cichoracearum* infections. *The Plant Journal, 44*, 516–529.

Koike, S. T., & Ochoa, O. E. (2007). Downy mildew caused by *Bremia lactucae* on strawflower (*Helichrysum bracteatum*) in California. *Plant Disease, 91*, 326.

Kuang, H., Ochoa, O. E., Nevo, E., & Michelmore, R. W. (2006). The disease resistance gene *Dm3* is infrequent in natural populations of *Lactuca serriola* due to deletions and frequent gene conversions at the RGC2 locus. *The Plant Journal, 47*, 38–48.

Kuang, H., Woo, S. S., Meyers, B. C., Nevo, E., & Michelmore, R. W. (2004). Multiple genetic processes result in heterogeneous rates of evolution within the major cluster disease resistance genes in lettuce. *The Plant Cell, 16*, 2870–2894.

Latijnhouwers, M., de Wit, P. J., & Govers, F. (2003). Oomycetes and fungi: Similar weaponry to attack plants. *Trends in Microbiology, 11*, 462–469.

Lebeda, A. (1986). Specificity of interactions between wild *Lactuca* species and *Bremia lactucae* isolates from *Lactuca serriola. Journal of Phytopathology, 117*, 54–64.

Lebeda, A. (2002). Occurrence and variation in virulence of *Bremia lactucae* in natural populations of *Lactuca serriola*. In P. T. N. Spencer-Phillips, U. Gisi, & A. Lebeda (Eds.) *Advances in downy mildew research* (pp. 179–183). Dordrecht: Kluwer.

Lebeda, A., & Boukema, I. W. (1991). Further investigation of the specificity of interactions between wild *Lactuca* spp. and *Bremia lactucae* isolates from *Lactuca serriola. Journal of Phytopathology, 133*, 57–64.

Lebeda, A., Doležalová, I., Feráková, V., & Astley, D. (2004). Geographical distribution of wild *Lactuca* species (Asteraceae, Lactuceae). *The Botanical Review, 70*, 328–356.

Lebeda, A., Doležalová, I., Křístková, E., Vinter, V., Vránová, O., Doležal, K., et al. (1999). Complex research of taxonomy and ecobiology of wild *Lactuca* spp. genetic resources. In A. Lebeda, & E. Křístková (Eds.) *Eucarpia leafy vegetables '99* (pp. 117–131). Olomouc (Czech Republic): Palacký University in Olomouc.

Lebeda, A., Luhová, L., Sedlářová, M., & Jančová, D. (2001a). The role of enzymes in plant–fungal pathogens interactions. *Journal of Plant Diseases and Protection, 108*, 89–111.

Lebeda, A., & Petrželová, I. (2004). Occurrence of race-specific resistance to *Bremia lactucae* in *Lactuca serriola* germplasm originating from four European countries. In J. Vollmann, H. Grausgruber, & P. Ruckenbauer (Eds.) *Genetic variation for plant breeding* (pp. (pp. 113–116)). Vienna, Austria: EUCARPIA & BOKU-University of Natural Resources and Applied Life Sciences.

Lebeda, A., & Petrželová, I. (2008). Screening for resistance to lettuce downy mildew (*Bremia lactucae*). In M. Miranda & A. Lebeda (Eds.), Disease resistance mass screening techniques for banana and other crops. IAEA, Vienna and Springer, Dordrecht, the Netherlands (in press).

Lebeda, A., Petrželová, I., & Maryška, Z. (2008). Structure and variation in the wild-plant pathosystem: *Lactuca serriola–Bremia lactucae. European Journal of Plant Pathology*, DOI 10.1007/s10658-008-9291-4.

Lebeda, A., & Pink, D. A. C. (1998). Histological aspects of the response of wild *Lactuca* spp. and their hybrids, with *L. sativa* to lettuce downy mildew (*Bremia lactucae*). *Plant Pathology, 47*, 723–736.

Lebeda, A., Pink, D. A. C., & Astley, D. (2002). Aspects of the interactions between wild *Lactuca* spp. and related genera and lettuce downy mildew (*Bremia lactucae*). In P. T. N. Spencer-Phillips, U. Gisi, & A. Lebeda (Eds.) *Advances in downy mildew research* (pp. 85–117). Dordrecht: Kluwer.

Lebeda, A., Pink, D. A. C., & Mieslerová, B. (2001b). Host–parasite specificity and defense variability in the *Lactuca* spp.–*Bremia lactucae* pathosystem. *Journal of Plant Pathology, 83*, 25–35.

Lebeda, A., & Reinink, K. (1991). Variation in the early development of *Bremia lactucae* on lettuce cultivars with different levels of field resistance. *Plant Pathology, 40*, 232–237.

Lebeda, A., & Reinink, K. (1994). Histological characterization of resistance in *Lactuca saligna* to lettuce downy mildew (*Bremia lactucae*). *Physiological and Molecular Plant Pathology, 44*, 125–139.

Lebeda, A., Ryder, E. J., Grube, R., Doležalová, I., & Křístková, E. (2007). Lettuce (Asteraceae; *Lactuca* spp.), Chapter 9. In R. Singh (Ed.) *Genetic resources, chromosome engineering, and crop improvement series, volume 3—vegetable crops* (pp. 377–472). Boca Raton, FL, USA: CRC.

Lebeda, A., & Schwinn, F. J. (1994). The downy mildews—an overview of recent research progress. *Journal of Plant Diseases and Protection, 101*, 225–254.

Lebeda, A., Sedlářová, M., Lynn, J., & Pink, D. A. C. (2006). Phenotypic and histological expression of different genetic backgrounds in interactions between lettuce, wild *Lactuca* spp., *L. sativa* × *L. serriola* hybrids and *Bremia lactucae*. *European Journal of Plant Pathology, 115*, 431–441.

Lebeda, A., & Syrovátko, P. (1988). Specificity of *Bremia lactucae* isolates from *Lactuca sativa* and some *Asteraceae* plants. *Acta Phytopathologica et Entomologica Hungarica, 23*, 39–48.

Lebeda, A., & Zinkernagel, V. (2003a). Evolution and distribution of virulence in the German population of *Bremia lactucae*. *Plant Pathology, 52*, 41–51.

Lebeda, A., & Zinkernagel, V. (2003b). Characterization of new highly virulent German isolates of *Bremia lactucae* and efficiency of resistance in wild *Lactuca* spp. germplasm. *Journal of Phytopathology, 151*, 274–282.

Lipka, V., Dittgen, J., Bednarek, P., Bhat, R., Wiermer, M., Stein, M., et al. (2005). Pre- and postinvasion defenses both contribute to nonhost resistance in *Arabidopsis*. *Science, 310*, 1180–1183.

Lucas, J. A., Hayter, J. B. R., & Crute, I. R. (1995). The downy mildews: Host specificity and pathogenesis. In K. Kohmoto, U. S. Singh, & R. P. Singh (Eds.) *Pathogenesis and host specificity in plant diseases* (pp. 217–238). Oxford: Pergamon.

MacLean, D. J., & Tommerup, I. C. (1979). Histology and physiology of compatibility between lettuce and the downy mildew fungus, *Bremia lactucae* Regel. *Physiological Plant Pathology, 14*, 291–312.

Mahajan, M., & Dhillon, M. (2003). Relation of leaf epidermal characteristics with susceptibility and resistance of potato (*Solanum tuberosum*) to late blight disease (*Phytophthora infestans*). *Indian Journal of Agricultural Sciences, 73*, 656–660.

Mansfield, J. W. (2005). Biophoton distress flares signal the onset of the hypersensitive reaction. *Trends in Plant Science, 10*, 307–309.

Mansfield, J. W., Bennett, M., Bestwick, C., & Woods-Tör, A. (1997). Phenotypic expression of gene-for-gene interaction involving fungal and bacterial pathogens: Variation from recognition to response. In I. R. Crute, E. B. Holub, & J. J. Burdon (Eds.) *The gene-for-gene relationship in plant parasite interactions* (pp. 265–291). Wallingford, UK: CABI.

Mauch-Mani, B. (2002). Host resistance to downy mildew diseases. In P. T. N. Spencer-Phillips, U. Gisi, & A. Lebeda (Eds.) *Advances in downy mildew research* (pp. 59–83). Dordrecht: Kluwer.

Mauch-Mani, B., & Slusarenko, A. J. (1996). Production of salicylic acid precursors is a major function of phenylalanine ammonia-lyase in the resistance of *Arabidopsis* to *Peronospora parasitica*. *Plant Cell, 8*, 203–212.

Michelmore, R. W., & Wong, J. W. (2008). Classical and molecular genetics of *Bremia lactucae*, lettuce downy mildew. *European Journal of Plant Pathology* DOI 10.1007/s10658-008-9305-2.

Michelmore, R. W., Ochoa, O. E., Truco, M. J., Grube, R., & Hayes, R. (2005). Breeding crisphead lettuce. In *California Lettuce Research Board Annual Report (April 1, 2004 through March 31, 2005)* (pp. 68–78). Salinas, CA, USA: California Lettuce Research Board.

Moriondo, M., Orlandini, S., Giuntoli, A., & Bindi, M. (2005). The effect of downy and powdery mildew on grapevine (*Vitis vinifera* L.) leaf gas exchange. *Journal of Phytopathology, 153*, 350–357.

Neill, S. J., Desikan, R., Clarke, A., Hurst, R. D., & Hancock, J. T. (2002). Hydrogen peroxide and nitric oxide as signalling molecules in plants. *Journal of Experimental Botany, 53*, 1237–1247.

Nordskog, B., Gadoury, D. M., Seem, R. C., & Hermansen, A. (2007). Impact of diurnal periodicity, temperature and light on sporulation of *Bremia lactucae*. *Phytopathology, 97*, 979–986.

Norwood, J. M., Crute, I. R., & Lebeda, A. (1981). The utilization of novel sources of resistance to *Bremia lactucae* from wild *Lactuca* species. *Euphytica, 30*, 659–668.

O'Connell, R. J., & Panstruga, R. (2006). Tête à tête inside a plant cell: Establishing compatibility between plants and biotrophic fungi and oomycetes. *New Phytologist, 171*, 699–718.

Pesacreta, T. C., & Hasenstein, K. H. (1999). The internal cuticle of *Cirsium horridulum* (Asteraceae) leaves. *American Journal of Botany, 86*, 923–928.

Petrželová, I., & Lebeda, A. (2004). Occurrence of *Bremia lactucae* in natural populations of *Lactuca serriola*. *Journal of Phytopathology, 152*, 391–398.

Petrželová, I., Lebeda, A., Nevo, E., & Beharav, A. (2007). Variation of response against *Bremia lactucae* in natural populations of *Lactuca saligna*. In A. Lebeda, & P. T. N. Spencer-Phillips (Eds.) *Advances in downy mildew research, vol. 3. Proceedings of the 2nd international downy mildews symposium* (pp. 169–173). Olomouc and Kostelec na Hané, Czech Republic: Palacký University in Olomouc and JOLA.

Petřivalský, M., Kočířová, J., Sedlářová, M., Piterková, J., Luhová, L., & Lebeda, A. (2007). On the role of nitric oxide in *Bremia lactucae* pathogenesis on *Lactuca sativa*. In A. Lebeda, & P. T. N. Spencer-Phillips (Eds.) *Advances in downy mildew research, vol. 3. Proceedings of the 2nd international downy mildews symposium* (pp. 175–184). Olomouc and Kostelec na Hané, Czech Republic: Palacký University in Olomouc and JOLA.

Pink, D. A. C. (2002). Strategies using genes for non-durable disease resistance. *Euphytica, 124*, 227–236.

Portz, D., Koch, E., & Slusarenko, A. (2008). Effects of garlic (*Allium sativum*) juice containing allicin on *Phytophthora infestans* and on downy mildew of cucumber caused by *Pseudoperonospora cubensis*. *European Journal of Plant Pathology* (this issue).

Prokopová, J., Nauš, J., Špundová, M., & Sedlářová, M. (2007). The effect of *meta*-topolin on optical parameters of lettuce leaf discs infected by *Bremia lactucae*. In A. Lebeda, & P. T. N. Spencer-Phillips (Eds.) *Advances in downy mildew research, vol. 3. Proceedings of the 2nd international downy mildews symposium* (pp. 73–78). Olomouc and Kostelec na Hané, Czech Republic: Palacký University in Olomouc and JOLA.

Rees, S., & Harborne, J. (1984). Flavonoids and other phenolics of *Cichorium* and related members of the Lactuceae (Compositae). *Botanical Journal of the Linnean Society, 89*, 313–319.

Restrepo, S., Myers, K. L., del Pozo, O., Martin, G. B., Hart, A. L., Buell, C. R., et al. (2005). Gene profiling of a compatible interaction between *Phytophthora infestans* and *Solanum tuberosum* suggests a role for carbonic anhydrase. *Molecular Plant–Microbe Interactions, 18*, 913–922.

Reuveni, R., Shimoni, M., & Crute, I. R. (1991). An association between high peroxidase activity in lettuce (*Lactuca sativa*) and field resistance do downy mildew (*Bremia lactucae*). *Journal of Phytopathology, 132*, 312–318.

Robatzek, S. (2007). Vesicle trafficking in plant immune responses. *Cellular Microbiology, 9*, 1–8.

Saltveita, M. E., Choia, Y. J., & Tomás-Barberán, F. A. (2005). Mono-carboxylic acids and their salts inhibit wound-induced phenolic accumulation in excised lettuce (*Lactuca sativa*) leaf tissue. *Physiologia Plantarum, 125*, 454–463.

Sargent, J. A. (1976). Germination of spores of *Bremia lactucae*. *Annals of Applied Biology, 84*, 290–294.

Sargent, J. A., & Payne, H. L. (1974). Effect of temperature on germination, viability and fine structure of conidia of *Bremia lactucae*. *Transactions of British Mycological Society, 63*, 509–518.

Sargent, J. A., Tommerup, I. C., & Ingram, D. S. (1973). The penetration of a susceptible lettuce variety by the downy mildew fungus, *Bremia lactucae* Regel. *Physiological Plant Pathology, 3*, 231–239.

Scharte, J., Schön, H., & Weis, E. (2005). Photosynthesis and carbohydrate metabolism in tobacco leaves during an incompatible interaction with *Phytophthora nicotianae*. *Plant, Cell and Environment, 28*, 1421–1435.

Schnabel, G., Strittmatter, G., & Noga, G. (1998). Changes in photosynthetic electron transport in potato cultivars with different field resistance after infection with *Phytophthora infestans*. *Journal of Phytopathology, 146*, 205–210.

Sedlářová, M., Binarová, P., & Lebeda, A. (2001a). Changes in microtubular alignment in *Lactuca* spp. (*Asteraceae*) epidermal cells during early stages of infection by *Bremia lactucae* (*Peronosporaceae*). *Phyton, 41*, 21–34.

Sedlářová, M., & Lebeda, A. (2001). Histochemical detection and role of phenolic compounds in defence response of *Lactuca* spp. to lettuce downy mildew (*Bremia lactucae*). *Journal of Phytopathology, 149*, 1–5.

Sedlářová, M., Lebeda, A., & Pink, D. A. C. (2001b). The early stages of interaction between effective and non-effective race-specific genes in *Lactuca sativa*, wild *Lactuca* spp. and *Bremia lactucae* (race NL16). *Journal of Plant Diseases and Protection, 108*, 477–489.

Sedlářová, M., Luhová, L., Petřivalský, M., & Lebeda, A. (2007a). Localisation and metabolism of reactive oxygen species during *Bremia lactucae* pathogenesis in *Lactuca sativa* and wild *Lactuca* spp. *Plant Physiology and Biochemistry, 45*, 607–616.

Sedlářová, M., Výtisková, M., Doležal, K., & Lebeda, A. (2007b). Pre-incubation with cytokinins delays chlorophyll degradation in *Lactuca* spp. tissues and reduces *Bremia lactucae* sporulation. In A. Lebeda, & P. T. N. Spencer-Phillips (Eds.) *Advances in downy mildew research, vol. 3. Proceedings of the 2nd international downy mildews symposium* (pp. 185–194). Olomouc and Kostelec na Hané, Czech Republic: Palacký University in Olomouc and JOLA.

Skidmore, D. I., & Ingram, D. S. (1985). Conidial morphology and the specialization of *Bremia lactucae* Regel (Peronosporaceae) on hosts in the family Compositae. *Botanical Journal of the Linnean Society, 91*, 503–522.

Spencer-Phillips, P. T. N. (1997). Function of fungal haustoria in epiphytic and endophytic infections. *Advances in Botanical Research, 24*, 309–333.

Takemoto, D., & Hardham, A. R. (2004). The cytoskeleton as a regulator and target of biotic interactions in plants. *Plant Physiology, 136*, 3864–3876.

Takemoto, D., Jones, D. A., & Hardham, A. R. (2003). GFP-tagging of cell components reveals the dynamics of subcellular re-organization in response to infection of *Arabidopsis* by oomycete pathogens. *The Plant Journal, 33*, 775–792.

Tang, X., Rolfe, S. A., & Scholes, J. D. (1996). The effect of *Albugo candida* (white blister rust) on the photosynthetic and carbohydrate metabolism of leaves of *Arabidopsis thaliana*. *Plant, Cell and Environment, 19*, 967–975.

Thines, M. (2007). Use of scanning electron microscopy in downy mildew systematics. In A. Lebeda, & P. T. N. Spencer-Phillips (Eds.) *Advances in downy mildew research, vol. 3. Proceedings of the 2nd international downy mildews symposium* (pp. 17–23). Olomouc and Kostelec na Hané, Czech Republic: Palacký University in Olomouc and JOLA.

Thines, M., Göker, M., Spring, O., & Oberwinkler, F. (2006). A revision of *Bremia graminicola*. *Mycological Research, 110*, 646–656.

Van Pelt-Heerschap, H., & Smit-Bakker, O. (1993). Cell-wall degrading enzymes synthesized by the obligate pathogen *Bremia lactucae*. In B. Fritig, & M. Legrand (Eds.) *Mechanisms of plant defense responses* (p. 82). Dordrecht: Kluwer.

Vieira, B. S., & Barreto, R. W. (2006). First record of *Bremia lactucae* infecting *Sonchus oleraceus* and *Sonchus asper* in Brazil and its infectivity to lettuce. *Journal of Phytopathology, 154*, 84–87.

Voglmayr, H. (2008). Progress and challenge in systematics of downy mildews and white blister rusts: New insights from genes and morphology. *European Journal of Plant Pathology* (this issue).

Voglmayr, H., Riethmuller, A., Göker, M., Weiss, M., & Oberwinkler, F. (2004). Phylogenetic relationships of *Plasmopara, Bremia* and other genera of downy mildew pathogens with pyriform haustoria based on Bayesian analysis of partial LSU rDNA sequence data. *Mycological Research, 108*, 1011–1024.

Walters, D., & McRoberts, N. (2006). Plants and biotrophs: A pivotal role for cytokinins. *Trends in Plant Science, 11*, 581–586.

Wasteneys, G. O. (2004). Progress in understanding the role of microtubules in plant cells. *Current Opinion in Plant Biology, 7*, 651–660.

Witsenboer, H., Vogel, J., & Michelmore, R. W. (1997). Identification, genetic localization, and allelic diversity of selectively amplified microsatellite polymorphic loci in lettuce and wild relatives (*Lactuca* spp.). *Genome, 40*, 923–936.

Woods, A. M., Fagg, J., & Mansfield, J. W. (1988). Fungal development and irreversible membrane damage in cells of *Lactuca sativa* undergoing the hypersensitive reaction to the downy mildew fungus *Bremia lactucae*. *Physiological and Molecular Plant Pathology, 32*, 483–498.

Zinkernagel, V. (1985). Investigations on the susceptibility and resistance of head lettuce (*Lactuca sativa*) to downy mildew (*Bremia lactucae*). 2. Light and electron-microscopic examinations of the host–parasite interface. *Journal of Phytopathology, 112*, 238–258.

Zinkernagel, V. (1986). Investigations on the susceptibility of head lettuce (*Lactuca sativa*) to downy mildew (*Bremia lactucae*). 3. Activities of peroxidase, catalase and polyphenoloxidase. *Journal of Phytopathology, 115*, 257–266.

Zinkernagel, V., & Bartscherer, H. C. (1978). The host–parasite interface in *Bremia lactucae* with *Lactuca sativa*. In *Book of Abstracts, 3rd International Congress of Plant Pathology* (p. 112). München, Germany.

Eur J Plant Pathol (2008) 122:91–109
DOI 10.1007/s10658-008-9286-1

Natural history of *Arabidopsis thaliana* and oomycete symbioses

Eric B. Holub

Received: 3 September 2007 / Accepted: 31 January 2008

Abstract Molecular ecology of plant–microbe interactions has immediate significance for filling a gap in knowledge between the laboratory discipline of molecular biology and the largely theoretical discipline of evolutionary ecology. Somewhere in between lies conservation biology, aimed at protection of habitats and the diversity of species housed within them. A seemingly insignificant wildflower called *Arabidopsis thaliana* has an important contribution to make in this endeavour. It has already transformed botanical research with deepening understanding of molecular processes within the species and across the Plant Kingdom; and has begun to revolutionize plant breeding by providing an invaluable catalogue of gene sequences that can be used to design the most precise molecular markers attainable for marker-assisted selection of valued traits. This review describes how *A. thaliana* and two of its natural biotrophic parasites could be seminal as a model for exploring the biogeography and molecular ecology of plant–microbe interactions, and specifically, for testing hypotheses proposed from the geographic mosaic theory of co-evolution.

E. B. Holub (✉)
Warwick-HRI, University of Warwick,
Wellesbourne CV35 9EF, UK
e-mail: eric.holub@warwick.ac.uk

Keywords *Hyaloperonospora parasitica* · *Albugo candida* · Downy mildew · White blister rust · Gene-for-gene · Innate immunity · Receptor-like proteins · Arms race · LRR · CATERPILLAR genes · *Boechera* · Geographic mosaic · Non-host resistance

Introduction

In the current age of molecular technology and big science questions, the practical aims for downy mildew research have remained essentially unchanged for more than a century. Investment in downy mildew research still seeks practical solutions for improved disease control, through better use of fungicides and to aid the breeding and deployment of downy mildew resistant crops.

Plant breeding, however, has been transformed over the past 20 years by molecular genetics, which has been providing knowledge of the genes that underlie natural variation in valued traits of species such as *Arabidopsis thaliana*, tomato and rice. Converting valued traits into knowledge of genes is an important step for genetic engineering, but more importantly, it provides the means to design the most precise molecular markers attainable for marker-assisted selection of valued traits in conventional breeding programmes (McCouch 2004). From the initial public investment in research, the knowledge gained has potential use across breeding efforts in different plant species. For instance, molecular genetics led to the discovery of numerous pathogen receptor-like genes that provide the molecular basis for downy mildew resistance in *A. thaliana* (Holub 2001, 2007). These major discoveries in turn have been instrumental in the early development of marker-assisted selection for disease resistance in crops (Aarts et al. 1998a; Botella et al. 1997; Michelmore and Wong 2008; Shen et al. 2002; Speulman et al. 1998).

In a similar process of scientific enquiry, conservation biology could also benefit significantly from the assimilation of molecular genetics research. Progress has been and will continue to be achieved in the protection of biological diversity without knowing much detail about the genetic variation within species. However, by analogy with the recent advances in plant breeding, knowledge about critical genes (e.g., that enable adaptability of a species) could provide useful indicators for improving the prospects for survival of animal and plant populations that are in most need of protection.

For example, the *geographic mosaic theory of co-evolution* has been proposed to explain how intimate interactions among multiple species will co-evolve across geographic landscapes and provides a major process in organising the earth's biodiversity (Thompson 1999a, b; Gomulkiewicz et al. 2007). The theory proposes a three-part hypothesis: (a) the species interactions will have different evolutionary trajectories in different populations thereby generating a *selection mosaic* of co-evolving traits among populations; (b) some of the interacting populations will be *hot spots* for reciprocal selection of the co-evolving traits, whereas *cold spots* will also be generated in which selection is unilateral or not occurring at all; and, (c) continual re-mixing of traits will occur across the geographic landscape (e.g., gene flow, genetic drift and local extinction of populations). Understanding the molecular basis of co-evolving traits will be essential for empirical testing of this tripartite hypothesis, for providing fundamental advances in evolutionary ecology, and ultimately for providing invaluable insight that will aid the more practical domain of conservation biology.

The purpose of this review is to describe how *A. thaliana* and two of its natural biotrophic parasites could be seminal as a model for exploring the biogeography and molecular ecology of plant–microbe interactions, and specifically, for testing hypotheses proposed from the geographic mosaic theory of coevolution. This subject is relatively young, but stands on the shoulders of important biologists whose contributions provide a foundation for this review.

Divergent enquiry from early revolutions in plant science

Natural history and molecular biology have developed as polar disciplines (Wilson 1994; see 'Molecular wars' chapter, pp. 218–237), determined by the environments in which the respective practitioners have generally explored their chosen subjects (field versus laboratory), and contrasting methodologies. They also differ markedly in the relative accessibility to amateur biologists and a public audience.

Molecular biology is rooted in a revolution of experimental science that began in the mid-1800s and led by European scientists including the botanists Anton DeBary in Germany and Harry Marshall Ward in England (Ayers 2005). Their dream was 'the Cause', a new approach to botany that could equal chemistry and physics as a rigorous, experimental science. They incorporated both field observations and laboratory experiments in their studies of plant–microbe interactions catalyzed by the invention of light microscopy but pre-dating genetics. Biochemistry and biophysics would eventually provide a strong basis for experimental botany, as with the whole of biology. However, Gregor Mendel's discovery of genetics from his seminal experiments with plants soon provided a fundamental mathematical rigour that is unique to biology. It was therefore fitting that the 2nd International Downy Mildew conference was held in central Europe, and specifically in Olomouc, Czech Republic where Mendel had been a student and near the monastery in Brno where he conducted his now famous experiments.

Genetics is the cornerstone of molecular biology, used to illuminate the cause and effect mechanisms of proteins within organisms under strict laboratory environments. In recent years, whole genome sequences have become publicly available from pioneering use of species like *A. thaliana* and *Phytophthora sojae*. And, this in turn has galvanized a Linnaean-like approach by an international community of molecular biologists to name and classify all of the proteins encoded within an organism, one representative organism at a time (Clark et al. 2007; Holub 2006, 2007; Kamoun 2007; Tör 2008; Tyler et al. 2006; Win et al. 2007). This in effect is providing a powerful online catalogue of genes that enables further comparisons to be made among species (so-called comparative genomics). Researchers will continue to pursue their experiments largely in the laboratory. However, molecular biology is increasingly being enriched by biometricians and mathematicians who are enabling extrapolation of what has been learned from biochemical mechanisms in model organisms into other less tractable species, and predictive modelling in the emerging field of *molecular systems biology* which aims to advance understanding of the cause and effect associations amongst molecules of cellular processes (Holub 2007; Jönsson et al. 2006; Prusinkiewicz et al. 2007).

Natural history, on the other hand, is transforming into evolutionary ecology in its modern research form (Thompson et al. 2001). This young cross-discipline is rooted in comparative methods and mathematical modelling as its cornerstones, and builds on ecology as the ultimate arena for 'systems biology' research (Wilson 1994). Evolutionary ecology has progressed from the early revolutions led by Karl Linnaeus (the founder of taxonomy) and Charles Darwin (the founder of evolutionary theory) into a molecular age of being able to, for example, reassess classifications of species through the inference of familial history (or phylogenetics) of species with comparative analyses of informative DNA, RNA and protein sequences.

Phylogenetic trees provide the structure for bold new ventures led by evolutionary biologists, ecologists and amateur taxonomists to assemble information on the earth's biodiversity in online encyclopaedias (www.eol.org; www.tolweb.org; www.nbn.org.uk). These publicly accessible libraries could eventually incorporate web links to vast amounts of genome-wide, DNA-based information from reference species and thereby stimulate growth of new research and educational tools that would join the polar disciplines together. This may require a new generation of cross-disciplinary biologists and creative educators.

At the moment, evolutionary ecology provides a powerful means for elucidating the specific and variable characteristics of habitats to which a species is unable to adapt and thereby render it vulnerable to local or global extinction. However, the molecular basis for the potential vulnerability of a species is poorly understood. Phrased as a more positive question: what naturally variable genes provide a species with 'genetic resilience', by enabling it to adapt in a changing environment or to survive in a diversity of circumstances across its geographic range of habitats? In a nutshell, this question encapsulates a large gap that exists between the disciplines of molecular systems biology and evolutionary ecology, and defines molecular genetics of natural variation as a key means for bridging the gap.

Converging the enquiry from polar disciplines within plant science

Answers to this fundamental question, aimed at improving habitat conservation and species protection, will benefit from a molecular genetic approach to understanding how natural variation within wild populations enables a species to survive in increasingly variable habitats. This will require linking of our genome-wide (genomic) knowledge of organisms with an equally in-depth knowledge of suitable habitats. This approach should at the very least provide an illustrative guide to the minimal levels of genetic diversity required to sustain the survival of a species in a natural habitat, and thereby provide genetic indicators (e.g., disease resistance traits, highlighted below) for the direction and scale of public investment required for species conservation.

The Plant Kingdom provides many opportunities to probe this question. We can see and experience the magnificence of biodiversity in vascular plants that currently exists across our planet, in vast habitats such as precious tropical rainforests or in small patches of aged meadow (Silvertown 2005). As sedentary autotrophs, plants are supreme providers of sustenance, indispensable generators of diverse micro-habitats for animals and microorganisms, and the backbone of terrestrial communities. They, along with their symbioses with microorganisms, make terrestrial life possible. Thus, understanding the genetic resilience of existing plant species is vital to human survival on Earth.

To address the question posed above, we will require genetically-based insights from three aspects of natural variation for a range of contrasting plant species representing different survival strategies. These insights include: 1. identifying the critical genes that are naturally variable within and among established populations of a species; 2. mapping the allelic variation in these genes onto a geographic distribution of established populations; and 3. determining the selective factors that vary among microhabitats and drive fluctuations in natural variation in a given species.

Establishing a genetic model for co-evolved plant–microbe interactions

Since a major selective characteristic in a plant's microhabitat is often likely to be biological (e.g., symbiotic associations with parasitic or mutualistic microorganisms; Thompson 1999b; Gomulkiewicz et al. 2007), the remainder of this review considers where the natural history of *A. thaliana* and its common oomycete symbionts currently stands as an early model which a new generation of experimental botanists could soon be using to explore fundamental genetically-based questions at the forefront of conservation biology.

Host specialization is a characteristic that corresponds with a high degree of speciation in fungus-like oomycetes that cause downy mildew and white rust diseases in a wide range of plant genera (Constantinescu and Fatehi 2002; Göker et al. 2004; Riethmüller et al. 2002; Voglmayr 2008; Voglmayr and Riethmüller 2006). Speciation of these plant parasites caught the attention of botanists before biology had been transformed by genetics (Lindau 1901; Gäumann 1918). Their reports included downy mildew and white rust in *A. thaliana*, which was rediscovered much later in 1988 by Professor Paul H. Williams (University of Wisconsin-Madison) who was seeking a potential model for research on the molecular genetics of plant–microbe interactions (Table 1).

A visiting scientist working with Williams, Eckhard Koch, had attempted to infect *A. thaliana* with downy mildew and white rust using parasite isolates collected from brassica species but with no success because, as we can conclude with hindsight, the chosen isolates were constrained by a high degree of host preference (Fig. 1a). Williams, however, informed by his reading of the earlier German reports, was successful in acquiring support from the US National Science Foundation for a solo trip to collect *A. thaliana* seed and parasite material in Europe. He visited Germany, France and England during May of 1988, and was successful in finding both downy mildew and white rust of *A. thaliana* in Kent (southeast England; Fig. 2) while visiting Professor Ian Crute, who had recently moved to a horticulture research institute at East Malling (now East Malling Research). Williams was unable to secure additional funding, but fortunately the seed of opportunity had been sown with both Koch (who found downy mildew on *A. thaliana* in Switzerland after completing his visit to UW-Madison) and Crute. They were each subsequently successful in establishing *A. thaliana* downy mildew and white rust as model pathosystems for molecular genetics research (Table 1).

With respect to the phylogenetics talks presented at the 2nd International Downy Mildew meeting by M. Thines and reviewed by Voglmayr (2008), I will refer to the two oomycete symbionts that Williams rediscovered in Europe on *A. thaliana* as *Hyaloperonospora arabidopsis* (Gäumann 1918; Göker et al. 2004; Rehmany et al. 2000; Constantinescu and Fatehi 2002; formerly *Peronospora parasitica* subsp. *arabidopsis*; abbreviated below as *HpA*) causing downy mildew, and *Albugo candida* subsp. *arabidopsis* (abbreviated below as *AcA*) causing white rust.

Current molecular knowledge of *A. thaliana*–oomycete symbioses has largely been constructed from genetic analyses of interactions amongst a small sample of host accessions and a highly variable but also relatively small collection of *HpA* isolates (Holub and Beynon 1997; Slusarenko and Schlaich 2003). The *HpA* isolate names (e.g., Emoy2) reflect their origins, predominantly from the UK, with two prefix letters that indicate the geographic source (As = Aspatria, Cumbria; Bi = Biggar, Scotland; Ca = Canterbury, Kent; Ed = Edinburgh, Scotland; Em = East Malling, Kent; Go = Godmersham, Kent; Hi = Hilliers arboretum, Hampshire; Ma = Maidstone, Kent; Nu = Nunnery Walks, Cumbria; Wa = Wageningen, Netherlands; We = Weiningen, Switzerland), and two suffix letters indicating an accession available from *A. thaliana* stock centres suitable for cultivation of the particular isolate (Co = Columbia; Ks = Keswick-1; La = Landsberg *erecta*; Nd = Niederzenz; Oy = Oystese; Wa = Wassilewskija). Most of the *HpA* isolates currently being used in research laboratories have been refined genetically by derivation from a single oospore and maintained subsequently by propagation of asexual inoculum (Holub and Beynon 1997).

Table 1 Two decades in the Linnaean genomics of *Arabidopsis thaliana*–oomycete natural history

Year	Milestone	References
1988	Paul Williams rediscovers downy mildew (DM) and white rust (WR) in European *At*, following an 80 year old trail.	Williams, PH (personal communication); Lindau 1901; Gäumann 1918.
1990	Natural variation of DM resistance in *At* is described as a model for molecular genetic investigation.	Koch and Slusarenko 1990.
1991	The DM isolate Emoy2 was borne from an oospore in a seedling of *At* 'Columbia'. This isolate would be used to establish genetics in the organism and provide the first reference genome of downy mildew parasites in 2007.	Holub 2006.
1995	Mutation of *NDR1* demonstrates that H.H. Flor's notion of disease resistance being conferred by single *R*genes can actually be a multi-genic process and involve common links in the signalling of defense against diverse pathogens. This established a precedent for using oomycete and bacterial pathogens of *At* in comparative laboratory experiments.	Century et al. 1995, 1997.
	Systemic acquired resistance to bacterial disease and DM in *At* is found to require salicylic acid.	Lawton et al. 1995.
	Natural variation in WR resistance provides a complementary model for molecular genetics of *At*–oomycete interactions.	Holub et al. 1995; Borhan et al. 2001.
1996	Mutation of *EDS1* demonstrates that species level barriers ('non-host' resistance) to biotrophic parasites can be amenable to mutation and genetic analysis. This gene and *PAD4*, which is also typically required for DM resistance in *At*, were found to encode lipase-like proteins.	Parker et al. 1996; Glazebrook et al. 1997; Falk et al. 1999; Jirage et al. 1999; Holub and Cooper 2004.
1997	The first DM resistance gene *RPP5* is cloned and found to encode a member of the previously described TIRNBS-LRR class of cytoplasmic receptor-like proteins.	Parker et al. 1997.
	Major *R*-gene clusters revealed on four chromosomes of *At* using a powerful combination of recombinant inbred *At* populations and DM isolates as physiological probes to map *RPP* loci.	Holub and Beynon 1997.
1998	The bacterial resistance genes *RPS4* and *RPS5* were cloned and demonstrated that non-host resistance in *At* can involve parasite recognition mediated by classic (receptor-like) R-genes. This established the use of a broadly virulent pathogen as a surrogate vector for transient expression to test the 'non-host' recognition of effector proteins. Interestingly, mutation of *RPS5* can interfere with recognition of some DM isolates.	Warren et al. 1998; Gassmann et al. 1999; Holub 2007; Rentel et al. 2008.
	R-like homologues provide a powerful class of molecular markers for map-based new oomycete resistance genes in *At* and in crops such as lettuce and potato.	Aarts et al. 1998a, b; Botella et al. 1997; Speulman et al. 1998.
	RPP8 is cloned, and found to be a CC-NBS-LRR gene (similar to the first bacterial resistance genes described in *At*) and subsequently associated with viral resistance.	McDowell et al. 1998; Cooley et al. 2000; Takahashi et al. 2002.
	The multi-copy locus *RPP1* contains several DM resistance genes (TIR-NBS-LRR subclass) that differ in specificity.	Botella et al. 1998.
	DM resistance genes vary in how they confer defense via different regulatory proteins.	Aarts et al. 1998a, b; Eulgem et al. 2004.
2000	Whole genome sequence of *At* is announced as a public resource. The 'Linnean genomics' age of plant biology is soon launched with ARABIDOPSIS 2010, an international community effort to ascribe a function for all of the genes in *At* by 2010.	AGI 2000; Holub 2007.
	A single DM resistance gene (*RPP7*) can confer accumulative (salicylic acid dependent and independent) defense responses.	McDowell et al. 2000; Tör et al. 2002; Eulgem et al. 2007.

Table 1 (continued)

Year	Milestone	References
	DM isolates collected from *At* appear to be phylogenetically distinct from brassica isolates: referred to hence as subsp. *arabidopsis* (*HpA*) or *brassica* (*HpB*).	Rehmany et al. 2000.
	RPP13 is cloned, encoding a protein homologous to RPP8 and providing the most extreme benchmark for allelic diversification of a receptor-like gene in *At*.	Bittner-Eddy et al. 2000; Rose et al. 2004.
2001	An *At-Phytophthora* model is described, providing *Agrobacterium*-mediated transformation of an oomycete for host–parasite research in *At*.	Roetschi et al. 2001.
2002	'Gene-for-gene' paradigm is established in the *At-HpA* pathosystem. An outcross of *HpA* enables genetic evidence for five independent *At*-recognizable effectors (*ATR1, ATR4, ATR5, ATR8* and *ATR13*) that correspond with different cloned DM resistance genes.	Gunn et al. 2002.
	DM3 is cloned from lettuce, and found to encode a TIRNBS-LRR protein similar to several *RPP* genes in *At*.	Shen et al. 2002.
	The DM parasite of crucifer species (previously *Peronospora parasitica*) is renamed as *Hyaloperonospora parasitica*.	Constantinescu and Fatehi 2002.
	SGT1b and *RAR1/PBS2* provide evidence for highly conserved regulators (also found in monocots) and the likely involvement of proteolysis in defense signalling.	Austin et al. 2002; Muskett et al. 2002; Tör et al. 2002; Tornero et al. 2002; Warren et al. 1999.
	A. thaliana-oomycete molecular ecology is launched.	Damgaard and Jensen 2002.
2004	The first *At*-recognized effector (*ATR13*) is cloned from *HpA* isolate Maks9 and found to encode a small secreted protein that exhibits a high degree of sequence variation amongst UK isolates of *HpA*.	Allen et al. 2004.
	An *EDS1*-independent WR resistance gene (*RAC1*) is cloned and found to encode a TIR-NBS-LRR protein.	Borhan et al. 2004.
	The broad spectrum late blight resistance gene *Rb* is cloned and found to encode a CC-NBS-LRR protein.	Song et al. 2004.
2005	Enhanced downy mildew resistance (DMR) mutants are described, and launch the genetic analyses of induced accessibility for oomycete parasites in *At*.	Van Damme et al. 2005; Holub 2006.
2006	The first draft sequence of two oomycete genomes (*Phytophthora sojae* and *Ph. ramorum*) is announced.	Tyler et al. 2006.
	ATR1 is cloned from *HpA* Emoy2 and ignites the 'Linnean genomics' age of oomycete effector biology by revealing a conserved RxLR motif also found in *ATR13*, effector proteins from *Phytophthora* species, and secreted proteins from the human malarial pathogen *Plasmodium*.	Rehmany et al. 2005; Holub 2007; Win et al. 2007.
2007	Draft genome sequence of *HpA* isolate Emoy2 is released by a *Phytophthora* consortium.	http://pmgn.vbi.vt.edu/

Noco2 is an exception, which was derived and has since been propagated from an original source of asexual inoculum.

Characterising the extent of phenotypic variation was an important step in establishing downy mildew and white rust of *A. thaliana* as model pathosystems for molecular genetics, following examples that Williams had established for assessing these diseases in vegetable brassica breeding (downy mildew example in Fig. 3a; Holub et al. 1994, 1995). Incompatibility with *HpA* isolates is common among *A. thaliana* accessions and much of the observed host resistance is *HpA*-isolate specific and simply inherited (Fig. 3b; Holub 2007). The naturally variable *HpA* isolates described above were therefore readily identified and differentiated according to their respective recognition by different combinations of downy mildew resistance (*RPP* = resistance to *Peronospora*

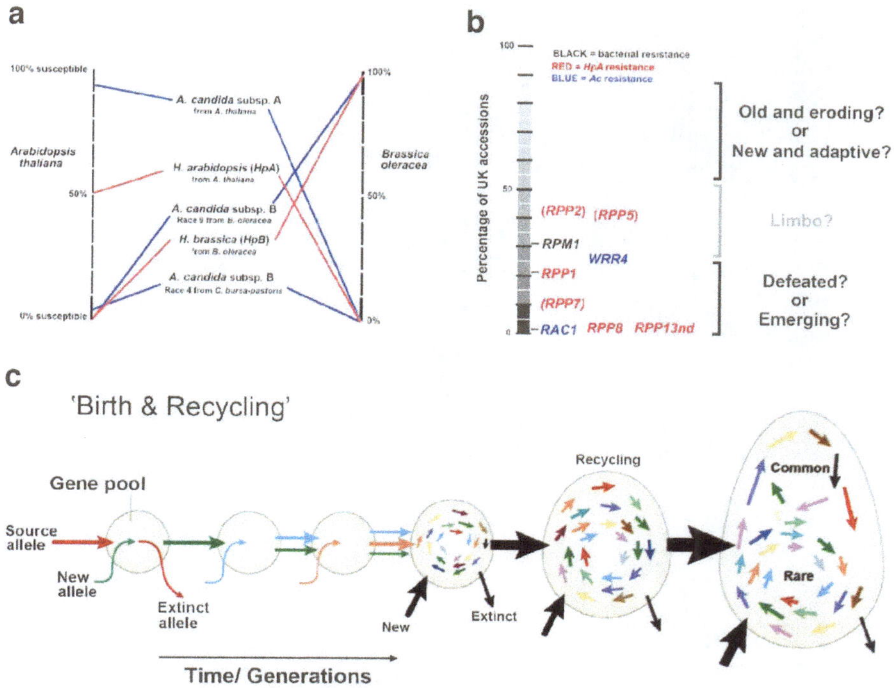

Fig. 1 *Arabidopsis thaliana* and its natural associations with parasitic oomycetes provide a rich and fascinating model for molecular and evolutionary ecology. *Arabidopsis thaliana* is a common host for two biotrophic oomycetes including *Hyaloperonospora arabidopsis* (*HpA*, causing downy mildew) and *Albugo candida* subsp. 'A' (*AcA*, causing white blister rust); and will also be exposed under field conditions to recurrent but foiled attacks from closely related parasites (e.g., *H. brassica* and *A. candida* subsp 'B') that proliferate on brassica crops and other wild crucifers (e.g., *Capsella bursa-pastoris*). **a** The relative degree of compatibility to different parasite subspecies is compared in *A. thaliana* and *B. oleracea*; the estimates are based on the percentage of susceptibility within diversity collections of each host species (Holub et al. 1995; Holub, unpublished). **b** Many naturally variable pathogen recognition proteins (encoded by so-called *R*-genes) from just a few standard accessions of *A. thaliana* are now known which confer resistance to specific isolates of *HpA* (*red labels*), *A. candida* (*blue*) or *Pseudomonas syringae* (*black*). All of the molecularly characterised examples are members of a large receptor-like 'NB-LRR' gene family (NB = *n*uceotide *b*inding site; LRR = *l*eucine *r*ich *r*epeat domain; Holub 2001). Preliminary analyses (Holub unpublished) indicate that the known functional proteins occur at a frequency of <50% in a UK diversity collection of 96 *A. thaliana* accessions, and many are potentially quite rare. Estimates for downy mildew and white rust resistance genes here have been based on cumbersome F_2 allelism testing (populations derived from out-crossing UK accessions to the standard accessions containing the known *R*-genes), except where *parentheses* indicate cruder estimates based only on using a single standard isolate to survey resistance in the 96 accessions. These are overestimates; however, they are sufficient to pose questions (*right margin*) concerning the natural history of each variant R-protein, and more theoretical consideration of the origins and fate of allelic variation within a plant species. **c** For example (modified from Holub 2001), attempts to illustrate how the very earliest means of host defence (the first 'source allele'), when an ancestral host–parasite relationship began to co-evolve, may have been replaced by a 'new allele' and presumably lost from the historical DNA-based record. Several rounds of this 'arms race' may have occurred before a final solution arose in vascular plants, with a highly adaptable capability of parasite recognition made possible by the accumulation in the gene pool of allelic variation in multiple copies of highly mutable NB-LRR proteins. These proteins enable individual plants, upon parasite detection, to rapidly elicit inducible and highly conserved (ancient) defence responses. It should be possible with population-level experiments to investigate whether recycling and 'common versus rare' variants of NB-LRR proteins actually occur in the molecular ecology of *A. thaliana*–parasite interactions

parasitica) genes in laboratory accessions of *A. thaliana* (Holub and Beynon 1997). This *HpA* collection provided the 'physiological probes' for rapid, iterative mapping of numerous *RPP* loci relative to the molecular markers already available in recombinant inbred populations of *A. thaliana* (Holub 1997). Targets for molecular characterisation were easily established (described below) on each of the five *A. thaliana* chromosomes that represented the breadth of phenotypic variation observed in inoculated cotyledons

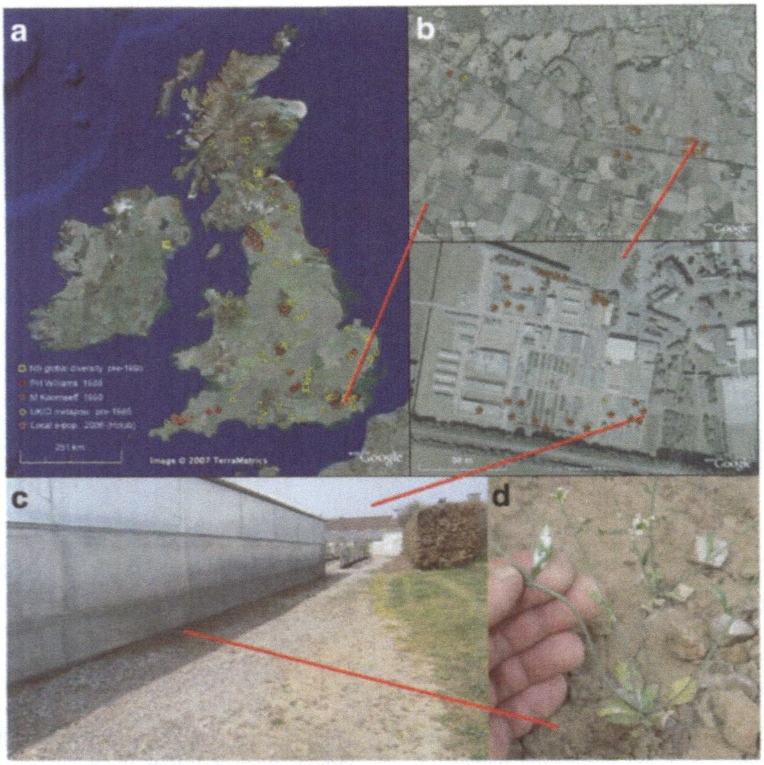

Fig. 2 The natural diversity and genomic history in *Arabidopsis thaliana* from the British Isles provides a unique resource for exploring molecular ecology of plants. **a** A large collection of *A. thaliana* has been assembled for a multi-national ARABIDOPSIS 2010 project which aims to generate a database of genome-wide, high density DNA sequencing from a global diversity collection of 1,152 accessions (Holub 2007). This collection includes all of the UK accessions currently available from public stock centres (Nordborg-Bergelson global diversity collection, and donations from P.H. Williams and M. Koorneeff) and a UK and Ireland diversity (UKID) collection assembled by E.B. Holub. This combined material provides a sample metapopulation of *A. thaliana* with a single accession representing the local population at each of ca. 150 locations. Extensive sampling of more than 100 established local populations (>10 years old; ca. 1,200 accessions total) was conducted in 2006, and this material was also included in the first phase (low density genotyping) of the 2010 project. **b** Satellite photographs of one heavily sampled region in southeast England, where in 1988 P.H. Williams observed downy mildew and white blister rust in *A. thaliana* and where the molecular genetics of these pathosystems began 2 years later in I.R Crute's laboratory at the horticulture research facility in East Malling, Kent (aerial view of buildings, lower panel). **c** Epidemics have occurred most years in established *A. thaliana* populations (>20 years old) that reside between polytunnels. **d** An example of white rust occurring on rosette (readily visible above as *bright yellow tissue*) and stem leaves of an individual plant

(Fig. 3b). It is important to note that natural variation exists in other characteristics of downy mildew or white rust resistance including: polygenic but isolate-specific in seedlings; broad spectrum in seedlings against the current *HpA*-collection; and adult leaf resistance. Although these further examples are ecologically important, they have yet to be characterised at the molecular level.

Natural variation in *A. thaliana–AcA* interactions is comparatively sparse in providing clear examples of genotype-specific resistance (Fig. 1a; Holub et al. 1995), so this closely related pathosystem was less attractive for the opening phase of molecular genetics research. As a consequence, this pathosystem remains undeveloped as a potentially excellent model for investigating such topics as species-level characteristics including host determinants of parasite speciation involving pathogen receptor-like genes (Fig. 1a; Rehmany et al. 2000; Borhan et al. 2008), and important aspects of compatible interactions including induced accessibility (pathogen-activated defence suppression) and source-sink dynamics affecting carbohydrate metabolism (Chou et al. 2000; Cooper et al. 2008; Holub and Cooper 2004; Tang et al. 1996).

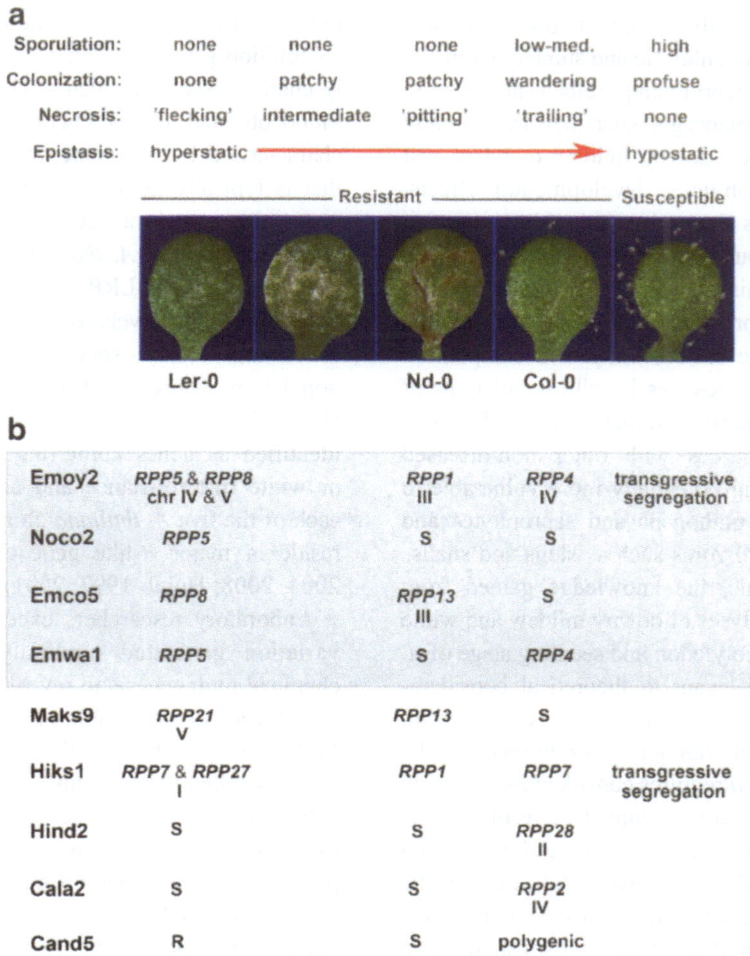

Fig. 3 Molecular genetics has revealed tremendous diversity in the ability of *Arabidopsis thaliana* to recognize genetic variants of *Hyaloperonospora arabidopsis* (*HpA*, causing downy mildew) under laboratory conditions. Research began two decades ago (see Table 1) with **a**, characterisation of phenotypic variation in the incompatibility of cotyledons with respect to the extent of host colonization and sporulation by the parasite, and the relative degree of visible host response (summary text above photographs). **b** These interaction phenotypes were observed among several standard accessions of *A. thaliana* (e.g., Columbia, Col-0; Landsberg *erecta*, Ler-0; and Niederzenz, Nd-0; labelled below photographs) following inoculations with a diverse collection of *HpA* isolates (nine examples listed in left hand column, below photos). One of the first *HpA* isolates collected, Emoy2, is broadly avirulent and was therefore chosen for an outcross to Maks9 to launch genetic analyses in *HpA* (Holub 2006), and consequently, also provided DNA for the first reference genome of downy mildews. The host × parasite table (*below photographs*) summarizes genetic analysis of downy mildew resistance in *A. thaliana* conferred by *RPP* genes (recognition of *Peronospora parasitica*, the previous name of *HpA*). Examples have been found on each of the five chromosomes of *A. thaliana*, and most of them have been molecularly characterised (*RPP21* and *RPP27* are exceptions). Five of them correspond with simply inherited avirulence (shown in *grey box*) in *HpA*, suggesting that the 'gene-for-gene' hypothesis is often valid in this pathosystem (Table 1). Susceptibility to some isolates (e.g., Emoy2 and Hiks1) was only observed in these experiments from transgressive segregation within host populations derived by intercrossing Col-0 with either Ler-0 or Nd-0

Only two UK isolates of *AcA* have thus far been named including the type isolate Aceml from East Malling, Kent and Acks1 from Keswick, Cumbria (Borhan et al. 2004; Holub et al. 1995).

From a laboratory perspective, downy mildew and white rust are most detrimental to wild-type *A. thaliana* when infections are initiated at the seedling stage in a susceptible host accession. Seedlings and juvenile rosettes can readily succumb to downy mildew or white rust, becoming chlorotic and stunted but plants are still capable of reproducing (albeit at severely reduced capacity). As plants get older, they are generally more able to 'outgrow' downy mildew or white rust infections with no obvious developmental effects, although the parasites themselves are able to survive endophytically (without symptoms) in older tissue of a compatible host. This suggests that under natural conditions, selection pressure by *HpA* and *AcA* will be most acute during the establishment and competition amongst early juvenile rosettes in wild populations of *A. thaliana*, because debilitated plants may be less able to compete for resources with other non-diseased plants as well as being potentially more vulnerable to attack by secondary pathogens and saprophytes and more attractive to herbivores such as slugs and snails.

Using this rationale, the knowledge gained from molecular genetic analyses of downy mildew and white rust resistance in the cotyledon and seedling stage of *A. thaliana* should be relevant to theoretical considerations of host–parasite co-evolution (Fig. 1c). However, endophytic survival of *AcA* and *HpA* throughout the life cycle of *A. thaliana* requires consideration in future studies in molecular ecology aimed at resolving the underlying selection drivers determining the observed distribution, epidemiology and maintenance of genetic variation in these symbiotic oomycetes. Similarly, host determinants of *HpA* and *AcA* endophytism should also be targeted for molecular investigation; for instance, transgressive segregation for natural variation in *A. thaliana* can generate recombinant plants which exhibit hyper-susceptibility to downy mildew and white rust (high levels of sporulation and severe stunting) following inoculation of mature rosettes (Holub, unpublished). Such recombinants can presumably arise in natural populations but are presumably poor competitors when disease is prevalent and perhaps also under seemingly disease-free conditions.

Two decades in the Linnaean genomics of *A. thaliana*

Molecular milestones from *A. thaliana*-oomycete pathology have added appreciably to the Linnaean-like effort to name and classify gene families (Table 1; Holub 2001, 2006, 2007). For example, naturally variable genes that control *HpA* isolate specific downy mildew resistance were among the first pathogen recognition genes (so-called *R*-genes) to be molecularly characterised from plants. These abundant genes in monocots and dicots encode receptor-like proteins characterised by a leucine-rich repeat (LRR) domain that is typically highly mutable, and most known *R*-genes also contain a nucleotide binding (NB) site. A single accession of *A. thaliana* will contain more than 100 full length NB-LRR genes, many of which exhibit unusually high levels of allelic variation amongst accessions of the species collected from natural populations (Borevitz et al. 2007; Clark et al. 2007; Holub 2001). At least 10 such genes have been identified as genes conferring either downy mildew or white rust resistance and are distributed amongst each of the five *A. thaliana* chromosomes; at least half reside in major *R*-like gene clusters (Borhan et al. 2004, 2008; Holub 1997, 2001).

Laboratory researchers excelled at using artificial variation generated randomly by irradiation and chemical mutagenesis to reveal other key components of induced defence (Table 1). These components (e.g., *EDS1, NDR1, RAR1/PBS2* and *SGT1b*) represent much smaller gene families more highly conserved than *R*-genes. Such genes have been shown to be essential for defence against a wide spectrum of pathogens, as first demonstrated by the discovery of *NDR1*, essential for resistance to plant pathogenic bacteria and downy mildew. Some of these genes have also been shown to represent essential and commonly-occurring components of induced defence across different families of vascular plants including monocots (e.g., *RAR1/PBS2* and *SGT1b*).

It is important to remember that all of these innate defence genes, as with *R*-genes, have been identified and defined functionally in an experimental context. Each gene was discoverable because they were non-lethal and non-redundant (unique functional copy) in the specific combination of host and parasite genotypes selected by investigators for mutagenesis experiments. For instance, null mutation of *sgt1b* exhibits full susceptibility to some *HpA* isolates (e.g., Cala2) in some genetic backgrounds of *A. thaliana* including the most popular laboratory accession Columbia (Tör et al. 2002). However, the same mutant exhibits residual resistance to other isolates of *HpA* and strong resistance to all bacterial isolates tested thus far. Similarly, the effect of *sgt1b* mutation can also be masked in other genetic backgrounds such as Landsberg *erecta* in an *RPP*-dependent manner (Tör et al. 2002). This effect may be due in some cases to redundant expression of a second copy of the gene (*SGT1A*) known to exist in Columbia *A. thaliana*, although this hypothesis is difficult to test decisively because double mutation of the gene pair (*sgt1a/ sgt1b*) appears to be lethal (Azevedo et al. 2006; Holt et al. 2005).

In other words, the currently known *R*-genes and other defence components were essentially 'low hanging fruit' for discovery because, despite fortuitous examples like *SGT1b*, functional redundancy and mutational lethality present major obstacles that limit the scope of an approach to the dissection of defence mechanisms in plants based purely on molecular genetics. This is true even in the case of excellent model organisms like *A. thaliana*. The science has nevertheless progressed a long way towards addressing Albert Ellingboe's challenge to identify the molecular 'pieces of the (disease resistance) puzzle' with genetics and mutational analyses (Ellingboe 1976, 2001; Holub 2006). Now however, complementary methods and techniques are required to extend knowledge incrementally from the low hanging fruit and fully reconstruct defence mechanisms in model systems (Glazebrook 2007; Kazan and Schenk 2007; Shen and Schulze-Lefert 2007; Staal and Dixelius 2007).

Nevertheless, for the purposes of this review, there is already sufficient molecular understanding of downy mildew resistance in *A. thaliana* to state what now seem to be three features underlying innate immunity in plants: (1) disease resistance typically involves the activation of an inducible defence process involving many other genes; (2) natural genetic variation in disease resistance is most prevalent at the level of receptor-like proteins (containing a highly mutable leucine-rich repeat domain), which acts as a trigger for the induced defence; and (3) exceptions to the above will arise from further investigation as a consequence of genetic background, plant tissue type and/or physiological age of tissue.

Linnaean genomics has also revolutionized our understanding of plant pathogenic oomycetes as a consequence of whole genome sequences that have become available for several *Phytophthora* species and *HpA* (Table 1). There is now a fervent interest in revealing so-called effector proteins from parasitic microorganisms that somehow affect biochemical responses in a potential plant host (Grant et al. 2006; Kamoun 2007; Tör 2008; Win et al. 2007). Microbial effectors that can be detected by matching R-protein activity were previously referred to in plant pathology literature as avirulence proteins or race-specific elicitors. Similarly, non-specific elicitors have come in recent years to be referred to as PAMPs (pathogen associated molecular patterns), an abbreviated term adopted from the research literature of innate immunity in animals. Many *ATR* (*A. thaliana* recognized) effectors were predicted in *HpA* following the genetic identification of numerous *R*-genes conferring downy mildew resistance in *A. thaliana* and classical genetic analysis of avirulence in *Bremia lactuca* (Holub and Beynon 1997; Michelmore and Wong 2008). However, confirmation of a 'gene-for-gene' correspondence has only recently been confirmed for five cloned *RPP* genes, followed swiftly by the molecular identification of the first two *ATR* effectors (Table 1; Fig. 3b; reviewed by Holub 2007).

Downy mildew research in *A. thaliana* has benefited enormously over nearly two decades from healthy competition and collaboration amongst several research groups. However, our current knowledge (summarized in Table 1) would not have been achieved so quickly and coincident with the precedents from molecular investigations of bacterial resistance in *A. thaliana* without the original catalytic fieldwork provided by P. H. Williams in 1988. Two downy mildew resistance genes (*RPP1* and *RPP5*) would probably have been discovered; however, *HpA* genetics and cloning of *HpA* effectors would still only be a theoretical prospect if Williams had not conducted his European collection trip. Similarly, the involvement of major defence components such as *EDS1, NDR1* and *SGT1b* in bacterial resistance would have been revealed but the parallel with downy mildew resistance would likely have come much later. Hence, Williams provided a crucial link between the founders of experimental botany and the modern laboratories of molecular plant pathologists.

Establishing an experimental basis for molecular ecology of plant–parasite interactions

Arabidopsis thaliana–oomycete symbioses could provide the basis for making the next crucial link, between recent laboratory advances in molecular biology of innate immunity in plants and genomics-based field experiments of molecular ecology (Holub 2007). Seven key resources will be: (1) access to microhabitats in the form of reasonably protected field locations that have been inhabited naturally by *A. thaliana*; (2) seed samples from these reference habitats collected from a high proportion of individuals at each location; (3) 'DNA finger-printing' of individuals derived from these initial seed lines involving genome-extensive sequence information to provide a base-line reference for future sampling; (4) access to comparable locations uninhabited by *A. thaliana* for hypothesis testing in future seed introduction experiments (including virgin sites which could provide suitable conditions for *A. thaliana*, and sites which are unsuitable for some unknown selective factor despite the initial appearance of the habitat); (5) satellite navigation (GPS) and photographic records enabling field biologists to accurately re-visit the marked populations; (6) public participation that must include land owners of reference sites, local and regional plant ecologists, and amateur naturalists; and (7) online access for researchers and educators to the public-sponsored data.

The best reference locations for field research will have been well-established (>10 years old, but ideally much older) natural populations of *A. thaliana*. In the British Isles for instance, *A. thaliana* is predominantly a winter annual (flowering May–June) throughout, but it can occur as a summer–autumn annual especially in wettest regions such as the Scottish highlands and the Hebrides (or even well-watered gardens with good slug control). The species relies largely on animals for seed dispersal to establish new populations, and human activity is particularly important so suitable locations for established populations will often be of man-made origin. Sites that support enduring natural populations of *A. thaliana* will include mural communities (described by Segal (1969) as vertical, such as the head, vertical face and foot of old stone or brick walls; and horizontal, such as cracks in urban pavements), ballast of disused and existing steam railway tracks, roadside verges, and cultivated areas such as historic gardens (Fig. 4).

All such habitats can be found throughout the British Isles, and even within regions such as the Lake District National Park that became a protected conservation area thanks to the investment and leadership of Beatrix Potter (Holub 2007; Thompson 2007). It will be fascinating to see whether the rich genomic history (DNA-based, whole genome) that is currently being accumulated from *A. thaliana* (Borevitz et al. 2007; Clark et al. 2007; Holub 2007) will enable us to resolve whether this species is a native in the British Isles or instead an archaeophyte (*i.e.*, pre-Medieval introduction by one or more human migrations) like its close relative *Capsella bursa-pastoris* (Preston et al. 2004).

Mural populations of *A. thaliana* will be particularly interesting for molecular ecology, especially where *A. thaliana* can cycle continuously in consecutive years in essentially the same spot (Fig. 4a and b). Human disturbance represents a major threat to the survival of these self-maintaining mural populations. The linear nature of old wall and railway habitats provides pertinent opportunities to assess the accumulation and maintenance of genetic variation. It should be possible to collect evidence for multiple migration events or out-cross pollination events, as well as for genetic differences that correlate with contrasting features of microhabitat (e.g., heavy shade vs. full sunlight along a wall extending several kilometre; different exposure levels to atmospheric pollutants), and for possibly predicting historical movement of genetic variants along a wall. Although *A. thaliana* is highly self-fertile, pollen dispersal and outcrossing may be more important than expected for movement and maintenance of natural variation within established populations (Bakker et al. 2006), especially on walls.

In a given region, the total available length of wall habitats will far exceed the natural distribution of *A. thaliana*, so with appropriate permission from property owners, there is excellent potential for experimental introductions such as reciprocal transplants of seed from genetically distinct populations or 'seeding' of apparently non-diseased populations with oospores of genetically defined *HpA* or *AcA* isolates. Defining the suitable environmental conditions for establishment and extended survival of *A. thaliana* in these natural habitats will, at the very least, be critical for judging the relative merits of experiments that are performed under controlled laboratory conditions, and whether sites chosen for 'common garden studies' (Mitchell-Olds and Schmitt 2006) are representative of habitat in a given geographic region.

In the British Isles, a good start has been made in assembling the key resources outlined above. A

◀**Fig. 4** *Arabidopsis thaliana* habitats are typically of man-made origin in the British Isles including: **a** in the cracks of pavements; **b** on the head, vertical face and foot of old stone or brick walls; **c** in the ballast of existing steam railway tracks, **d** persisting in disused sections of railways; **e** in grassland meadows and roadside verges; and **f** in cultivated settings (e.g., a flower bed in the Cambridge University Botanical garden). The resident populations are typically well established in these locations; for example, 'X' literally marks the spot in **a** where three to five plants reside annually at this location on a street in Stratford-on-Avon, and similarly in **b** for the position on top of a drystone wall in the Lake District. White blister rust (*Albugo candida* subsp. A) is the disease most commonly visible during late spring in UK habitats (example shown in **b**, *right panel*); however, downy mildew (*Hyaloperonospora arabidopsis*) can also occur frequently during the juvenile rosette stage (October–March). *Yellow arrowheads* indicate the location of *A. thaliana* plants in each of these habitats

starter set of *A. thaliana* accessions has been available from public stock centres, and this has been extended with a metapopulation sample that was assembled over 5 years between 1988 and 1993 (Fig. 2a). The local population at each of the locations is represented by a single accession (seed from one individual). In 2006, GPS marking was initiated of field sites that were inhabited with established populations of *A. thaliana*, and seed was collected from at least five individuals per population (80% or more of individuals in each population; ca. 1,200 samples in total). Many of the sites had been sampled in previous forays, including wall populations in southeast England that P.H. Williams had visited in 1988. In most cases, populations were found in the same site as in previous forays indicating an age of at least 15–20 years. Such walls are often more than 100 years old. Many of the newly identified sites from 2006 were subsequently revisited in 2007 to confirm that the populations were also recurrent. Unfortunately, given the timing and logistics of handling seed samples, it was not possible to thoroughly assess disease and collect pathogen samples. The sole purpose was to accumulate potential sites of scientific interest and reference host material. Nonetheless, *AcA* was observed in ca. 70% of the established populations in both the spring of 2006 and 2007 (e.g., Figs. 2d and 4b), whereas *HpA* was only found in a few locations (<2%). These observations at least identify sites for future host–parasite investigations.

For DNA fingerprinting, the complete collection of *A. thaliana* from the British Isles shown in Fig. 2a has been incorporated into a major international project led by US scientists, including sampling from established populations in 2006 (Holub 2007). This project is aimed at providing laboratory researchers with a global diversity collection of 1,152 accessions, a large database of molecular markers (<10 kb spacing), and linkage analysis tools as a research capability for the mapping of genes that underlie natural phenotypic variation in *A. thaliana*. The first phase of the project, however, was to genotype more than 5,000 accessions (mostly European and USA samples) at 149 genome-wide loci containing SNP (single nucleotide polymorphism) or indel (small insertion or deletions) sequence variation. This relatively low-cost dataset provides a means for choosing accessions with unique genotypes that represent global diversity in an elite collection that will undergo the further high-density genotyping. All of the British accessions have been genotyped in the first phase, and ca. 20% are likely to be advanced for inclusion in the elite diversity collection. The low-density genotyping should provide sufficient information for assessing the frequency and distribution of genetic variation in UK *A. thaliana* at different geographic scales and in contrasting habitats, and for choosing prime diseased and non-diseased sites for future field research.

Access to *A. thaliana* seed stocks and DNA information will be possible via the North American and European stock centres (www.biosci.ohio-state.edu/pcmb/Facilities/abrc/abrchome.htm; www.arabidopsis.info) and the central information website (www.arabidopsis.org). The UK National Biodiversity Network (www.nbn.org.uk) provides an exciting precedent for international efforts to release biodiversity information online, and will therefore provide a superb means for releasing information about reference field sites for *A. thaliana* in the British Isles.

Pump-priming molecular pathology in natural populations of *A. thaliana*

Although the seven key resources described above are at various stages of development, it is already possible to envisage how the molecular ecology of *A. thaliana*–parasite interactions will begin to unfold.

For instance, the conventional method of using standard parasite isolates (diagnostic for different *R*-genes) to assess wild accessions for compatibility phenotypes indicates that most of the currently known *R*-genes, conferring bacterial or oomycete resistance, occur across the UK at a frequency well below 50%. This may suggest that the resistance alleles are transient, and are either newly-emergent alleles that have yet to provide a selective advantage under natural conditions, or are much older alleles that have been persistently defeated by a virulent pathogen across the species distribution (Fig. 1b). Alternatively, they may be relatively old alleles that have been recycled by undergoing fluctuations in frequency due to changes in virulence of the pathogen population (Fig. 1c; Holub 2001).

These hypotheses need testing by spatial and temporal experiments in populations of *A. thaliana* and should be extended to include other ecologically relevant examples of natural variation (e.g., polygenic and adult plant resistance). Seedling assays with *HpA* and *AcA* isolates provide a simple and high-throughput means of generating phenotypic data (Fig. 5); however the results obtained will lack precision for genotyping purposes because uncharacterised R-protein/avirulent effector interactions may occur in the sampled material which can confound interpretations from the observed phenotypes. Additional data generated from bacterial assays may eventually be possible, using a broadly virulent isolate of *Pseudomonas syringae* as genetically engineered physiological probes to predictably deliver different avirulent effectors from downy mildew and white rust parasites (Holub 1997, 2007; Rentel et al. 2008).

The next step will be to generate phenotypic data using the existing UK seed stocks and remaining elite global diversity collection, to supplement the rich DNA-based dataset. Once the prime reference locations have been identified, the major investigation can commence with experiments designed to monitor changes in known *R*-genes and oomycete effectors. Low-cost genotyping and PCR-based assays that can be used to detect specific alleles will be instrumental as case studies in aiding progress, especially at fine levels of spatial distribution where populations may have, for example, extended along a kilometre or more of wall and are exposed to variable physical and biological conditions of microhabitat.

Concluding remarks

The natural history of *A. thaliana*–oomycete interactions was the topic of a reviews published a decade

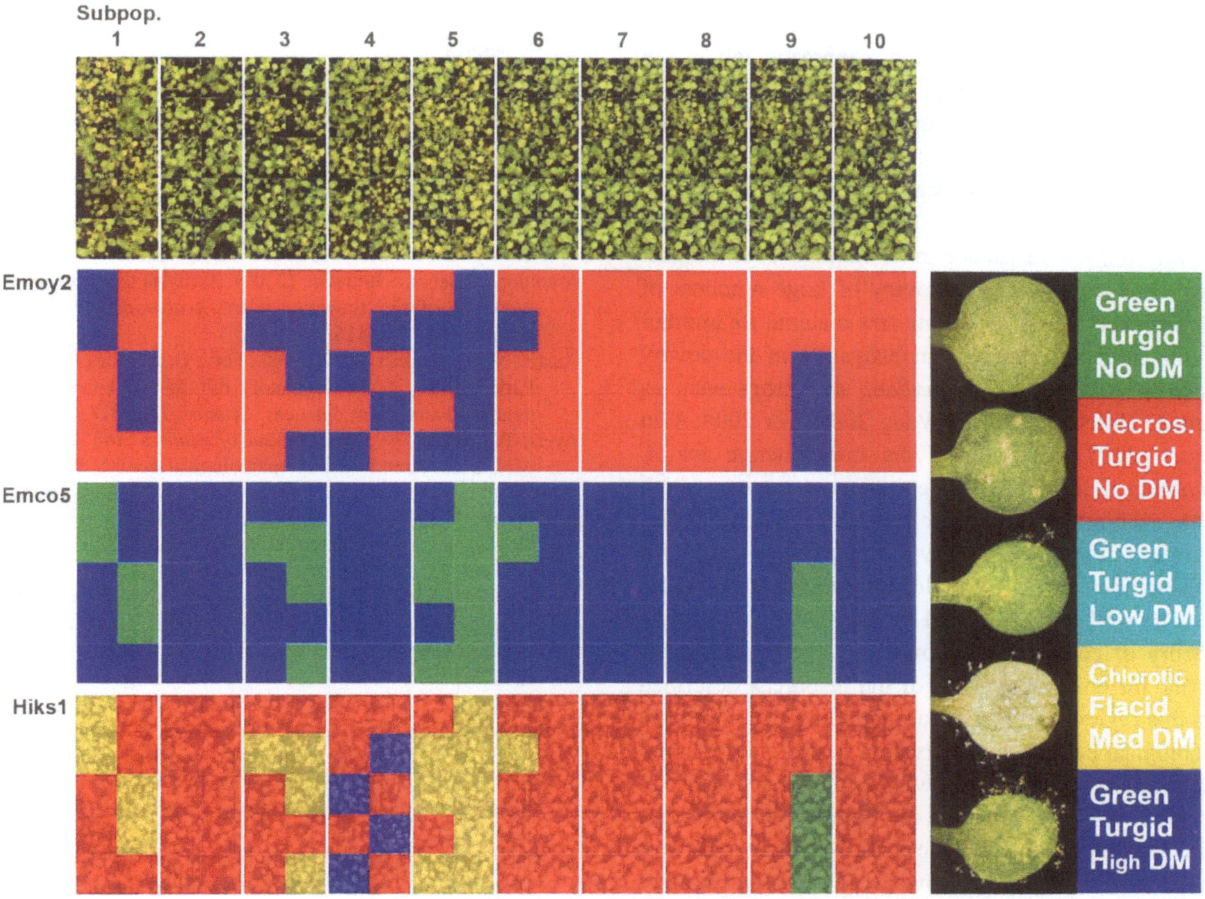

Fig. 5 *Arabidopsis thaliana* from the grounds of a horticulture research site in southeast England (located at East Malling Research) provides one important example of a well-established wild population (at least 25 years old) for investigating the molecular ecology of downy mildew. The reference isolate of *Hyaloperonospora arabidopsis* (*HpA*-Emoy2) was derived in 1991 from oospore-infested leaf tissue, collected from *A. thaliana* plants that were growing naturally outside a polytunnel at this location (Holub 2006; see Fig. 2c). Seed samples were subsequently collected in 1993, from 200 plants that were distributed across the research site, and each seed line was increased by self-pollination to establish larger individual seed stocks of this sample population for use in laboratory experiments. In a pilot experiment, 100 seed lines (arranged as illustrated by top row of seedling photographs; divided in 10 subpopulations of 10 lines in each) were inoculated as 7 day-old seedlings with different *HpA* isolates including Emoy2, Emoc5 (also from the same East Malling site), and Hiks1 (from Hampshire, ca. 250 km west of East Malling). Interaction phenotypes (*far right*, *vertical panel* of cotyledon photographs, along with brief description and colour coding) were recorded for each combination of seed line and *HpA* isolate. The pattern of colour coding summarizes the observed phenotype from each combination (seed line × isolate) across the host population. The *red* phenotype is specifically indicative of downy mildew resistance mediated by *RPP1* on chromosome 3 (recognizing both Emoy2 and Hiks1), and appears to occur at a high frequency throughout this population. The only alternative phenotype observed with Emoy2 was full susceptibility, whereas two alternative resistance phenotypes were observed with Hiks1; the *RPP* gene(s) conferring these examples of resistance have yet to be determined. Interestingly, Emco5 inoculations revealed an opposite pattern of susceptibility and resistance compared with Emoy2. In this case, the GREEN or full immunity phenotype may be indicative of *RPP13* which is known to recognize Emco5 and is closely linked to *RPP1*; however, this too has yet to be determined). Nonetheless, these preliminary data indicate that two of the most important *RPP* genes from previous laboratory research most likely occur in this natural population of *A. thaliana* as well as the matching avirulence effectors within the resident *HpA* population

ago (Holub 1997; Holub and Beynon 1997). However, the emphasis then was on laboratory research and the 'hardcore' molecular biology of innate immunity in plants. Molecular ecology of plant–parasite interactions at that time was an aspiration for a small community of biologists. Times have clearly changed with the complete genome sequence of the first *A. thaliana* accession, rich DNA-sequence data from another 20 *A. thaliana* accessions (Borevitz et al. 2007; Clark et al. 2007), and a dramatic fall in the cost of DNA sequencing. The aspiration has progressed from plausibility to a very real and exciting opportunity.

Much of the infrastructure is falling into place, behind major advances in the molecular genetics of downy mildew resistance in *A. thaliana* over the past two decades, and the discovery of large numbers of oomycete effectors providing raw material for another complementary round of investigation in the downy mildew parasites of *A. thaliana* and crops such as lettuce (Michelmore and Wong 2008; Tör 2008; Win et al. 2007). In parallel, the infrastructure for *A. thaliana* field biology is beginning to emerge, which has exciting potential for advancements in understanding the evolutionary context of a range of developmental and physiological processes in plants.

The well-developed and timely geographic mosaic theory of co-evolution provides the essential framework for constructive use of the resources described in this review. Mathematical modelling that combines ecological parameters (to explain spatial and temporal changes in populations) with evolutionary genetics (natural selection acting on multiple loci of interacting species) represents an enormous challenge for this exciting research field. However, a cross-disciplinary 'systems biology' approach is inevitable. Somehow, all of the public investment in laboratory research of *A. thaliana* will not make sense without further practical applications in plant breeding, and equally, without concerted contributions from a new generation of field botanists who are committed to aiding conservation biology. I would hope that the likes of A. DeBary and H.M. Ward would agree.

Acknowledgements The author is grateful to Robin Allaby, Simon Bright, Ian Crute, Greg Gilbert, Cindy Morris and Caroline Young for critical reading of the manuscript; to Joy Bergelson for the opportunity to join the 2010 association mapping project; and for generous support for related research from the UK Biotechnology and Biological Sciences Research Council and the Gatsby Charitable Trust.

References

Aarts, M. G. M., Hekkert, B. L., Holub, E. B., Beynon, J. L., Stiekema, W. J., & Pereira, A. (1998a). Identification of R-gene homologous DNA fragments genetically linked to disease resistance loci in *Arabidopsis thaliana*. *Molecular Plant–Microbe Interactions, 11*, 251–258.

Aarts, N., Metz, M., Holub, E. B., Staskawicz, B. J., Daniels, M. J., & Parker, J. E. (1998b). Different requirements for *EDS1* and *NDR1* by disease resistance genes define at least two *R* gene-mediated signalling pathways in *Arabidopsis*. *Proceedings of the National Academy of Science, USA, 95*, 10306–10311.

Allen, R. L., Bittner-Eddy, P. D., Grenville-Briggs, L. J., Meitz, J. C., Rehmany, A. P., Rose, L. E., et al. (2004). Host–parasite coevolutionary conflict between *Arabidopsis* and downy mildew. *Science, 306*, 1957–1960.

Arabidopsis Genome Initiative (2000). Analysis of the genome sequence of the flowering plant *Arabidopsis thaliana*. *Nature, 408*, 796–815.

Austin, M. J., Muskett, P., Kahn, K., Feys, B. J., Jones, J. D., & Parker, J. E. (2002). Regulatory role of SGT1 in early R gene-mediated plant defences. *Science, 15*, 2077–2080.

Ayers, P. (2005). *The fungal thread of death* p. 168. St Paul, MN, USA: American Phytopathological Society Press.

Azevedo, C., Betsuyaku, S., Peart, J., Takahashi, A., Noël, L., Sadanandom, A., et al. (2006). Role of SGT1 in resistance protein accumulation in plant immunity. *The EMBO Journal, 25*, 2007–2016.

Bakker, E. G., Stahl, E. A., Toomajian, C., Nordborg, M., Krietman, M., & Bergelson, J. (2006). Distribution of genetic variation within and among local populations of *Arabidopsis thaliana* over its species range. *Molecular Ecology, 15*, 1405–1418.

Bittner-Eddy, P. D., Crute, I. R., Holub, E. B., & Beynon, J. L. (2000). *RPP13* is a simple locus in *Arabidopsis* for alleles that specify downy mildew resistance to different avirulence determinants in *Peronospora parasitica*. *The Plant Journal, 21*, 177–188.

Borevitz, J. O., Hazen, S. P., Michael, T. P., Morris, G. P., Baxter, I. R., et al. (2007). Genome-wide patterns of single-feature polymorphism in *Arabidopsis thaliana*. *Proceedings of the National Academy of Science, USA, 104*, 12057–12062.

Borhan, M. H., Brose, E., Beynon, J. L., & Holub, E. B. (2001). White rust (*Albugo candida*) resistance loci on three *Arabidopsis* chromosomes are closely linked to downy mildew (*Peronospora parasitica*) resistance loci. *Molecular Plant Pathology, 2*, 87–95.

Borhan, M. H., Gunn, N., Cooper, A., Gulden, S., Tör, M., Rimmer, S. R., et al. (2008). *WRR4* encodes a TIR-NB-LRR protein that confers broad spectrum white rust resistance in *Arabidopsis thaliana* to four physiological races of *Albugo candida*. *Molecular Plant–Microbe Interactions*, (in press).

Borhan, M. H., Holub, E. B., Beynon, J. L., Rozwadowski, K., & Rimmer, S. R. (2004). The *Arabidopsis* TIR-NB-LRR gene *RAC1* confers resistance to *Albugo candida* (white rust) and is dependent on *EDS1* but not *PAD4*. *Molecular Plant–Microbe Interactions, 17*, 711–719.

Botella, M. A., Coleman, M. J., Hughes, D. E., Nishimura, M. T., Jones, J. D., & Somerville, S. C. (1997). Map positions of 47 *Arabidopsis* sequences with sequence similarity to disease resistance genes. *The Plant Journal, 12*, 1197–1211.

Botella, M. A., Parker, J. E., Frost, L. N., Bittner-Eddy, P. D., Beynon, J. L., Holub, E. B., et al. (1998). Three genes of the *Arabidopsis RPP1* complex locus recognize distinct

Peronospora parasitica (downy mildew) avirulence determinants. *The Plant Cell, 10*, 1847–1860.

Century, K. S., Holub, E. B., & Staskawicz, B. J. (1995). *NDR1,* a locus of *Arabidopsis thaliana* that is required for disease resistance to both a bacterial and a fungal pathogen. *Proceedings of the National Academy of Science, USA, 92*, 6597–6601.

Century, K., Shapiro, A. D., Repetti, P. P., Dahlbeck, D., Holub, E. B., & Staskawicz, B. J. (1997). *NDR1*, a pathogen induced component required for *Arabidopsis* disease resistance. *Science, 278*, 1963–1965.

Chou, H.-M., Bundock, N., Rolfe, S. A., & Scholes, J. D. (2000). Infection of *Arabidopsis thaliana* leaves with *Albugo candida* (white blister rust) causes a reprogramming of host metabolism. *Molecular Plant Pathology, 1*, 99–113.

Clark, R. M., Schweikert, G., Toomajian, C., Ossowski, S., Zeller, G., et al. (2007). Common sequence diversity in *Arabidopsis thaliana*. *Science, 317*, 338–342.

Cooley, M. B., Pathirana, S., Wu, H. J., Kachroo, P., & Klessig, D. F. (2000). Members of the *Arabidopsis HRT/RPP8* family of resistance genes confer resistance to both viral and oomycete pathogens. *The Plant Cell, 12*, 663–676.

Cooper, A. J., Latunde-Dada, A. O, Woods-Tör, A., Lynn, J., Lucas, J. A., Crute, I. R., et al. (2008). Basic compatibility of *Albugo candida* in *Arabidopsis thaliana* and *Brassica juncea* causes broad-spectrum suppression of innate immunity. *Molecular Plant–Microbe Interactions*, (in press).

Constantinescu, O., & Fatehi, J. (2002). *Peronospora*-like fungi (Chromista, Peronosporales) parasitic on Brassicaceae and related hosts. *Nova Hedwigia, 74*, 291–338.

Damgaard, C., & Jensen, B. D. (2002). Disease resistance in *Arabidopsis thaliana* increases the competitive ability and the predicted probability of long-term ecological success under disease pressure. *Oikos, 98*, 459–466.

Ellingboe, A. H. (1976). Genetics of host–parasite interactions. In R. Heitefuss, & P. H. Williams (Eds.) *Physiological plant pathology (Encyclopedia of plant physiology, new series, vol 4* (pp. 761–778). Berlin: Springer.

Ellingboe, A. H. (2001). Plant–pathogen interactions: Genetic and comparative analyses. *European Journal of Plant Pathology, 107*, 79–84.

Eulgem, T., Wang, X. J., Tsuchiya, T., Beasley, B., Tör, M., Zhu, T., et al. (2007). EDM2 is a novel component of *RPP7*-dependent disease resistance in *Arabidopsis* that controls *RPP7* transcript levels. *The Plant Journal, 49*, 829–839.

Eulgem, T., Weigman, V. J., Chang, H. S., McDowell, J. M., Holub, E. B., Glazebrook, J., et al. (2004). Three genetically separable *R* gene signaling pathways converge to regulate a largely overlapping transcriptome. *Plant Physiology, 135*, 1129–1144.

Falk, A., Feys, B. J., Frost, L. N., Jones, J. D., Daniels, M. J., & Parker, J. E. (1999). *EDS1,* an essential component of R gene-mediated disease resistance in *Arabidopsis* has homology to eukaryotic lipases. *Proceedings of the National Academy of Science, USA, 96*, 3292–3297.

Gassmann, W., Hinsch, M. E., & Staskawicz, B. J. (1999). The *Arabidopsis RPS4* bacterial-resistance gene is a member of the TIR-NBS-LRR family of disease-resistance genes. *The Plant Journal, 20*, 265–277.

Gäumann, E. (1918). Ueber die Formen der *Peronospora parasitica* (Pers.) Fries. *Beih. Bot. Zentralblatt, 35*(1. Abt.), 395–533.

Glazebrook, J. (2007). Use of microarry analysis to dissect the plant defence response. *Methods of Molecular Biology, 354*, 121–130.

Glazebrook, J., Zook, M., Mert, F., Kagan, I., Rogers, E. E., Crute, I. R., et al. (1997). Phytoalexin-deficient mutants of *Arabidopsis* reveal that *PAD4* encodes a regulatory factor and that four *PAD* genes contribute to downy mildew resistance. *Genetics, 146*, 381–392.

Göker, M., Riethmüller, A., Voglmayr, H., Weiss, M., & Oberwinkler, F. (2004). Phylogeny of *Hyaloperonospora* based on nuclear ribosomal internal transcribed spacer sequences. *Mycological Progress, 3*, 83–94.

Gomulkiewicz, R., Drown, D. M., Dybdahl, M. F., Godsoe, W., Nuismer, S. L., Pepin, K. M., et al. (2007). Dos and don'ts of testing the geographic mosaic theory of coevolution. *Heredity, 98*, 249–258.

Grant, S. R., Fisher, E. J., Chang, J. H., Mole, B. M., & Dangl, J. L. (2006). Subterfuge and manipulation: Type III effector proteins of phytopathogenic bacteria. *Annual Review of Microbiology, 60*, 425–449.

Gunn, N., Byrne, J., & Holub, E. B. (2002). Outcrossing of two homothallic isolates of *Peronospora parasitica* and segregation of avirulence matching six resistance loci in *Arabidopsis thaliana*. In P. T. N. Spencer-Phillips, U. Gisi, & A. Lebeda (Eds.) *Advances in downy mildew research* (pp. 185–188). Dordrecht: Kluwer Academic Publishers.

Holt, B. F., Belkhadir, Y., & Dangl, J. L. (2005). Antagonistic control of disease resistance protein stability in the plant immune system. *Science, 309*, 929–932.

Holub, E. B. (1997). Organization of resistance genes in *Arabidopsis*. In I. R. Crute, E. B Holub, & J. J. Burdon (Eds.) *The gene-for-gene relationship in plant parasite interactions* (pp. 5–26). Wallingford, UK: CAB International.

Holub, E. B. (2001). The Arms race is ancient history in *Arabidopsis*, the wildflower. *Nature Reviews Genetics, 2*, 516–527.

Holub, E. B. (2006). Evolution of symbioses between plants and filamentous parasites. *Current Opinion of Plant Biology, 9*, 397–405.

Holub, E. B. (2007). Natural variation in innate immunity of a pioneer species. *Current Opinion of Plant Biology, 10*, 415–424.

Holub, E. B., & Beynon, J. L. (1997). Symbiology of mouse-ear cress (*Arabidopsis thaliana*) and oomycetes. *Advances in Botanical Research, 24*, 227–273.

Holub, E. B., Beynon, J. L., & Crute, I. R. (1994). Phenotypic and genotypic characterisation of interactions between isolates of *Peronospora parasitica* and accessions of *Arabidopsis thaliana*. *Molecular Plant–Microbe Interactions, 7*, 223–239.

Holub, E. B., Brose, E., Tör, M., Clay, C., Crute, I. R., & Beynon, J. L. (1995). Phenotypic_and genotypic variation in the interaction between *Arabidopsis thaliana* and *Albugo candida*. *Molecular Plant–Microbe Interactions, 8*, 916–928.

Holub, E. B., & Cooper, A. (2004). Matrix, reinvention in plants: How genetics is unveiling secrets from non-host disease resistance. *Trends Plant Science, 9*, 211–214.

Jirage, D., Tootle, T. L., Reuber, T. L., Frost, L. N., Feys, B. J., Parker, J. E., et al. (1999). *Arabidopsis thaliana PAD4* encodes a lipase-like gene that is important for salicylic acid signaling. *Proceedings of the National Academy of Science, USA, 96*, 13583–13588.

Jönsson, H., Heisler, M. G., Shapiro, B. E., Meyerowitz, E. M., & Mjolsness, E. (2006). An auxin-driven polarized transport model for phyllotaxis. *Proceedings National Academy of Science, USA, 103*, 1633–1638.

Kamoun, S. (2007). Groovy times: Filamentous pathogen effectors revealed. *Current Opinion of Plant Biology, 10*, 358–365 (Aug issue).

Kazan, K., & Schenk, P. M. (2007). Genomics in induced resistance. In D. Walters, A. Newton, & G. Lyon (Eds.) *Induced resistance for plant defence* (pp. 31–64). Oxford: Blackwell.

Koch, E., & Slusarenko, A. (1990). *Arabidopsis* is susceptible to infection by a downy mildew fungus. *The Plant Cell, 2*, 437–445.

Lawton, K., Weymann, K., Friedrich, L., Vernooij, B., Uknes, S., & Ryals, J. (1995). Systemic acquired resistance in *Arabidopsis* requires salicylic acid but not ethylene. *Molecular Plant–Microbe Interactions, 8*, 863–870.

Lindau, G. (1901). *Hilfsbuch für das Sammein Parasitischer Pilze*. Berlin: Bornträger Verlag.

McCouch, S. (2004). Diversifying selection in plant breeding. *PLoS Biology, 2*, 1507–1512.

McDowell, J. M., Cuzick, A., Can, C., Beynon, J. L., Dangl, J. D., & Holub, E. B. (2000). Downy mildew (*Peronospora parasitica*) resistance genes in *Arabidopsis* vary in functional requirements for *NDR1*, *EDS1*, *NPR1*, and salicylic acid accumulation. *The Plant Journal, 22*, 523–529.

McDowell, J. M., Dhandaydham, M., Long, T. A., Aarts, M. G. M., Goff, S., Holub, E. B., et al. (1998). Intragenic recombination and diversifying selection contribute to the evolution of downy mildew resistance at the *RPP8* locus of *Arabidopsis*. *The Plant Cell, 10*, 1861–1874.

Michelmore, R. W. & Wong, J. W. (2008). Classical and molecular genetics of *Bremia lactucae*, lettuce downy mildew. *European Journal of Plant Pathology*, (this issue).

Mitchell-Olds, T., & Schmitt, J. (2006). Genetic mechanisms and evolutionary significance of natural variation in *Arabidopsis*. *Nature, 441*, 264–271.

Muskett, P. R., Kahn, K., Austin, M. J., Moisan, L. J., Sadanandom, A., Shirasu, K., et al. (2002). *Arabidopsis* RAR1 exerts rate-limiting control of R gene-mediated defences against multiple pathogens. *The Plant Cell, 14*, 979–992.

Parker, J. E., Coleman, M. J., Szado, V., Frost, L. N., Schmidt, R., van der Biezen, E. A., et al. (1997). The *Arabidopsis* downy mildew resistance gene *RPP5* shares similarity to the Toll and interleukin-1 receptors with *N* and *L6*. *The Plant Cell, 9*, 879–894.

Parker, J. E., Holub, E. B., Frost, L. N., Falk, A., Gunn, N. D., & Daniels, M. J. (1996). Characterisation of *eds1*, a mutation in *Arabidopsis* suppressing resistance to *Peronospora parasitica* specified by several different *RPP* genes. *The Plant Cell, 8*, 2033–2046.

Preston, C. D., Pearman, D. A., & Hall, A. R. (2004). Archaeophytes in Britain. *Botanical Journal of the Linnean Society, 145*, 257–294.

Prusinkiewicz, P., Erasmus, Y., Lane, B., Harder, L. D., & Coen, E. (2007). Evolution and development of inflorescence architectures. *Science, 316*, 1452–1456.

Rehmany, A. P., Gordon, A., Rose, L. E., Allen, R. L., Armstrong, M. R., Whisson, S. C., et al. (2005). Differential recognition of highly divergent downy mildew avirulence gene alleles by *RPP1* resistance genes from two *Arabidopsis* lines. *The Plant Cell, 17*, 1839–1850.

Rehmany, A. P., Lynn, J. R., Tör, M., Holub, E. B., & Beynon, J. L. (2000). A comparison of *Peronospora parasitica* (downy mildew) isolates from *Arabidopsis thaliana* and *Brassica oleracea* using amplified fragment length polymorphism and internal transcribed spacer 1 sequence analyses. *Fungal Genetics & Biology, 30*, 95–103.

Rentel, M. C., Leonelli, L., Dahlbeck, D., Zhao, B., & Staskawicz, B. J. (2008). Recognition of the *Hyaloperonospora parasitica* effector ATR13 triggers resistance against oomycete, bacterial, and viral pathogens. *Proceedings of the National Academy of Science, USA, 105*, 1091–1096.

Riethmüller, A., Voglmayer, H., Göker, M., Weiß, M., & Oberwinkler, F. (2002). Phylogenetic relationships of the downy mildew (Peronosporales) and related groups based on nuclear large subunit ribosomal DNA sequences. *Mycologia, 94*, 834–849.

Rose, L. E., Bittner-Eddy, P. D., Holub, E. B., Langley, C. H., Michelmore, R. W., & Beynon, J. L. (2004). The maintenance of extreme amino acid diversity at the disease resistance gene, *RPP13*, in *Arabidopsis*. *Genetics, 166*, 1517–1527.

Roetschi, A., Si-Ammour, A., Belbahri, L., Mauch, F., & Mauch-Mani, B. (2001). Characterisation of an *Arabidopsis–Phytophthora* pathosystem: Resistance requires a functional *PAD2* gene and is independent of salicylic acid, ethylene and jasmonic acid signalling. *The Plant Journal, 28*, 293–305.

Segal, S. (1969). *Ecological notes on wall vegetation* p. 326. The Hague, Netherlands: Dr W. Junk.

Shen, K. A., Chin, D. B., Arroyo-Garcia, R., Ochoa, O. E., Lavelle, D. O., Wroblewski, T., et al. (2002). *Dm3* is one member of a large constitutively expressed family of nucleotide binding site-leucine-rich repeat encoding genes. *Molecular Plant–Microbe Interactions, 15*, 251–261.

Shen, Q. H., & Schulze-Lefert, P. (2007). Rumble in the nuclear jungle: Compartmentalization, trafficking and nuclear action of plant immune receptors. *EMBO Journal, 26*, 4293–4301.

Silvertown, J. (2005). *Demons in Eden, the paradox of plant diversity*. Chicago, IL, USA: University of Chicago Press (ISBN: 0-226-75771-4).

Slusarenko, A. J., & Schlaich, N. L. (2003). Downy mildew of *Arabidopsis thaliana* caused by *Hyaloperonospora parasitica* (formerly *Peronospora parasitica*). *Molecular Plant Pathology, 4*, 159–170.

Song, J., Bradeen, J. M., Naess, S. K., Raasch, J. A., Wielgus, S. M., Haberlach, G. T., et al. (2004). Gene *RB* cloned from *Solanum bulbocastanum* confers broad spectrum resistance to potato late blight. *Proceedings of the National Academy of Science, USA, 100*, 9128–9133.

Speulman, E., Bouchez, D., Holub, E., & Beynon, J. (1998). Disease resistance gene homologs correlate with disease resistance loci of *Arabidopsis thaliana*. *The Plant Journal, 14*, 467–474.

Staal, J., & Dixelius, C. (2007). Tracing the ancient origins of plant innate immunity. *Trends in Plant Science*, *12*, 334–342.

Takahashi, H., Miller, J., Nozaki, Y., Takeda, M., Shah, J., Hase, S., et al. (2002). *RCY1*, an *Arabidopsis thaliana RPP8/HRT* family resistance gene, conferring resistance to *Cucumber mosaic virus* requires salicylic acid, ethylene and a novel signal transduction mechanism. *The Plant Journal*, *32*, 655–667.

Tang, X., Rolfe, S. A., & Scholes, J. D. (1996). The effect of *Albugo candida* (white blister rust) on the photosynthetic and carbohydrate metabolism of leaves of *Arabidopsis thaliana*. *Plant, Cell & Environment*, *19*, 967–975.

Thompson, J. N. (1999a). Specific hypotheses on the geographic mosaic of coevolution. *The American Naturalist*, *153*, S1–S14.

Thompson, J. N. (1999b). The evolution of species interactions. *Science*, *284*, 2116–2118.

Thompson, K. S. (2007). Beatrix Potter, conservationist. *American Scientist*, *95*, 210–212.

Thompson, J. N., Reichman, O. J., Morin, P., Polis, G. A., Power, M. E., Sterner, R. W., et al. (2001). Frontiers of ecology. *BioScience*, *51*, 15–24.

Tör, M. (2008). Tapping into molecular conversation between oomycete plant pathogens and their hosts. *European Journal of Plant Pathology*, (this issue).

Tör, M., Gordon, P., Cuzick, A., Eulgem, T., Sinapidou, E., Mert-Türk, F., et al. (2002). *Arabidopsis SGT1b* is required for defence signaling conferred by several downy mildew resistance genes. *The Plant Cell*, *14*, 993–1003.

Tornero, P., Merritt, P., Sadanandom, A., Shirasu, K., Innes, R. W., & Dangl, J. L. (2002). RAR1 and NDR1 contribute quantitatively to disease resistance in *Arabidopsis*, and their relative contributions are dependent on the *R* gene assayed. *The Plant Cell*, *14*, 1005–1015.

Tyler, B. M., Tripathy, S., Zhang, X., Dehal, P., Jiang, R. H., et al. (2006). Phytophthora genome sequences uncover evolutionary origins and mechanisms of pathogenesis. *Science*, *313*, 1261–1266.

Van Damme, M., Andel, A., Hulbers, R. P., Panstruga, R., Weisbeek, P. J., & Van den Ackerveken, G. (2005). Identification of *Arabidopsis* loci required for susceptibility to the downy mildew pathogen *Hyaloperonospora parasitica*. *Molecular Plant–Microbe Interactions*, *18*, 583–592.

Voglmayr, H. (2008). Progress and challenge in systematics of downy mildews and white blister rusts: new insights from genes and morphology. *European Journal of Plant Pathology*, (this issue).

Voglmayr, H., & Riethmüller, A. (2006). Phylogenetic relationships of *Albugo* species (white blister rusts) based on LSU rDNA sequence and oospore data. *Mycological Research*, *110*, 75–85.

Warren, R. F., Henk, A., Mowery, P., Holub, E. B., & Innes, R. W. (1998). A mutation within the leucine rich repeat domain of the *Arabidopsis* disease resistance gene *RPS5* partially suppresses multiple bacterial and downy mildew resistance. *The Plant Cell*, *10*, 1439–1452.

Warren, R. F., Merritt, P. M., Holub, E., & Innes, R. W. (1999). Identification of three putative signal transduction genes involved in *R* gene-specified disease resistance in *Arabidopsis*. *Genetics*, *152*, 401–412.

Wilson, E. O. (1994). *Naturalist* p. 380. Washington D.C., USA: Island Press.

Win, J., Morgan, W., Bos, J., Krasileva, K. V., Cano, L. M., Chaparro-Garcia, A., et al. (2007). Adaptive evolution has targeted the C-terminal domain of the RXLR effectors of plant pathogenic oomycetes. *The Plant Cell*, *19*, 2349–2369.

Eur J Plant Pathol (2008) 122:111–126
DOI 10.1007/s10658-008-9289-y

Comparative epidemiology of zoosporic plant pathogens

Mike J. Jeger · Marco Pautasso

Received: 4 October 2007 / Accepted: 4 February 2008

Abstract Loss of zoospores has happened independently several times in different phylogenic lines and has, it is claimed, no major phylogenetic significance. But whether or not, how, and under which conditions plant pathogens retain the ability to produce motile asexual spores has fundamental importance from an ecological and epidemiological perspective. Recent molecular investigations of the early evolution of fungi and oomycetes are shedding light on the issue of zoospore loss in organisms able to cause plant diseases. Zoospore loss may have accompanied the development of new forms of dispersal adapted to the terrestrial environment, or the simplification processes which often follow the shift to parasitic or biotrophic life-forms. In this review we consider hybridisation events between *Phytophthora* species, long distance dispersal of oomycetes, sporangia and zoospore survival, direct and indirect infection processes and newly observed sporulating structures. These aspects are all relevant features for an understanding of the epidemiology of zoosporic plant pathogens. Disease management should not be based on the presumption that the zoosporic stage is a weak link in the life cycle. Oomycete plant pathogens show remarkable flexibility in their life cycles and ability to adapt to changing environmental circumstances.

Keywords Fungal phylogeny · Landscape pathology · Pathogen evolution · Plant epidemiology · Zoosporangia

M. J. Jeger (✉) · M. Pautasso
Division of Biology, Imperial College London,
Silwood Park Campus,
Ascot SL5 7PY, UK
e-mail: m.jeger@ic.ac.uk

Introduction

Plant diseases are the outcome of the interaction of plants with a variety of pathogenic organisms in a disease-conducing environment. Many important plant pathogens are zoosporic, i.e. with motile asexual spores. Zoosporic plant pathogens cause significant crop losses worldwide and are the object of a substantial amount of epidemiological research. In our use of the term 'zoosporic plant pathogens' we include both zoosporic fungi and oomycetes. Although we wish to avoid becoming entangled in a systematics debate, modern molecular phylogenetic studies must be at the very heart of any attempt to discuss the comparative epidemiology (Kranz 1980, 2003) of plant pathogens in relation to the evolutionary loss of zoospores, a feature present in both true fungi and oomycetes. Zoospores, which are single-nucleated, formed in sporangia, and motile in aqueous environments, are however a key feature in the life cycle of many plant pathogens. They have been thought to be a weak link, as zoospores have no cell walls, which makes them particularly vulnerable and transient (Lange and Olson 1983; Stanghellini 1997). Here, we aim at a selective review of relevant literature, focusing on a limited number of case studies that we believe provide insights in the issue of zoospore function and loss in plant pathogens. We then move on to discuss the epidemiological and ecological implications for sustainable plant disease management. We emphasize plant pathogens within the Straminipila[1], mostly oomycetes, but refer also to the true fungi with zoospores when appropriate. Zoospores are common in oomycetes and less common in the true fungi (Hardham et al. 1994; Lebeda and Schwinn 1994; Judelson and Blanco 2005).

Phylogenetic and epidemiological significance of zoospore loss

According to Dick (2002), "loss of the zoospore and therefore flagellation is a feature of both the Peronosporales and Sclerosporales and has minor phylogenetic significance." If the term 'fungus' is considered to be an essentially physiological concept and not a taxonomic one, then several independent phylogenetic lines of fungi have evolved (and lost) flagella (Dick 1997). We would not dismiss such a lack of phylogenetic significance for zoospore loss, but argue here that loss of flagellation and motility must have considerable significance from an epidemiological point of view. This follows from the differences in dispersal potential, infection processes and survival between pathogens with or without zoospores.

In spite of this, the ability to produce zoospores varies amongst different groups of oosporic plant pathogens. For example, it is usual in *Albugo* (e.g. Whipps and Cooke 1978), variable in *Plasmopara* (e.g. Kast and Stark-Urnau 1999), environmentally-dependent in *Phytophthora* (e.g. Judelson and Blanco 2005), and lost in *Hyaloperonospora* (e.g. Slusarenko and Schlaich 2003) and *Peronosclerospora* (e.g. Jeger et al. 1998). Whenever organisms have evolved to occupy niches in which their pre-existing complexity might have been superfluous, there has been the potential for features not contributing to the fitness of the species to be lost abruptly or over a period of time.

Whether zoosporic loss happened during punctuated events or over longer periods of time can only be the subject of speculation; given the paucity of the fossil record for fungi and oomycetes alike, the important point is that antagonistic interactions may inherently lead towards simplification—once one organism becomes dependent on another for its sustenance it may discard features previously required as a free-living organism. Parasitism, for instance, is often accompanied by morphological simplification involving, in the system we are interested here, the evolution of sporangia originally water-dependent and producing zoospores into sporangiophores producing directly germinating conidia (Brasier and Hansen 1992).

There may be here a conceptual connection with the argument that pathogens with higher genetic diversity and thus evolutionary potential pose a greater risk to plant populations, other things being equal, as these pathogens will be more likely than those with less genetic diversity to overcome the defence apparatus of their host(s) (McDonald and Linde 2002). Host specialization may on the one hand lead to genetic impoverishment, as the pathogen no longer needs the ability to infect various hosts, and can thus discard the machinery upon which it relied to successfully infect that host diversity; on the other hand, host specialization may also lead to the creation of new pathogen genetic diversity due to species-specific evolutionary arms races between host and pathogen (Clay and Kover 1996).

For species of the genus *Phytophthora*, both specialization to a single host and general aggressiveness towards a wide range of hosts are observed (Brasier and Hansen 1992; Hardham 2007). For example, *P. cinnamomi* affects several tree, shrub and herbaceous species in the Jarrah forest of South-Western Australia (e.g. Shearer et al. 2007). A similar wide range of potential and actual hosts is found with *P. ramorum* (e.g. Rizzo et al. 2005). Conversely, *P. sojae* (e.g. Tyler 2007), *P. ilicis* (Coyier 1981) and *P. porri* (Smilde et al. 1995) are all examples of Phytophthoras which are specialized to a single host or to a taxonomically related group of hosts. This host specialization implies a distinct co-evolution of attack

[1] Alternatively, Stramenopila: spelling of the taxon and of its various derivatives urgently needs standardization. Analysis of 32 publications since 2000 breaks down into 20 using the spelling above and 12 using the alternative. In some cases both taxon spellings are given as keywords (Money et al. 2004; Honda et al. 2007).

and defence in these pathosystems. Zoospore loss seems not to be dependent on whether or not a certain *Phytophthora* has undergone host specialization, but rather on environmental conditions.

Increasing numbers of molecular studies are elucidating the early evolution of various groups of plant pathogens, including the true fungi (James et al. 2006a) and oomycetes (Göker et al. 2004; Tyler et al. 2006). Assembling the fungal tree of life (Bruns 2006) also provides insights on the issue of zoospore loss in organisms able to cause plant diseases. The ancestors of fungi are believed to have been simple aquatic forms with flagellated spores (James et al. 2006a). Also the earliest fungi were aquatic and lacked aerial spore dispersal. The traditional view is then that a monophyletic core developed producing zoospores (phylum Chytridiomycota, with the exception of *Hyaloraphidium curvatum*, where the presence of flagella has never been reported; Ustinova et al. 2000). As opposed to that, loss of zoospores was generally thought to have happened in the Zygomycota, with the exception of the single-flagellated *Olpidium* (Lange and Olson 1976), which has now been reclassified (James et al. 2006b). However, recent molecular work based on a six-gene phylogeny suggests that the Chytridiomycota are not monophyletic, and that at least four independent events of zoospore loss can be traced back in the kingdom Fungi (James et al. 2006a).

This surge of molecular activity is not just relevant for the production of a more accurate phylogeny (Tyler et al. 2006; Göker et al. 2007), but also for applied epidemiology, as zoosporic fungi can act as vectors of plant viruses (e.g. Teakle 1983; Adams 1991; Campbell 1996; Rochon et al. 2004), although suspicions that oomycetes may be implicated in virus transmission, e.g. *Lagena radicicola* and flame chlorosis of cereals (Haber et al. 1991), have not been confirmed. But before dealing with the ecological and epidemiological implications of zoospore loss in oomycetes, we briefly discuss potential explanations for such an evolutionary development and some case studies.

Explanations for the loss of zoospores

Loss of flagellated spores is believed to have been concurrent with the development of new mechanisms of spore production and dispersal (James et al. 2006a). When fungi moved on to the terrestrial environment, some of them shed their 'ancient baggage' which had made them successful in water, and focused on new means of dispersal, more adapted to the new life in periodically water-poor environments. For example, in the Peronosporales, *Hyaloperonospora parasitica* has no zoosporic stage in its life cycle, and this has been related to its independence from the aqueous environment (Slusarenko and Schlaich 2003).

Alternatively, zoospore loss may have accompanied the development of parasitism and biotrophy. An example is *Peronospora*, which is thought to derive from a *Phytophthora* that lost the ability to produce zoospores and became an obligate biotroph (Cooke et al. 2000). There is a wide spectrum of angiosperm hosts that is parasitised by the morphologically 'advanced' (i.e. lacking zoospores) genus *Peronospora*. For species-specific parasitic interactions, it has been claimed that suppression and inhibition are likely to be less important than attraction and growth stimulation (Dick 2002).

There are many examples of zoosporic loss of plant pathogens in relation to the presence or absence of humidity in their typical environment. Prime case studies are tropical graminaceous downy mildews of sorghum and pearl millet (Jeger et al. 1998; Fig. 1). On the one hand, *Sclerospora graminicola* produces zoospores and affects pearl millet, which is generally found in regions with higher temperatures and lower rainfall than sorghum. Sorghum is affected by *Peronosclerospora sorghi*, which does not produce zoospores in spite of sorghum growing in regions of higher humidity than those where pearl millet is cultivated. Given that flagellated zoospores are propagules for dispersal in the presence of humidity, it is perhaps counter-intuitive that *S. graminicola* should have kept zoospores whilst *P. sorghi* should have lost them. Conversely, it can be argued that zoospores are even more important in an arid environment where water is available only rarely and needs to be used efficiently.

There are recent examples where plant pathogens have made a rapid transition to a new environment. Turf grass rapid blight disease has recently emerged as a terrestrial plant pathogen (Olsen 2007). It was first observed in California in 1995 and was subsequently associated with high salinity irrigation in

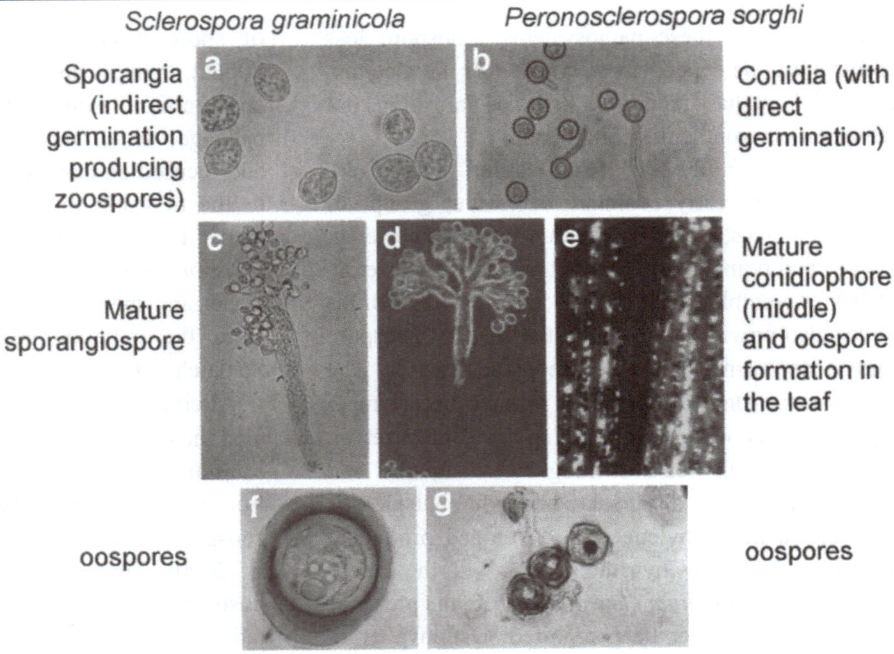

Fig. 1 Sexual and asexual phases of *Sclerospora graminicola* **a**, **c**, **f** and *Peronosclerospora sorghi* **b**, **d**, **e**, **g** (from Jeger et al. 1998, with kind permission of Blackwell)

water and golf courses. Preliminary diagnosis identified the pathogen as a species of the *Labyrinthula* genus, which is associated with the marine environment. For example, *L. zosterae* causes marine grass wasting disease (Olsen et al. 2003). The pathogen (Fig. 2) was then aptly named as *Labyrinthula terrestis* sp. nov. (Bigelow et al. 2005), as it is the first observation of this type of organism (a straminipile; Leander and Porter 2001) on land plants. It is considered to have originated from a single infected population and to share a recent common ancestor with other labyrinthulids (Craven et al. 2005). *Labyrinthula terrestris* builds digitate colonies in an extracellular network produced by specialized organelles called bothrosomes and uses these networks to move rapidly (Stowell et al. 2005). The disease has spread onto golf courses in Arizona and nine other US states; there has been a first report of a *Labyrinthula* sp. on amenity turf grass in the UK (Entwistle et al. 2006). In many Labyrinthulid species there is an absence of zoospore production, although biflagellated zoospores are clearly described (Amon and Perkins 1968; Perkins 1973; Amon 1978). Perhaps the formation of the extracellular network enables the local but rapid movement of somatic cells analogous to the swimming of zoospores.

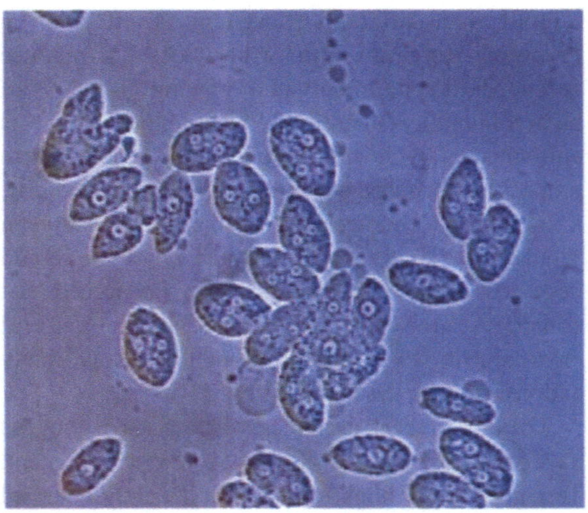

Fig. 2 Vegetative cells of *Labyrinthula terrestris* illustrating longitudinal cell division (photo, D. Bigelow, with kind permission of American Phytopathological Society)

Pythium species are root-infecting oomycetes closely related to Phytophthoras (Brasier and Hansen 1992; Deacon and Donaldson 1993). They are characterized by flexibility in their life cycle. Oospores can either germinate directly or produce cysts via sporangia and zoospores. Zoospores can also be produced by sporangiophora on infected seedlings (van West et al. 2003). Some species, e.g. *P. glomeratum* from soil, are reported to produce no sporangia or zoospores (Paul 2003) but as a rule *Pythium* species do have the ability to undergo zoosporogenesis (Walker and van West 2007). Other species, such as *P. helicoides*, are reported to produce only sporangia and zoospores in ebb-and-flow culture systems (Kageyama et al. 2003; Fig. 3). Some related oomycetes, e.g. *Saprolegnia* species, are able to release a new secondary zoospore after encystment of a primary zoospore. The secondary zoospore is the better swimming spore (Walker and van West 2007). Thus *Pythium* and related species such as *Aphanomyces* show remarkable flexibility in their life cycles and the ability to respond and adapt to changing environmental conditions.

Epidemiological and ecological implications

Zoospore loss has been reported widely in plant pathogens, but it is important to relate this knowledge to its potential epidemiological implications and to its relevance for disease management (Jeger 2004; Madden 2006). We discuss here hybridisation events for Phytophthoras, long distance dispersal for tobacco blue mold, the relation of sporangia and zoospore release with pathogen survival, infection processes (direct and indirect germination), sporulating structures in *Phytophthora ramorum,* integrating life cycles in *P. syringae*, and epidemic modelling in *P. infestans*.

Hybridisation events

The advent of molecular phylogenetics has revealed the potential for interspecific hybridisation of many plant pathogens (Schardl and Craven 2003). Hybrids may create devastating disease on both cultivated and wild plants (Olsen and Stenlid 2002) and have the potential to jump on new host species or to increase their virulence on traditionally infected hosts. For *Phytophthora*, the occurrence of multiple species in the rhizosphere of individual nursery plants can enhance the evolution and emergence of new tree diseases (Brasier and Jung 2003). Natural hybrids of *P. nicotianae* and *P. cactorum* have been observed in glasshouse hydroponic systems (Bonants et al. 2000). Similarly, there are reports of interspecific crosses between *Phytophthora sojae* and *P. vignae* (May et al. 2003) and of nuclear hybrids from protoplasts of *P. parasitica* and *P. capsici* followed by completion of the parasexual cycle (Gu and Ko 2000). In vitro fusion of zoospores of *P. nicotianae* and *P. capsici* has been achieved (Érsek et al. 1995; English et al. 1999).

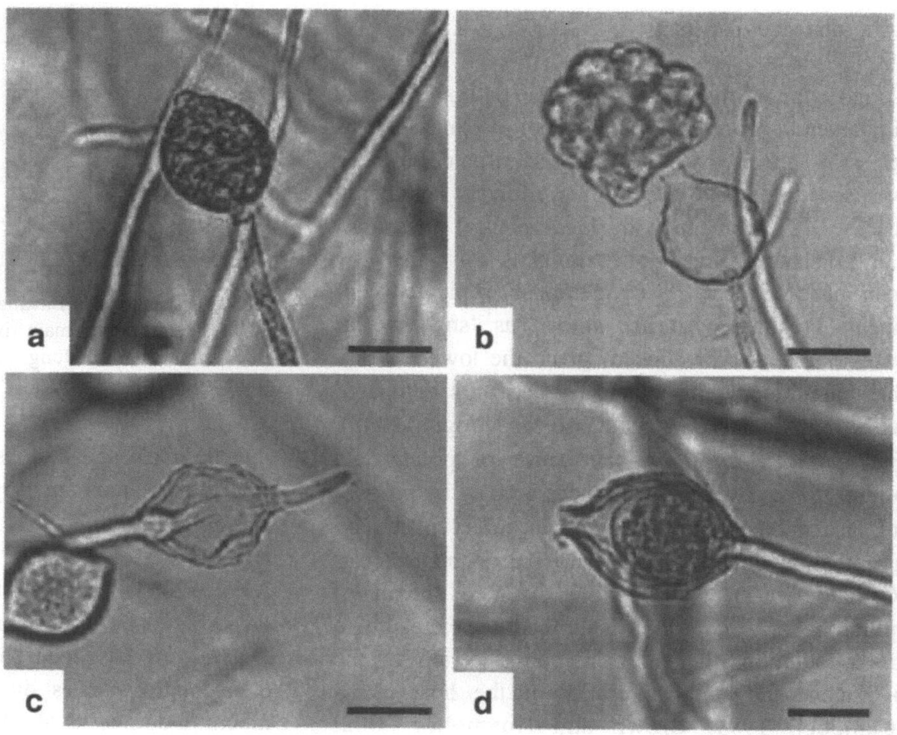

Fig. 3 Morphology and germination mode of group P of *Pythium* (scale bars = 20 μm). **a** Papillate sporangium, **b** zoospore formation in a vesicle originating from a sporangium, **c** hyphae proliferating from the base of the sporangium, **d** a sporangium proliferating from inside an old sporangial wall (from Kageyama et al. 2003, with kind permission of Blackwell)

There has been much less work done with downy mildews although genetic recombination through the parasexual cycle has been demonstrated in *Plasmopara halstedii* (Spring and Zipper 2006).

The emergence and spread of the hybrid alder *Phytophthora* is a good example of the potential of hybridization events to create new pathosystems (Brasier et al. 1995, 2004). Extensive field surveys of riparian and plantation alder in Bavaria (Germany) have revealed that symptoms were widespread on the majority of river courses and one third of plantation stands (Jung and Blaschke 2004; Fig. 4; see also Gibbs et al. 1999 for Britain, and Streito et al. 2002 and Thoirain et al. 2007 for France). The source of inoculum was traced back to young infected alder plantations at sites that drain into the river system. Rootstocks of alder plants might have been infected in nurseries, possibly due to the presence of disease propagules in irrigation water. The subsequent direct spread of zoospores from infected plantations (during seasonal flooding or waterlogged sites) to older and naturally regenerating trees, as well as to river catchments and riparian alders, can be seen as an example of disease spread at the landscape level along a physical network (Holdenrieder et al. 2004; Jeger et al. 2007).

Long-distance dispersal

Long-distance dispersal of plant pathogens is a fundamental process in the dynamic of plant epidemics, as it enables disease to jump from patch to patch of susceptible hosts, overcoming efforts at containing disease development with local control measures. Long-distance spread of pathogens is helped by man-made connectivity of previously separated continents creating what are known as 'small-world' networks, and is of concern given the lower disease threshold of epidemics in such networks compared with regular lattices (Pautasso and Jeger 2008). *Phytophthora infestans,* the cause of potato late blight, moves over long distances aerially by producing asexual sporangia which can infect plants by germinating directly or by releasing zoospores (e.g. Ristaino 2002). Natural long-distance spread of sporangia of *P. infestans* is limited by exposure to UVB radiation, the short infectious period of the pathogen, and rapid mortality of the host plants (Campbell 1999; Brown and Hovmøller 2002;

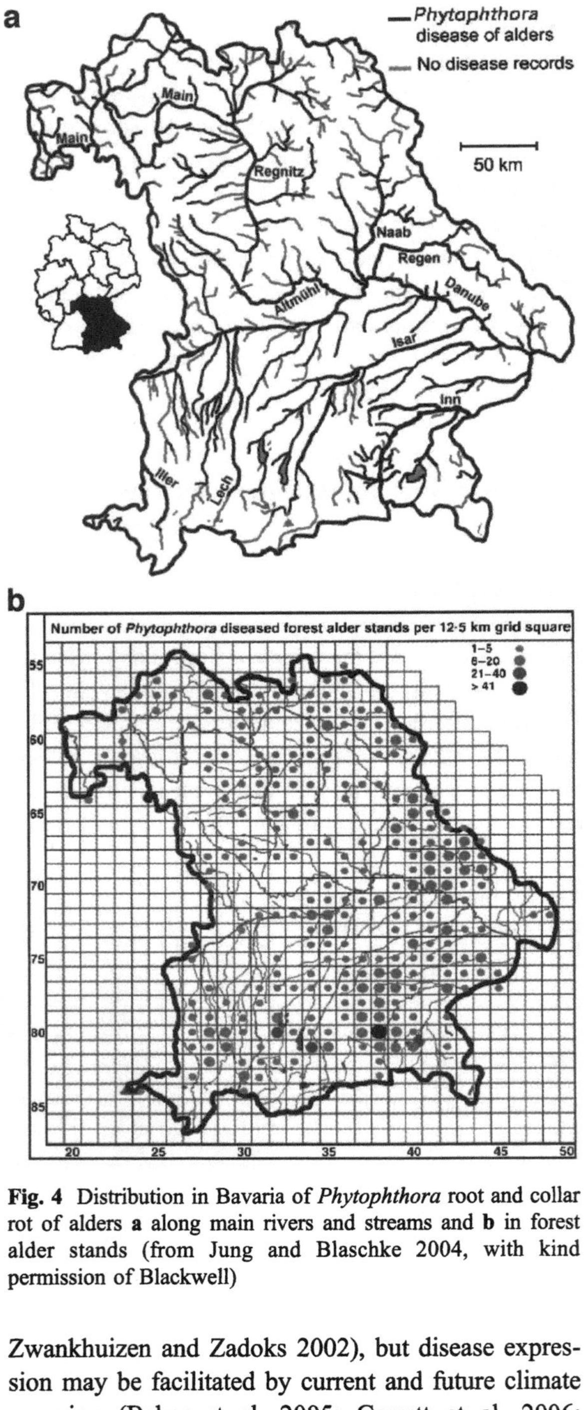

Fig. 4 Distribution in Bavaria of *Phytophthora* root and collar rot of alders **a** along main rivers and streams and **b** in forest alder stands (from Jung and Blaschke 2004, with kind permission of Blackwell)

Zwankhuizen and Zadoks 2002), but disease expression may be facilitated by current and future climate warming (Baker et al. 2005; Garrett et al. 2006; Hannukkala et al. 2007; Jeger and Pautasso 2008). Aylor (2003) assessed the critical gap width for dispersal to be approximately 35–50 km. However, *P. infestans* has been shown to spread rapidly and over long distances due to movement of infected tubers (Goodwin et al. 1998).

Long-distance dispersal of tobacco blue mold (*Peronospora tabacina*) is another example of the potential for plant pathogens to spread and act over vast regions. Each year, blue mold advances in a wave from the southern-most tobacco-growing regions to the northern-most ones in the eastern USA (Aylor 1999). This is consistent with the observed low rates of genetic diversity in this pathogen throughout the USA (Sukno et al. 2002). Calculated rates of advance range from 9 to 18 km per day. Aylor (2003) estimated the critical gap width for disease spread to be 10^2 km for dispersal under full sun and 10^3 km under cloud cover, depending on spore density. The effects on disease spread of the mode of dispersal of inoculum, with particular attention to Phytophthoras, was summarized by Ristaino and Gumpertz (2000). In general, although flagellated spores have epidemiological relevance, the presence of absence of zoospores does not necessarily have an impact on dispersal, particularly for foliar pathogens.

Survival

The occurrence of full sun or cloud cover is an important variable in plant epidemics, as it can affect the survival of spores. Some chytrids have the ability to survive periodic drying and high summer temperatures typical of cropping soils (Gleason et al. 2004). There are many examples of the influence of environmental conditions on oomycetes, both above ground and below. Solar radiation is the dominant factor determining survival of sporangia of *Bremia lactucae* in California. Infection by sporangia that have survived a day is only likely on cloudy days or shaded leaves (Wu et al. 2000, 2005). However, there is a lower ability of zoospores of *P. infestans* to survive under the cool temperatures which favour their development. Sporangia that do not form zoospores under conditions favourable for formation may be specially adapted for survival in the absence of a host (Porter and Johnson 2004). Release of zoospores from sporangia of *Plasmopora viticola* occurred for at least seven days if free water was available (Kast and Stark-Urnau 1999). Many sporangia of *P. viticola* do not survive during clear daylight periods following their production. However, with overcast conditions for 12–24 h, 50% still released zoospores (Kennelly et al. 2007). The formation of sporangia in *P. viticola* has been shown to be photosensitive, with a prolonged period of dark as a necessary condition (Rumbolz et al. 2002). Assessment of survival abilities in soil, and hence the influence of edaphic factors, depends on the techniques used. Assays for detecting and quantifying surviving *P. capsici* in soil differed in efficacy according to propagule type: oospores, mycelial fragments, sporangia and zoospores. Zoospore inoculum was detected at 10 propagules per gram (ppg) of soil, whereas sporangia were detected at 1 ppg (Larkin et al. 1995).

Infection processes

Host targeting is a fundamental strategy for zoosporic plant pathogens to successfully infect their hosts (Tyler 2002). This is true both in aquatic and terrestrial environments. Zoospore chemotaxis was observed in mangrove strains of *Halophytophthora vesicula* (Leano et al. 1998). However, no evidence for this phenomenon was obtained for *Pythium porphyrae* parasitising the red alga *Porphyra yezoensis* (Uppalapati et al. 2001). For terrestrial pathosystems, it is known that host factors can influence the development of *Plasmopara viticola* by (1) accelerating the release of zoospores from mature sporangia, (2) coordinating the morphogenesis of the germ tube, and (3) directing zoospores to stomata (Kiefer et al. 2002). Similar evidence for host-mediation of zoospore development was reported for *Phytophthora infestans* infecting *Solanum phureja* (Oyarzun et al. 2004). However, *Pythium oligandrum* zoospores are not attracted to hyphae of their fungal host, but if encysted on hyphae show a significant germ-tube emergence towards the host (Madsen et al. 1995).

Direct germination of conidia may be an advantage in some cases. Conidia of *Peronospora rubi* germinate and infect most commonly through direct penetration or enter through stomata (Williamson et al. 1995). Conidia of *Peronospora parasitica* enter through the stigma, ovary wall and establish in the ovary enabling embryo infection and seed transmission (Jang and Safeeulla 1990). Direct germination exists in *Phytophthora drechsleri*, where sporangia are stimulated by microbial interaction in soil. With indirect germination zoospore infectivity may be suppressed (Hardy and Sivasithamparam 1991). A study on the effect of the biocontrol bacterium *Burkholderia cepacia* on *Pythium aphanidermatum* indicated that although antibiosis was the main mechanism involved in suppression there was some contribution of competition for zoospore homing compounds (Heungens and Parke 2000). This effect was not apparent against *Aphanomyces euteiches* zoospores.

Many studies have shown that temperature has an important effect on zoospore infection. For example, heat stress (40°C rather than 25 to 35°C) enhanced the severity of root rot caused by *Phytophthora cryptogea* on container-grown *Chrisanthemum* (MacDonald 1991). Also for *P. cryptogea* on *Lycopersicon esculentum*, enhanced temperature (above 25°C) was ineffective to counter established infection in summer-grown plants (Kennedy and Pegg 1990). Together with wetness duration, higher day temperature was found to be associated with increasing incidence and severity of *P. cactorum* on apple and pear fruits (Grove and Boal 1991). However, citrus root colonization by *P. citrophthora* and *P. parasitica* was shown to be restricted or limited above a certain temperature threshold (27 and 33°C, respectively). A similar result was obtained for early infection of *Vitis vinifera* by *Plasmopara viticola* in Western Australia (Williams et al. 2007). In general, the effect of temperature on disease severity caused by zoosporic plant pathogens will depend not only on the temperature preferences of the pathogens, but also on the temperature threshold at which they will tend to switch from zoospore to sporangial infection (Judelson and Blanco 2005), and will be confounded by other factors such as inoculum density and plant age (Raftoyannis and Dick 2002).

Sporulation structures

In *Phytophthora ramorum*, the causal agent of sudden oak death and ramorum dieback of many shrubby species (Rizzo et al. 2005), sporangia and zoospores are the elements driving the observed disease epidemic. Moralejo et al. (2006) observed structures termed sporangiomata on susceptible woody species. This is the first description of stromata produced by a *Phytophthora* species, and may be a significant environmental adaptation in *P. ramorum*. In particular, adaxial positioning suggests adaptation for rain-splash dispersal. Moreover, sub-epidermal positioning of the stroma may in part protect from desiccation or solar radiation and clustering of sporangia may contribute to moisture retention. Oversummering survival structures may provide a way to avoid the challenge posed by the Mediterranean climate in the current region of outbreak, as well as in other regions with potentially susceptible hosts (Moralejo et al. 2006).

Integrating life cycles and predictive models

Oospore germination and zoospore infection in *Phytophthora syringae* also pose a challenge to understanding disease epidemiology and management. *Phytophthora syringae* persists as oospores in fallen apple leaves. Oospores germinate by giving rise to one or two sporangia and, when free water is available, each sporangia produces 20 to 30 motile spores. Undehised sporangia may germinate to create a secondary sporangium which may produce zoospores or give rise to a tertiary sporangium, potentially an important adaptation providing flexibility in response to variable environmental conditions. One open question in this pathosystem is the long-term viability of ungerminated zoospores. Harris and Xu (2003) found that infection of fruit depended mainly on sufficient rain being available to keep soil moist for at least 2–3 days (oospore germination) and wetness periods of at least 4 h (zoospore infection; Fig. 5).

Typically mechanistic and/or forecasting models should take account of zoospore behaviour, because in many cases this factor seems to be essential in understanding and predicting epidemic development. Examples of various predictive models where zoospore activity could significantly improve forecasting involve outbreaks of *Phytophthora infestans* (Johnson et al. 1996; Aylor et al. 2001; Bourgeois et al. 2004; Andrade-Piedra et al. 2005; Powell et al. 2005).

Disease management

Other than resistance breeding and sanitation, disease management for zoosporic plant pathogens has relied heavily on chemical control, and the emergence of resistance has been observed repeatedly. Apparently, the cost of fungicides used against *Phytophthora infestans* on *Solanum tuberosum* accounts worldwide for approximately 25% of the total sum spent on fungicides on all crops (Erwin and Ribeiro 1996). In

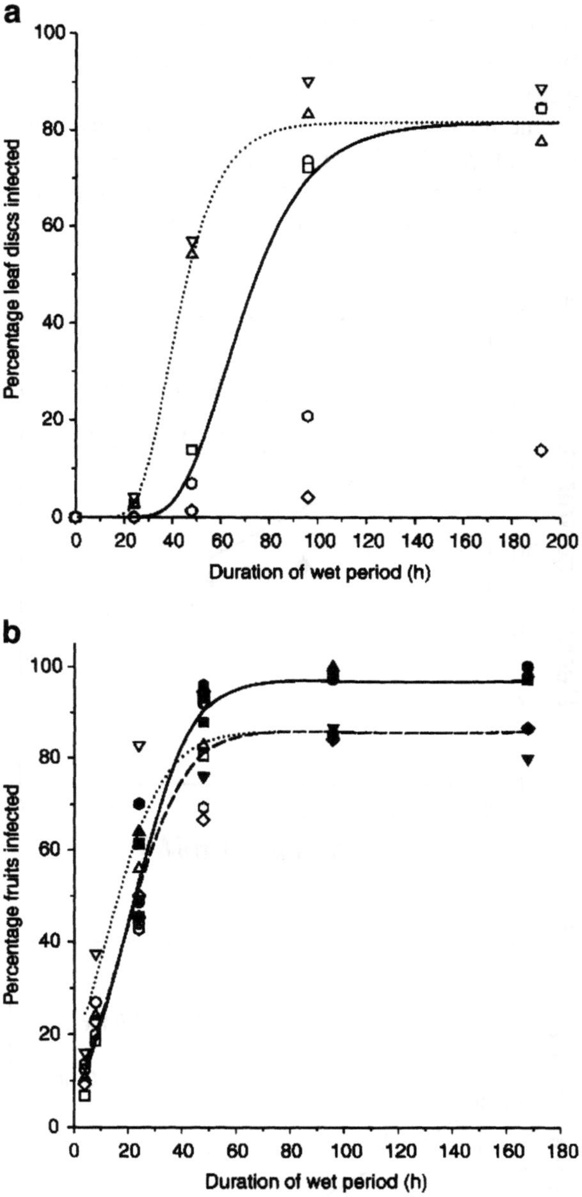

Fig. 5 Observed and predicted percentage of **a** *Phytophthora syringae* oospore activation, estimated as the percentage of infected leaf discs, and **b** apple fruits infected by zoospores of *P. syringae*, in relation to temperature and duration of wet period; *circle* 10°C, *square* 12°C, *triangle* 14°C, *inverted triangle* 16°C, *diamond* 18°C, *hexagon* 20°C; **a** *solid line* 10 and 12°C, *dashed line* 14 and 16°C; no models can be fitted to data at 18 and 20°C; and **b** *solid line* 10, 12 and 14°C, *dotted line* 16°C, *dashed line* 18 and 20°C (from Harris and Xu 2003, with kind permission of Blackwell)

many cases the effects of oomycete fungicides have been tracked through the various stages of zoospore formation, release, motility, cyst formation, germination, and infection (e.g. Mitani et al. 2001; Reuveni 2003) and similarly for plant extracts (Rohner et al. 2004), secondary metabolites (Shimai et al. 2002) and mineral supplementation (Xu and Morris 1998). In relatively few studies has the relative effect of control of sporangia/conidia and zoospores been directly compared.

In a comprehensive study the response of *Plasmopara halstedii* to anti-oomycete fungicides varied during ontogeny defined in terms of 13 developmental stages of the pathogen (Viranyi and Oros 1991). A principal component analysis of responses formed two main groupings with same separation of sporangial and zoosporic responses in one of the two groups. Famoxadone used against *P. infestans* and *Plasmopara halstedii* inhibited zoospore release and caused lysis of zoospores. Higher doses were required to inhibit direct germination (Andrieu et al. 2001). In *P. infestans* zoospore encystment and cyst germination were highly sensitive to dimethomorph; direct sporangial germination less so (Stein and Kirk 2003). Multi-drug resistant isolates of *P. infestans* significantly reduced sporulation and sporangial germination but not differentiation into zoospores (Ziogas et al. 2006). Tomato treated with PGPR, and BABA for induced systemic protection had reduced germination of *P. infestans* sporangia and zoospores with a marginally greater effect on sporangia (Yan et al. 2002; Fig. 6). Both direct and indirect germination of sporangia of *P. infestans* were suppressed by a range of calcium-modulating treatments, marginally greater for indirect germination (Hill et al. 1998; Fig. 7).

From a biological control point of view, a different line of work has built on the discovery that biosurfactants produced by the bacterium *Pseudomonas aeruginosa* were an effective way to protect hydroponic plant specimens inoculated with four species of *Pythium* and *Phytophthora parasitica* (Stanghellini and Miller 1997). In order to achieve long-term sustainability, strategies alternative to pesticides are needed for the management of zoosporic plant pathogens (Hoitink and Boehm 1999; Martin and Loper 1999; Paulitz and Belanger 2001; Hong and Moorman 2005). This research showed that rhamnolipids (Nitschke et al. 2005) produced by bacteria or directly applied to plants are able to lyse the membranes of zoospores (e.g. Kim et al. 2000; Maier and Soberon-Chavez 2000; see also Tomlinson and Faithfull 1979). Subsequent work showed that fluorescent pseudomonads colonizing the rhizosphere are

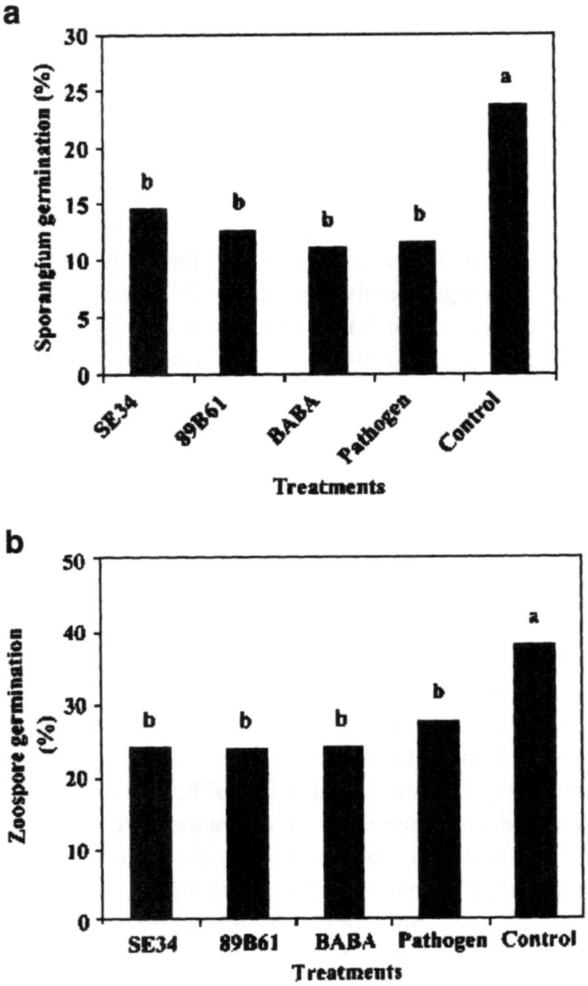

Fig. 6 Percent germination of **a** sporangia and **b** zoospores of *Phytophthora infestans* on tomato leaves induced with plant growth-promoting rhizobacteria (PGPR) strains SE34 and 89B61, β-amino butyric acid (BABA), and pathogen. Data are means of two experiments (from Yan et al. 2002, with kind permission of American Phytopathological Society)

able both to elicit systemic defence response in plants and to affect the pathogenicity of zoosporic plant pathogens (Haas and Defago 2005). The potential of the approach has been confirmed empirically in various pathosystems (e.g. *Phytophthora capsici* on *Capsicum annuum*; Ristaino and Johnston 1999; Nielsen et al. 2006; *Albugo occidentalis* on *Spinacia oleracea*; Irish et al. 2002; *Pythium aphanidermatum* on *Cucumis sativus*; Folman et al. 2004; *Phytophthora cryptogea* on *Cicorium intybus*; De Jonghe et al. 2005; *Phytophthora infestans* on *Solanum tuberosum*; Lozoya-Saldana et al. 2006; *Pythium aphanidermatum* or *Phytophthora* spp. on *Lycopersicon esculentum*; Calvo-Bado et al. 2006; Sharma et al. 2007). Widespread adoption is dependent on economic circumstances in different crop production systems.

One often overlooked management strategy is the effect of spatial and temporal mixtures of resistant and susceptible species or varieties on diseases. Devoting different fields to different crops and rotating crops

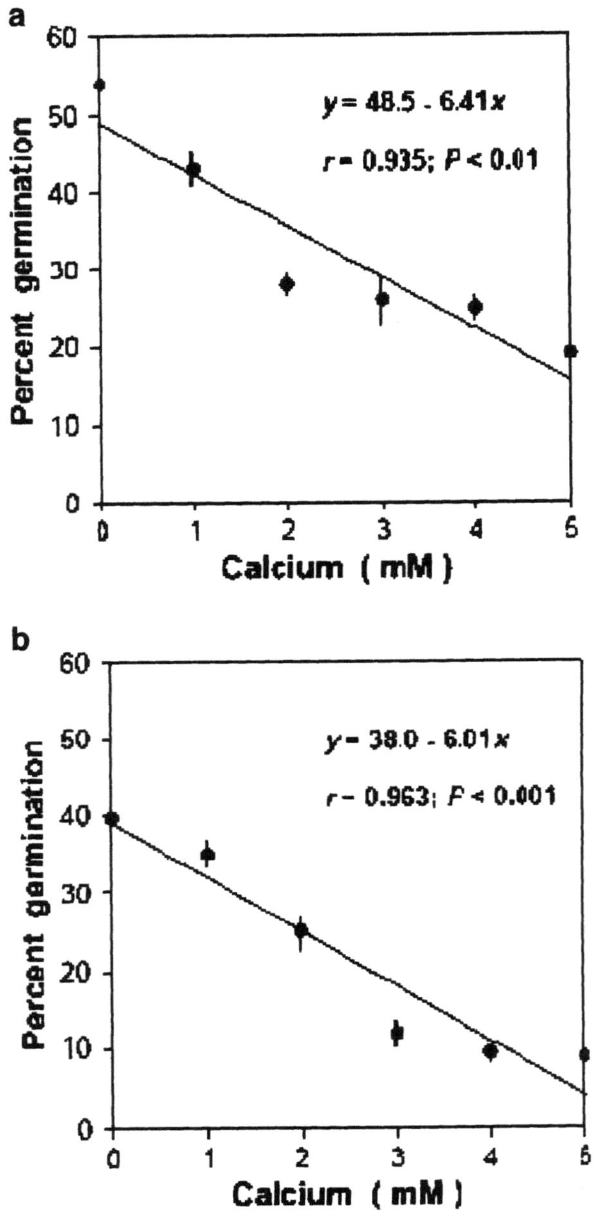

Fig. 7 Effect of [Ca^{2+}] on sporangial germination by **a** hyphal outgrowth (20°C) and **b** zoospore release (12°C). Data points are means ± SE of three replicates, based on counts of 100 sporangia in each replicate (from Hill et al. 1998, with kind permission of Kluwer)

from year to year is a traditional agricultural practice which makes sense also as a control strategy for zoosporic plant pathogens. Indeed, monocultures grown year after year in the same soil are often remarkably susceptible to disease, as exemplified by potato late blight. A study of the effect of mixtures of *Solanum tuberosum* varieties with differing levels of susceptibility to *P. infestans* showed that mixtures of an immune or near immune variety substantially reduced disease on susceptible ones (Phillips et al. 2005). That host diversity can reduce potato blight severity has been now shown repeatedly, although with varying degrees (Garrett and Mundt 2000; Garrett et al. 2001; Andrivon et al. 2003; Pilet et al. 2006). It is likely that the mechanisms underlying these findings involve sporangial dispersal, as immune plants constitute a physical barrier and reduce the overall density of susceptible individuals in a field (Burdon and Chilvers 1982; Keesing et al. 2006; see also Jactel and Brockerhoff 2007). At a landscape level, a similar protective mechanism could be implemented for sudden oak death. In this emerging pathosystem, connectivity of woodland patches is playing a key role in the spread of *Phytophthora ramorum* and forests could be managed so as to decrease connectivity of susceptible hosts (such as bay laurel) by increasing the diversity of resistant understory species (Condeso and Meentemeyer 2007). In tropical forests, *Phytophthora* and *Pythium* species have been suggested as contributing to the high tree diversity by producing density-dependent mortality of seedlings close to parent trees (e.g. Packer and Clay 2000; Hood et al. 2004; Pautasso et al. 2005; Bell et al. 2006; Augspurger and Wilkinson 2007).

Conclusions

Loss of flagellated cells, zoospores, has occurred independently in different phylogenetic lineages. No single explanation is apparent for these evolutionary losses. The case studies discussed in this review suggest that it would be an oversimplification to view lack of zoospores as progressing from free-living aquatic to parasitic terrestrial organisms. Indeed, oomycetes show remarkable flexibility (and redundancy) in 'spore' structure and function in relation to their environment. Zoospores have perhaps mistakenly been seen as the weak link in pathogen life cycles. Evidence from disease management studies on the best targets for control interventions is inconclusive and needs further comparative analysis.

Acknowledgement This review is partly based on an invited talk at the Downy Mildews Second International Symposium, 2–6 July 2007, Olomouc, Czech Republic. Many thanks to Sandra Denman, Ottmar Holdenrieder, Geert Kessel, Alan Slusarenko, Laetitia Willocquet, and two anonymous reviewers for helpful comments and encouragement in approaching the topic.

References

Adams, M. J. (1991). Transmission of plant viruses by fungi. *Annals of Applied Biology, 118*, 479–492.

Amon, J. P. (1978). Method for obtaining sporulating *Labyrinthula*. *Mycologia, 70*, 1297–1299.

Amon, J. P., & Perkins, F. O. (1968). Structure of *Labyrinthula* sp. zoospores. *The Journal of Eukaryotic Microbiology, 15*, 543–546.

Andrade-Piedra, J. L., Hijmans, R. J., Forbes, G. A., Fry, W. E., & Nelson, R. J. (2005). Simulation of potato late blight in the Andes. I: Modification and parameterization of the lateblight model. *Phytopathology, 95*, 1191–1199.

Andrieu, N., Jaworska, G., Genet, J.-L., & Bompeix, G. (2001). Biological mode of action of Famoxadone on *Plasmopara viticola* and *Phytophthora infestans*. *Crop Protection, 20*, 253–260.

Andrivon, D., Lucas, J. M., & Ellisseche, D. (2003). Development of natural late blight epidemics in pure and mixed plots of potato cultivars with different levels of partial resistance. *Plant Pathology, 52*, 586–594.

Augspurger, C. K., & Wilkinson, H. T. (2007). Host specificity of pathogenic *Pythium* species: implications for tree species diversity. *Biotropica, 39*, 702–708.

Aylor, D. E. (1999). Biophysical scaling and the passive dispersal of fungus spores: Relationship to integrated pest management strategies. *Agricultural and Forest Meteorology, 97*, 275–292.

Aylor, D. E. (2003). Spread of plant disease on a continental scale: Role of aerial dispersal of pathogens. *Ecology, 84*, 1989–1997.

Aylor, D. E., Fry, W. E., Mayton, H., & Andrade-Piedra, J. L. (2001). Quantifying the rate of release and escape of *Phytophthora infestans* sporangia from a potato canopy. *Phytopathology, 91*, 1189–1196.

Baker, K. M., Kirk, W. W., Stein, J. M., & Andresen, J. A. (2005). Climatic trends and potato late blight risk in the upper Great Lakes Region. *Horttechnology, 15*, 510–518.

Bell, T., Freckleton, R. P., & Lewis, O. T. (2006). Plant pathogens drive density-dependent seedling mortality in a tropical tree. *Ecology Letters, 9*, 569–574.

Bigelow, D. M., Olsen, M. W., & Gilbertson, R. L. (2005). *Labyrinthula terrestris* sp nov., a new pathogen of turf grass. *Mycologia, 97*, 185–190.

Bonants, P. J. M., Hagenaar-de Weerdt, M., Man in't Veld, W. A., & Baayen, R. P. (2000). Molecular characterization of natural hybrids of *Phytophthora nicotianae* and *P. cactorum*. *Phytopathology, 90*, 867–874.

Bourgeois, G., Bourque, A., & Deaudelin, G. (2004). Modelling the impact of climate change on disease incidence: A bioclimatic challenge. *Canadian Journal of Plant Pathology, 26*, 284–290.

Brasier, C. M., & Hansen, E. M. (1992). Evolutionary biology of *Phytophthora*. Part II: Phylogeny, speciation, and population structure. *Annual Review of Phytopathology, 30*, 173–200.

Brasier, C. M., & Jung, T. (2003). Progress in understanding Phytophthora diseases of trees in Europe. In J. A. McComb, G. E. St J. Hardy, & I. Tommerup (Eds.) *Phytophthora in forests and natural ecosystems. Proceedings of the second international meeting of IUFRO working party 0.02.09, Albany, Western Australia, 2001* (pp. 4–18). Perth, Australia: Murdoch University.

Brasier, C. M., Kirk, S. A., Delcan, J., Cooke, D. E. L., Jung, T., & Man in't Veld, W. A. (2004). *Phytophthora alni* sp. nov. and its variants: designation of emerging heteroploid hybrid pathogens spreading on Alnus trees. *Mycological Research, 108*, 1172–1184.

Brasier, C. M., Rose, J., & Gibbs, J. N. (1995). An unusual *Phytophthora* associated with widespread alder mortality in Britain. *Plant Pathology, 44*, 999–1007.

Brown, J. K. M., & Hovmøller, M. S. (2002). Aerial dispersal of pathogens on the global and continental scales and its impact on plant disease. *Science, 297*, 537–541.

Bruns, T. (2006). A kingdom revised. *Nature, 43*, 758–761.

Burdon, J. J., & Chilvers, G. A. (1982). Host density as a factor in plant disease ecology. *Annual Review of Phytopathology, 20*, 143–166.

Calvo-Bado, L. A., Petch, G., Parsons, N. R., Morgan, J. A. W., Pettitt, T. R., et al. (2006). Microbial community responses associated with the development of oomycete plant pathogens on tomato roots in soilless growing systems. *Journal of Applied Microbiology, 100*, 1194–1207.

Campbell, C. L. (1999). The importance of dispersal mechanisms in the epidemiology of *Phytophthora* blights and downy mildews on crop plants. *Ecosystem Health, 5*, 146–157.

Campbell, R. N. (1996). Fungal transmission of plant viruses. *Annual Review of Phytopathology, 34*, 87–108.

Clay, K., & Kover, P. X. (1996). The Red Queen Hypothesis and plant/pathogen interactions. *Annual Review of Phytopathology, 34*, 29–50.

Condeso, T. E., & Meentemeyer, R. K. (2007). Effects of landscape heterogeneity on the emerging forest disease sudden oak death. *Journal of Ecology, 95*, 364–375.

Cooke, D. E. L., Drenth, A., Duncan, J. M., Wagels, G., & Brasier, C. M. (2000). A molecular phylogeny of *Phytophthora* and related oomycetes. *Fungal Genetics and Biology, 30*, 17–32.

Coyier, D. L. (1981). Control of *Phytophthora ilicis* on English holly with sprays or soil drenches. *Phytopathology, 71*, 868–868.

Craven, K. D., Peterson, P. D., Windham, D. E., Mitchell, T. K., & Martin, S. B. (2005). Molecular identification of the turf grass rapid blight pathogen. *Mycologia, 97*, 160–166.

Deacon, J. W., & Donaldson, S. P. (1993). Molecular recognition in the homing responses of zoosporic fungi, with special reference to *Pythium* and *Phytophthora*. *Mycological Research, 97*, 1153–1171.

De Jonghe, K., De Dobbelaere, I., Sarrazyn, R., & Hofte, M. (2005). Control of *Phytophthora cryptogea* in the hydroponic forcing of witloof chicory with the rhamnolipid-based biosurfactant formulation PRO1. *Plant Pathology, 54*, 219–226.

Dick, M. W. (1997). Fungi, flagella and phylogeny. *Mycological Research, 101*, 385–394.

Dick, M. W. (2002). Towards an understanding of the evolution of the downy mildews. In P. T. N. Spencer-Phillips (Ed.) *Advances in downy mildew research* (pp. 1–57). Dordrecht: Kluwer.

English, J. T., Laday, M., Bakonyi, J., Schoelz, J. E., & Érsek, T. (1999). Phenotypic and molecular characterization of species hybrids derived from induced fusion of zoospores of *Phytophthora capsici* and *Phytophthora nicotianae*. *Mycological Research, 103*, 1003–1008.

Entwistle, C. A., Olsen, M. W., & Bigelow, D. M. (2006). First report of a *Labyrinthula* spp. causing rapid blight of *Agrostis capillaris* and *Poa annua* on amenity turfgrass in the UK. *Plant Pathology, 55*, 306.

Érsek, T., English, J. T., & Schoelz, J. E. (1995). Creation of species hybrids of *Phytophthora* with modified host ranges by zoospore fusion. *Phytopathology, 85*, 1343–1347.

Erwin, D. C., & Ribeiro, O. K. (1996). Introduction to the genus *Phytophthora*. In D. C. Erwin, & O. K. Ribeiro (Eds.) *Phytophthora disease worldwide* pp. 1–7. American Phytopathological Society: St. Paul.

Folman, L. B., De Klein, M. J. E. M., Postma, J., & van Veen, J. A. (2004). Production of antifungal compounds by *Lysobacter enzymogenes* isolate 3.1T8 under different conditions in relation to its efficacy as a biocontrol agent of *Pythium aphanidermatum* in cucumber. *Biological Control, 31*, 145–154.

Garrett, K. A., Dendy, S. P., Frank, E. E., Rouse, M. N., & Travers, S. E. (2006). Climate change effects on plant disease: genomes to ecosystems. *Annual Review of Phytopathology, 44*, 489–509.

Garrett, K. A., & Mundt, C. C. (2000). Host diversity can reduce potato late blight severity for focal and general patterns of primary inoculum. *Phytopathology, 90*, 1307–1312.

Garrett, K. A., Nelson, R. J., Mundt, C. C., Chacon, G., Jaramillo, R. E., & Forbes, G. A. (2001). The effects of host diversity and other management components on epidemics of potato late blight in the humid highland tropics. *Phytopathology, 91*, 993–1000.

Gibbs, J. N., Lipscombe, M. A., & Peace, A. J. (1999). The impact of *Phytophthora* disease on riparian populations of common alder (*Alnus glutinosa*) in southern Britain. *European Journal of Forest Pathology, 29*, 39–50.

Gleason, F. H., Letcher, P. M., & McGee, P. A. (2004). Some *Chytridiomycota* in soil recover from drying and high temperatures. *Mycological Research, 108*, 583–589.

Göker, M., Riethmüller, A., Voglmayr, H., Weiss, M., & Oberwinkler, F. (2004). Phylogeny of *Hyaloperonospora* based on nuclear ribosomal internal transcribed spacer sequences. *Mycological Progress, 3*, 83–94.

Göker, M., Voglmayr, H., Riethmüller, A., & Oberwinkler, F. (2007). How do obligate parasites evolve? A multi-gene phylogenetic analysis of downy mildews. *Fungal Genetics and Biology, 44*, 105–122.

Goodwin, S. B., Smart, C. D., Sandrock, R. W., Deahl, K. L., Punja, Z. K., & Fry, W. E. (1998). Genetic change within populations of *Phytophthora infestans* in the United States and Canada during 1994 to 1996: role of migration and recombination. *Phytopathology, 88*, 939–949.

Grove, G. G., & Boal, R. J. (1991). Influence of temperature and wetness duration on infection of immature apple and pear fruit by *Phytophthora cactorum*. *Phytopathology, 81*, 1465–1471.

Gu, Y. H., & Ko, W. H. (2000). Segregation following interspecific transfer of isolated nuclei between *Phytophthora parasitica* and *P. capsici*. *Canadian Journal of Microbiology, 46*, 410–416.

Haas, D., & Defago, G. (2005). Biological control of soil-borne pathogens by fluorescent pseudomonads. *Nature Reviews Microbiology, 3*, 307–319.

Haber, S., Barr, D. J. S., & Platford, R. G. (1991). Observations on the distribution of flame chlorosis in Manitoba and its association with certain zoosporic fungi and the intensive cultivation of cereals. *Canadian Journal of Plant Pathology, 13*, 241–246.

Hannukkala, A. O., Kaukoranta, T., Lehtinen, A., & Rahkonen, A. (2007). Late-blight epidemics on potato in Finland, 1933–2002; increased and earlier occurrence of epidemics associated with climate change and lack of rotation. *Plant Pathology, 56*, 167–176.

Hardham, A. R. (2007). Cell biology of plant–oomycete interactions. *Cellular Microbiology, 9*, 31–39.

Hardham, A. R., Cahill, D. M., Cope, M., Gabor, B. K., Gubler, F., & Hyde, G. J. (1994). Cell surface antigens of *Phytophthora* spores—biological and taxonomic characterization. *Protoplasma, 181*, 213–232.

Hardy, G. E. St J., & Sivasithamparam, K. (1991). Sporangial responses do not reflect microbial suppression of *Phytophthora drechsleri* in composted Eucalyptus bark mix. *Soil Biology and Biochemistry, 23*, 757–765.

Harris, D. C., & Xu, X. M. (2003). Conditions for infection of apple by *Phytophthora syringae*. *Journal of Phytopathology, 151*, 190–194.

Heungens, K., & Parke, J. L. (2000). Zoospore homing and infection events: Effects of the biocontrol bacterium *Burkholderia cepacia* AMMDR1 on two oomycete pathogens of pea (*Pisum sativum* L.). *Applied and Environmental Microbiology, 66*, 5192–5200.

Hill, A. E., Grayson, D. E., & Deacon, J. W. (1998). Suppressed germination and early death of *Phytophthora infestans* sporangia caused by pectin, inorganic phosphate, ion chelators and calcium-modulating treatments. *European Journal of Plant Pathology, 104*, 367–376.

Hoitink, H. A. J., & Boehm, M. J. (1999). Biocontrol within the context of soil microbial communities: A substrate-dependent phenomenon. *Annual Review of Phytopathology, 37*, 427–446.

Holdenrieder, O., Pautasso, M., Weisberg, P. J., & Lonsdale, D. (2004). Tree diseases and landscape processes: The challenge of landscape pathology. *Trends in Ecology & Evolution, 19*, 446–452.

Honda, D., Shono, T., Kimura, K., Fujita, S., Iseki, M., Makino, Y., et al. (2007). Homologs of the sexually induced gene 1 (sig1) product constitute the stramenopile mastigonemes. *Protist, 158*, 77–88.

Hong, C. X., & Moorman, G. W. (2005). Plant pathogens in irrigation water: Challenges and opportunities. *Critical Reviews in Plant Sciences, 24*, 189–208.

Hood, L. A., Swaine, M. D., & Mason, P. A. (2004). The influence of spatial patterns of damping-off disease and arbuscular mycorrhizal colonization on tree seedling establishment in Ghanaian tropical forest soil. *Journal of Ecology, 92*, 816–823.

Irish, B. M., Correll, J. C., & Morelock, T. E. (2002). The effect of synthetic surfactants on disease severity of white rust on spinach. *Plant Disease, 86*, 791–796.

Jactel, H., & Brockerhoff, E. G. (2007). Tree diversity reduces herbivory by forest insects. *Ecology Letters, 10*, 835–848.

James, T. Y., et al. (2006a). Reconstructing the early evolution of Fungi using a six-gene phylogeny. *Nature, 443*, 818–822.

James, T. Y., Letcher, P. M., Longcore, J. E., Mozley-Standridge, S. E., Porter, D., Powell, M. J., et al. (2006b). A molecular phylogeny of the flagellated fungi (Chytridiomycota) and description of a new phylum (Blastocladiomycota). *Mycologia, 98*, 860–871.

Jang, P., & Safeeulla, K. M. (1990). Modes of entry, establishment and seed transmission of *Peronospora parasitica* in radish. *Proceedings of the Indian Academy of Sciences—Plant Sciences, 100*, 369–373.

Jeger, M. J. (2004). Analysis of disease progress as a basis for evaluating disease management practices. *Annual Review of Phytopathology, 42*, 61–82.

Jeger, M. J., Gilijamse, E., Bock, C. H., & Frinking, H. D. (1998). The epidemiology, variability and control of the downy mildews of pearl millet and sorghum, with particular reference to Africa. *Plant Pathology, 47*, 544–569.

Jeger, M. J., & Pautasso, M. (2008). Plant disease and global change—The importance of long-term data sets. *New Phytologist, 177*, 8–11.

Jeger, M. J., Pautasso, M., Holdenrieder, O., & Shaw, M. W. (2007). Modelling disease spread and control in networks: implications for plant sciences. *New Phytologist, 174*, 279–297.

Johnson, D. A., Alldredge, J. R., & Vakoch, D. L. (1996). Potato late blight forecasting models for the semiarid environment of south-central Washington. *Phytopathology, 86*, 480–484.

Judelson, H. S., & Blanco, F. A. (2005). The spores of *Phytophthora*: weapons of the plant destroyer. *Nature Reviews Microbiology, 3*, 47–58.

Jung, T., & Blaschke, M. (2004). *Phytophthora* root and collar rot of alders in Bavaria: distribution, modes of spread and possible management strategies. *Plant Pathology, 53*, 197–208.

Kageyama, K., Suzuki, M., Priyatmojo, A., Oto, Y., Ishiguro, K., Suga, H., et al. (2003). Characterization and identification of asexual strains of *Pythium* associated with root rot of rose in Japan. *Journal of Phytopathology, 151*, 485–491.

Kast, W. K., & Stark-Urnau, M. (1999). Survival of sporangia from *Plasmopara viticola*, the downy mildew of grapevine. *Vitis, 38*, 185–186.

Keesing, F., Holt, R. D., & Ostfeld, R. S. (2006). Effects of species diversity on disease risk. *Ecology Letters*, *9*, 485–498.

Kennedy, R., & Pegg, G. F. (1990). *Phytophthora cryptogea* root rot of tomato in rockwool nutrient culture. 2. Effect of root zone temperature on infection, sporulation and symptom development. *Annals of Applied Biology*, *117*, 537–551.

Kennelly, M. M., Gadoury, D. M., Wilcox, W. F., Magarey, P. A., & Seem, R. C. (2007). Primary infection, lesion productivity, and survival of sporangia in the grapevine downy mildew pathogen *Plasmopara viticola*. *Phytopathology*, *97*, 512–522.

Kiefer, B., Riemann, M., Buche, C., Kassemeyer, H. H., & Nick, P. (2002). The host guides morphogenesis and stomatal targeting in the grapevine pathogen *Plasmopara viticola*. *Planta*, *215*, 387–393.

Kim, B. S., Lee, J. Y., & Hwang, B. K. (2000). In vivo control and in vitro antifungal activity of rhamnolipid B, a glycolipid antibiotic, against *Phytophthora capsici* and *Colletotrichum orbiculare*. *Pest Management Science*, *56*, 1029–1035.

Kranz, J. (1980). Comparative epidemiology: an evaluation of scope, concepts and methods. In J. Palti, & J. Kranz (Eds.) *Comparative epidemiology. A tool for better disease management* (pp. 18–28). Centre for Agricultural Publishing and Documentation: Wageningen.

Kranz, J. (2003). *Comparative epidemiology of plant diseases*. Springer: Berlin.

Lange, L., & Olson, L. W. (1976). Flagellar apparatus and striated rhizoplast of zoospore of *Olpidium brassicae*. *Protoplasma*, *89*, 339–351.

Lange, L., & Olson, L. W. (1983). The fungal zoospore. Its structure and biological significance. In S. T. Buczacki (Ed.) *Zoosporic plant pathogens* (pp. 1–42). Academic: London.

Larkin, R. P., Ristaino, J. B., & Campbell, C. L. (1995). Detection and quantification of *Phytophthora capsici* in soil. *Phytopathology*, *85*, 1057–1063.

Leander, C. A., & Porter, D. (2001). The Labyrinthulomycota is comprised of three distinct lineages. *Mycologia*, *93*, 459–464.

Leano, E. M., Vrijmoed, L. L. P., & Jones, E. B. G. (1998). Zoospore chemotaxis of two mangrove strains of *Halophytophthora vesicula* from Mai Po, Hong Kong. *Mycologia*, *90*, 1001–1008.

Lebeda, A., & Schwinn, F. J. (1994). The downy mildews—An overview of recent research progress. *Journal of Plant Diseases and Protection*, *101*, 225–254.

Lozoya-Saldana, H., Coyote-Palma, M. H., Ferrera-Cerrato, R., & Lara-Hernandez, M. E. (2006). Microbial antagonism against *Phytophthora infestans* (Mont) de Bary. *Agrociencia*, *40*, 491–499.

MacDonald, J. D. (1991). Heat stress enhances *Phytophthora* root rot severity in container-grown Chrysanthemums. *Journal of the American Society for Horticultural Science*, *116*, 36–41.

Madden, L. V. (2006). Botanical epidemiology: Some key advances and its continuing role in disease management. *European Journal of Plant Pathology*, *115*, 3–23.

Madsen, A. M., Robinson, H. L., & Deacon, J. W. (1995). Behaviour of zoospore cysts of the mycoparasite *Pythium oligandrum* in relation to their potential for biocontrol of plant pathogens. *Mycological Research*, *99*, 1417–1424.

Maier, R. M., & Soberon-Chavez, G. (2000). *Pseudomonas aeruginosa* rhamnolipids: biosynthesis and potential applications. *Applied Microbiology and Biotechnology*, *54*, 625–633.

Martin, F. N., & Loper, J. E. (1999). Soilborne plant diseases caused by *Pythium* spp: Ecology, epidemiology, and prospects for biological control. *Critical Reviews in Plant Sciences*, *18*, 111–181.

May, K. J., Drenth, A., & Irwin, J. A. G. (2003). Interspecific hybrids between the homothallic *Phytophthora sojae* and *Phytophthora vignae*. *Australasian Plant Pathology*, *32*, 353–359.

McDonald, B. A., & Linde, C. (2002). Pathogen population genetics, evolutionary potential, and durable resistance. *Annual Review of Phytopathology*, *40*, 349–379.

Mitani, S., Araki, S., Yamaguchi, T., Takii, Y., Ohshima, T., & Matsuo, N. (2001). Antifungal activity of the novel fungicide cyazofamid against *Phytophthora infestans* and other plant pathogenic fungi in vitro. *Pesticide Biochemistry and Physiology*, *70*, 92–99.

Money, N. P., Davis, C. M., & Ravishankar, J. P. (2004). Biomechanical evidence for convergent evolution of the invasive growth process among fungi and oomycete water molds. *Fungal Genetics and Biology*, *41*, 872–876.

Moralejo, E., Puig, M., García, J. A., & Descals, E. (2006). Stromata, sporangiomata and chlamydosori of *Phytophthora ramorum* on inoculated Mediterranean woody plants. *Mycological Research*, *110*, 1323–1332.

Nielsen, C. J., Ferrin, D. M., & Stanghellini, M. E. (2006). Efficacy of biosurfactants in the management of *Phytophthora capsici* on pepper in recirculating hydroponic systems. *Canadian Journal of Plant Pathology*, *28*, 450–460.

Nitschke, M., Costa, S. G. V. A. O., & Contiero, J. (2005). Rhamnolipid surfactants: An update on the general aspects of these remarkable biomolecules. *Biotechnology Progress*, *21*, 1593–1600.

Olsen, M. W. (2007). *Labyrinthula terrestris*: a new pathogen of cool-season turfgrasses. *Molecular Plant Pathology*, *8*, 817–820.

Olsen, M. W., Bigelow, D. M., & Gilbertson, R. L. (2003). First report of a *Labyrinthula* sp. causing rapid blight disease of rough bluegrass and perennial ryegrass. *Plant Disease*, *87*, 1267.

Olsen, A., & Stenlid, J. (2002). Pathogenic fungal species hybrids infecting plants. *Microbes and Infection*, *4*, 1353–1359.

Oyarzun, P. J., Yanez, J., & Forbes, G. A. (2004). Evidence for host mediation of preinfection stages of *Phytophthora infestans* on the leaf surface of *Solanum phureja*. *Journal of Phytopathology*, *152*, 651–657.

Packer, A., & Clay, K. (2000). Soil pathogens and spatial patterns of seedling mortality in a temperate tree. *Nature*, *404*, 278–281.

Paul, B. (2003). *Pythium glomeratum*, a new species isolated from agricultural soil taken in north-eastern France, its ITS region and its comparison with related species. *FEMS Microbiology Letters*, *225*, 47–52.

Paulitz, T. C., & Belanger, R. R. (2001). Biological control in greenhouse systems. *Annual Review of Phytopathology*, *39*, 103–133.

Pautasso, M., Holdenrieder, O., & Stenlid, J. (2005). Susceptibility to fungal pathogens of forests differing in tree diversity. In M. Scherer-Lorenzen, Ch. Koerner, & D. Schulze (Eds.) *Forest diversity and function: temperate and boreal systems* (pp. 263–289). Berlin, Germany: Springer.

Pautasso, M., & Jeger, M. J. (2008). Epidemic threshold and network structure: The interplay of probability of transmission and of persistence in small-size directed networks. *Ecological Complexity*, *5*, 1–8.

Perkins, F. O. (1973). A new species of marine labyrinthulid *Labyrinthuloides yorkensis* gen. nov. spec. nov.—Cytology and fine structure. *Archiv für Mikrobiologie*, *90*, 1–17.

Phillips, S. L., Shaw, M. W., & Wolfe, M. S. (2005). The effect of potato variety mixtures on epidemics of late blight in relation to plot size and level of resistance. *Annals of Applied Biology*, *147*, 245–252.

Pilet, F., Chacon, G., Forbes, G. A., & Andrivon, D. (2006). Protection of susceptible potato cultivars against late blight in mixtures increases with decreasing disease pressure. *Phytopathology*, *96*, 777–783.

Porter, L. D., & Johnson, D. A. (2004). Survival of *Phytophthora infestans* in surface water. *Phytopathology*, *94*, 380–387.

Powell, J. A., Slapničar, I., & van der Werf, W. (2005). Epidemic spread of a lesion-forming plant pathogen—Analysis of a mechanistic model with infinite age structure. *Linear Algebra and its Applications*, *398*, 117–140.

Raftoyannis, Y., & Dic, M. W. (2002). Effects of inoculum density, plant age and temperature on disease severity caused by pythiaceous fungi on several plants. *Phytoparasitica*, *30*, 67–76.

Reuveni, M. (2003). Activity of the new fungicide benthiavalicarb against *Plasmopara viticola* and its efficacy in controlling downy mildew in grapevines. *European Journal of Plant Pathology*, *109*, 243–251.

Ristaino, J. B. (2002). Tracking historic migrations of the Irish potato famine pathogen, *Phytophthora infestans*. *Microbes and Infection*, *4*, 1369–1377.

Ristaino, J. B., & Gumpertz, M. L. (2000). New frontiers in the study of dispersal and spatial analysis of epidemics caused by species in the genus *Phytophthora*. *Annual Review of Phytopathology*, *38*, 541–576.

Ristaino, J. B., & Johnston, S. A. (1999). Ecologically based approaches to management of *Phytophthora* blight on bell pepper. *Plant Disease*, *83*, 1080–1089.

Rizzo, D. M., Garbelotto, M., & Hansen, E. A. (2005). *Phytophthora ramorum*: Integrative research and management of an emerging pathogen in California and Oregon forests. *Annual Review of Phytopathology*, *43*, 309–335.

Rochon, D'. A., Kakani, K., Robbins, M., & Reade, R. (2004). Molecular aspects of plant virus transmission by olpidium and plasmodiophorid vectors. *Annual Review of Phytopathology*, *42*, 211–41.

Rohner, E., Carabet, A., & Buchenauer, H. (2004). Effectiveness of plant extracts of *Paeonia suffruticosa* and *Hedera helix* against diseases caused by *Phytophthora infestans* in tomato and *Pseudoperonospora cubensis* in cucumber. *Journal of Plant Diseases and Protection*, *111*, 83–95.

Rumbolz, J., Wirtz, S., Kassemeyer, H. H., Guggenheim, R., Schafer, E., & Buche, C. (2002). Sporulation of *Plasmopara viticola*: differentiation and light regulation. *Plant Biology*, *4*, 413–422.

Schardl, C. L., & Craven, K. D. (2003). Interspecific hybridization in plant-associated fungi and oomycetes: A review. *Molecular Ecology*, *12*, 2861–2873.

Sharma, A., Wray, V., & Johri, B. N. (2007). Rhizosphere *Pseudomonas* sp. strains reduce occurrence of pre- and post-emergence damping-off in chile and tomato in Central Himalayan region. *Archives of Microbiology*, *187*, 321–335.

Shearer, B. L., Crane, C. E., Barrett, S., & Cochrane, A. (2007). *Phytophthora cinnamomi* invasion, a major threatening process to conservation of flora diversity in the South-west Botanical Province of Western Australia. *Australian Journal of Botany*, *55*, 225–238.

Shimai, T., Islam, M. T., Fukushi, Y., Hashidoko, Yo., Yokosawa, R., & Tahara, S. (2002). Nicotinamide and structurally related compounds show halting activity against zoospores of the phytopathogenic fungus *Aphanomyces cochlioides*. *Zeitschrift fur Naturforschung C*, *57*, 323–331.

Slusarenko, A. J., & Schlaich, N. L. (2003). Downy mildew of *Arabidopsis thaliana* caused by *Hyaloperonospora parasitica* (formerly *Peronospora parasitica*). *Molecular Plant Pathology*, *4*, 159–170.

Smilde, W. D., Vannes, M., & Reinink, K. (1995). Resistance to *Phytophthora porri* in leek and some of its wild relatives. *Euphytica*, *83*, 131–138.

Spring, O., & Zipper, R. (2006). Evidence for asexual genetic recombination in sunflower downy mildew, *Plasmopara halstedii*. *Mycological Research*, *110*, 657–663.

Stanghellini, M. (1997). Inert components: Are they really so? *Phytoparasitica*, *25*, S81–S86.

Stanghellini, M. E., & Miller, R. M. (1997). Biosurfactans: Their identity and potential efficacy in the biological control of zoosporic plant pathogens. *Plant Disease*, *81*, 4–12.

Stein, J. M., & Kirk, W. W. (2003). Variations in the sensitivity of *Phytophthora infestans* isolates from different genetic backgrounds to dimethomorph. *Plant Disease*, *87*, 1283–1289.

Stowell, L. J., Martin, S. B., Olsen, M., Bigelow, D., Kohout, M., Peterson, P. D., Camberato, J., & Gelernter, W.D. (2005). *Rapid blight: A new plant disease. APSnet feature story.* Retrieved July 2007 from http://apsnet.org/online/feature/rapid/.

Streito, J. C., Legrand, P., Tabary, F., & De Villartay, G. J. (2002). *Phytophthora* disease of alder (*Alnus glutinosa*) in France: Investigations between 1995 and 1999. *Forest Pathology*, *32*, 179–191.

Sukno, S. A., Taylor, A. M., & Farman, M. L. (2002). Genetic uniformity among isolates of *Peronospora tabacina*, the tobacco blue mold pathogen. *Phytopathology*, *92*, 1236–1244.

Teakle, D. S. (1983). Zoosporic fungi and viruses. Double trouble. In S. T. Buczacki (Ed.) *Zoosporic plant pathogens* (pp. 233–248). Academic: London.

Thoirain, B., Husson, C., & Marcais, B. (2007). Risk factors for the *Phytophthora*-induced decline of alder in northeastern France. *Phytopathology, 97*, 99–105.

Tomlinson, J. A., & Faithfull, E. M. (1979). Effects of fungicides and surfactants on the zoospores of *Olpidium brassicae*. *Annals of Applied Biology, 93*, 13–19.

Tyler, B. M. (2002). Molecular basis of recognition between *Phytophthora* pathogens and their hosts. *Annual Review of Phytopathology, 40*, 137–167.

Tyler, B. M. (2007). *Phytophthora sojae*: Root rot pathogen of soybean and model oomycete. *Molecular Plant Pathology, 8*, 1–8.

Tyler, B. M., et al. (2006). *Phytophthora* genome sequences uncover evolutionary origins and mechanisms of pathogenesis. *Science, 313*, 1261–1266.

Uppalapati, S. R., Kerwin, J. L., & Fujita, Y. (2001). Epifluorescence and scanning electron microscopy of host–pathogen interactions between *Pythium porphyrae* (Peronosporales, Oomycota) and *Porphyra yezoensis* (Bangiales, Rhodophyta). *Botanica Marina, 44*, 139–145.

Ustinova, I., Krienitz, L., & Huss, V. A. R. (2000). *Hyaloraphidium curvatum* is not a green alga, but a lower fungus; *Amoebidium parasiticum* is not a fungus, but a member of the DRIPs. *Protist, 151*, 253–262.

van West, P., Appiah, A. A., & Gow, N. A. R. (2003). Advances in research on oomycete root pathogens. *Physiological and Molecular Plant Pathology, 62*, 99–113.

Viranyi, F., & Oros, G. (1991). Developmental stage response to fungicides of *Plasmopara halstedii* (Sunflower downy mildew). *Mycological Research, 95*, 199–205.

Walker, C. A., & van West, P. (2007). Zoospore development in the oomycetes. *Fungal Biology Reviews, 21*, 10–18.

Whipps, J. M., & Cooke, R. C. (1978). Behaviour of zoosporangia and zoospores of *Albugo tragopogonis* in relation to infection of *Senecio squalidus*. *Transactions of the British Mycological Society, 71*, 121–127.

Williams, M. G., Magarey, P. A., & Sivasithamparam, K. (2007). Effect of temperature and light intensity on early infection behaviour of a Western Australian isolate of *Plasmopara viticola*, the downy mildew pathogen of grapevine. *Australasian Plant Pathology, 36*, 325–331.

Williamson, B., Breese, W. A., & Shattock, R. C. (1995). A histological study of downy mildew (*Peronospora rubi*) infection of leaves, flowers and developing fruits of Tummelberry and other *Rubus* spp. *Mycological Research, 99*, 1311–1316.

Wu, B. M., Subbarao, K. V., & van Bruggen, A. H. C. (2000). Factors affecting the survival of *Bremia lactucae* sporangia deposited on lettuce leaves. *Phytopathology, 90*, 827–833.

Wu, B. M., Subbarao, K. V., & van Bruggen, A. H. C. (2005). Analyses of the relationships between lettuce downy mildew and weather variables using geographic information system techniques. *Plant Disease, 89*, 90–96.

Xu, C., & Morris, P. F. (1998). External calcium controls the development strategy of *Phytophthora sojae* cysts. *Mycologia, 90*, 269–275.

Yan, Z., Reddy, M. S., Ryu, C.-M., McInroy, J. A., Wilson, M., & Kloepper, J. W. (2002). Induced systemic protection against tomato late blight elicited by plant growth-promoting rhizobacteria. *Phytopathology, 92*, 1329–1333.

Ziogas, B. N., Markoglou, A. N., Theodosiou, D. I., Anagnostou, A., & Boutopoulou, S. (2006). A high multi-drug resistance to chemically unrelated oomycete fungicides in *Phytophthora infestans*. *European Journal of Plant Pathology, 115*, 283–292.

Zwankhuizen, M. J., & Zadoks, J. C. (2002). *Phytophthora infestans*'s 10-year truce with Holland: a long-term analysis of potato late-blight epidemics in the Netherlands. *Plant Pathology, 51*, 413–423.

Eur J Plant Pathol (2008) 122:127–146
DOI 10.1007/s10658-008-9291-4

Structure and variation in the wild-plant pathosystem: *Lactuca serriola–Bremia lactucae*

Aleš Lebeda · Irena Petrželová · Zbyněk Maryška

Received: 18 August 2007 / Accepted: 14 February 2008

Abstract Over the past decade, extensive research on the wild-plant pathosystem, *Lactuca serriola* (prickly lettuce)–*Bremia lactucae* (lettuce downy mildew), has been conducted in the Czech Republic. Studies focused on pathogen occurrence and distribution, host range, variation in symptom expression and disease severity, interactions of *B. lactucae* with another fungal species (*Golovinomyces cichoracearum*) on *L. serriola*, variation in resistance within natural populations of *L. serriola*, the structure and dynamics of virulence within populations of *B. lactucae*, sexual reproduction of *B. lactucae*, and a comparison of virulence structure and changes in *B. lactucae* populations occurring in wild (*L. serriola*) and crop (*L. sativa*) pathosystems. The incidence of *B. lactucae* on naturally growing *L. serriola* and other Asteraceae was recorded. *Lactuca serriola* was the most commonly occurring host species, followed by *Sonchus oleraceus*. Over the duration of these studies, the incidence of *B. lactucae* in *L. serriola* populations varied between 45–87%. Disease incidence and disease prevalence were partly related to the size, density and different habitats of *L. serriola* populations. In addition to *B. lactucae* infection, infection by the lettuce powdery mildew fungus (*Golovinomyces cichoracearum*) was quite common, including co-infection. Variation in resistance to *B. lactucae* was studied by using ten isolates (NL and BL races with known virulence patterns) at a metapopulation level, i.e. 250 *L. serriola* samples representing 16 populations from the Czech Republic (CZ). Our comparisons revealed broad variation in host resistance among host populations and also intrapopulation variability. In the CZ populations, 45 resistance phenotypes were recorded, the most frequent were race-specific reaction patterns. Structural and temporal changes in virulence variation of *B. lactucae* populations on *L. serriola* were studied during 1998–2005. Altogether, 313 isolates of *B. lactucae* originating from the Czech Republic were examined for the presence of 32 virulence factors (v-factors), and 93 different virulence phenotypes (v-phenotypes) were recorded. A study of v-factor frequency showed that common v-factors in *B. lactucae* populations match some of the race-specific resistance genes/factors (*Dm* genes or R-factors) originating from *L. serriola*. The highest frequency was recorded by v-factors v7, v11, v15–17, and v24–30. In contrast, v-factors (e.g. v1–4, 6, and 10) matching *Dm* genes originating from *L. sativa* were very rare. This demonstrates the close adaptation of *B. lactucae* virulence to the host (*L. serriola*) genetic background. Temporal changes in virulence frequencies over the period were recorded. In many v-factors (v11, v14, v16, and v25–28), fluctuations were

A. Lebeda (✉) · I. Petrželová · Z. Maryška
Faculty of Science, Department of Botany,
Palacký University in Olomouc,
Šlechtitelů 11, 783 71 Olomouc, Czech Republic
e-mail: ales.lebeda@upol.cz

observed, some (v14 and v17) shifting to higher frequencies, and others (v5/8 and v23) decreasing. The occurrence of mating types was studied (1997–1999) in a set of 59 *B. lactucae* isolates. Both compatibility types (B1 and B2) were recorded; however the majority of the isolates (96%) were type B2. A comparative study of *B. lactucae* virulence variation between the wild (*L. serriola*) and crop (*L. sativa*) pathosystems showed major differences. Migration and gene flow between both pathosystems and the potential danger of wild *B. lactucae* populations for cultivated lettuce are discussed. This paper summarizes comprehensive and unique research on an oomycete pathogen (*B. lactucae*) that is shared between a crop (lettuce, *L. sativa*) and its close wild relative (prickly lettuce, *L. serriola*). The data demonstrate clear evidence about race-specific interactions, variation and changes in virulence, and coevolutionary relationships in the wild pathosystem *L. serriola–B. lactucae*. Conclusions contribute to the broadening and better understanding of gene-for-gene systems in natural host–pathogen populations and their relationships to crop pathosystems.

Keywords Disease incidence · Disease prevalence · Gene flow · Gene-for-gene · Host range · Intra- and inter-population variability · Lettuce downy mildew · Lettuce powdery mildew · Metapopulation · Migration · Natural plant communities · Prickly lettuce · Race-specific resistance · Virulence structure · Wild- and crop-pathosystems

Introduction

Plant pathogens play a substantial role in the structure, dynamics, and evolution of natural plant communities. They may cause increased mortality, reduced fitness of individual plants, or dramatic shifts in the structure or composition of plant populations and communities. However, they may also help to maintain plant species diversity, and enhance the genetic diversity and structure of host populations (Gilbert 2002). The first detailed studies focusing on wild-plant pathosystem structure and function were published in the 1980s (Burdon 1987; Dinoor and Eshed 1984), and research on host–pathogen interactions in natural communities is a rapidly growing area of investigation in plant pathology (Burdon et al. 2006). Most of these studies have focused on interactions between host plants and plant parasitic fungi. Investigations of interactions between host plants and oomycetes are still very rare (Lebeda 2002; Lebeda and Schwinn 1994), with one of the most extensively studied of such pathosystems being *Lactuca* spp.–*Bremia lactucae* (Lebeda et al. 2002, 2007c).

Bremia lactucae (lettuce downy mildew) is an oomycete pathogen of cultivated lettuce (*Lactuca sativa*) and many other species of Asteraceae (Lebeda et al. 2002) that is distributed worldwide. The most common wild host species of this pathogen is *Lactuca serriola* (prickly lettuce), and it can also be frequently found on *Sonchus* species (Lebeda et al. 2002; Lebeda and Petrželová 2004a; Lebeda and Syrovátko 1988; Petrželová and Lebeda 2004b). However, it is well documented that *B. lactucae* is highly host-specific and mostly limited to a single plant genus (Crute and Dixon 1981; Lebeda and Syrovátko 1988). Thus, except for certain *Lactuca* species, weedy growing Asteraceae cannot serve as a source of inoculum for cultivated lettuce (Lebeda and Syrovátko 1988) and vice versa.

The interaction between *L. sativa* and *L. serriola* and *B. lactucae* generally conforms to a gene-for-gene relationship (Crute 1992a, b), in which resistance is determined by dominant *Dm* resistance genes (or R-factors) in the hosts, matched by dominant avirulence factors in the pathogens (Hammond-Kosack and Jones 1997). Detailed analyses of the genetics of these host–parasite interactions (Crute and Johnson 1976; Farrara et al. 1987) made it possible to interpret the variability of virulence in *B. lactucae* individuals and populations in terms of virulence factors (v-factors) and virulence phenotypes (v-phenotypes; Lebeda 1981, 1982).

Several different mechanisms of resistance to *B. lactucae* have been identified in cultivated and wild lettuce (Lebeda et al. 2001a). Most of the resistance is considered to be race-specific (Lebeda et al. 2002, 2007b). This type of resistance has a big disadvantage as it does not provide durable protection against lettuce downy mildew and the introduction of new resistant cultivars is often followed by the appearance of new virulent pathogen races (Lebeda and Schwinn 1994; Lebeda and Zinkernagel 2003a). During the last few decades, lettuce resistance breeding has focused on searching for and utilizing novel sources of resistance to *B. lactucae* from wild *Lactuca* species (Lebeda et al. 2002, 2007b). However, these new resistances could be quickly overcome by *B. lactucae* isolates from wild pathosystems (Lebeda 2002; Lebeda et al. 2002).

During the last few decades, studies of host resistance, variation and distribution of *B. lactucae* virulence phenotypes have focused on the population level and only on cultivated lettuce (*L. sativa*; e.g. Crute 1987; Lebeda and Zinkernagel 2003a). So far, there have been no studies of interactions between *Lactuca* spp.–*B. lactucae* in natural populations, especially from the viewpoint of host resistance, pathogen virulence, and their temporal and spatial dynamics (Lebeda 2002). In the Czech Republic, studies of the wild *L. serriola*–*B. lactucae* pathosystem were initiated at the beginning of the 1980s (Lebeda 1984, 1986; Lebeda and Boukema 1991; Lebeda and Syrovátko 1988). However, more detailed research focusing on the structure, spatial and temporal changes in this pathosystem, including interactions with the crop (*L. sativa*) pathosystem and coevolutionary studies, began only recently (Lebeda 2002; Lebeda et al. 2002; Lebeda and Petrželová 2004a, b; Petrželová and Lebeda 2003, 2004a, b, c).

In populations of *B. lactucae*, sexual reproduction has an important role in genetic recombination (Michelmore 1981) and is considered to be the major source of virulence variation (Crute 1992b; Lebeda and Schwinn 1994). *Bremia lactucae* is predominantly heterothallic, and two sexual compatibility types (mating types), designated B1 and B2, have been described (Michelmore 1981). Lebeda and Schwinn (1994) documented sexual reproduction in populations of *B. lactucae* occurring on lettuce (*L. sativa*), but studies focussing on pathogen isolates from *L. serriola* have been more limited (Lebeda and Blok 1990). These reports documented both mating types of *B. lactucae*.

The aim of this paper is to describe and analyze patterns of variation in interactions between naturally growing *L. serriola* populations and *B. lactucae*. This report includes both previously published and new data (collected between 1998 and 2006) about the structure and dynamics of this pathosystem from the viewpoint of host range, disease distribution and severity, and the occurrence of various types of symptoms, variation and spatial distribution of host resistance and pathogen virulence, temporal dynamics and microevolutionary shifts in *B. lactucae* populations on naturally growing populations of *L. serriola*. Of particular interest are the interactions between the wild- and crop (*Lactuca sativa*) pathosystems. Coincidence of *B. lactucae* with *Golovinomyces cichoracearum* and its potential for competitive interactions is also considered.

Host range of *B. lactucae* in natural populations of Asteraceae plants

The natural incidence of *B. lactucae* on wild Asteraceae species was surveyed in two main areas of the Czech Republic (Fig. 1) during the period 1999–2006, with the main focus on populations of weedy *L. serriola* populations and associated Asteraceae plants. Field surveys usually took place between May and early September. Whenever possible, locations were visited repeatedly during the growing season. During the course of this study, *B. lactucae* was recorded on eight Asteraceae species (Table 1). It is evident that, in the Czech Republic, the most common host species of *B. lactucae* are *L. serriola* and *Sonchus oleraceus*; however sparse occurrence was also observed on *Arctium tomentosum*, *Carduus crispus*, *Cirsium arvense*, *Lapsana communis*, *Sonchus arvensis* and *Sonchus asper* (Table 1).

Bremia lactucae is an obligate biotrophic pathogen with a broad host range within the Asteraceae. On the lettuce crop (*L. sativa*), *B. lactucae* has a worldwide distribution (e.g. Achar 1996; Crute 1987; Datnoff et al. 1994; Lebeda 1979; Lebeda and Zinkernagel 2003a; Marlatt 1974; Sharaf et al. 2007; Trimboli and Crute 1983). It has also been recorded on more than 200 other Asteraceae species from about 40 genera of the tribes Lactuceae, Cynareae and Arctotideae (Crute and Dixon 1981; Koike and Ochoa 2007; Lebeda et al. 2002). However, information about the natural distribution and patterns of variation of *B. lactucae* populations on wild composites is very rare (Lebeda and Syrovátko 1988). Recently, *B. lactucae* was noted as a common disease on *L. serriola* in The Netherlands (Hooftman et al. 2007); however, only sporadic occurrence has been recorded in other European countries, such as Austria, France, Germany and Switzerland (Lebeda et al. 2001b), and Slovenia and Sweden (Doležalová et al. 2001). Detailed data on the distribution of *B. lactucae* within Europe are lacking. Only in the Czech Republic has

Fig. 1 Areas and locations (*by dots*) in the Czech Republic surveyed for the natural distribution of *Bremia lactucae* in the period 1998–2006

the natural occurrence of *B. lactucae* been studied more intensively during the last decade, with *L. serriola* and *Sonchus* species (especially *S. oleraceus*) found as its most common hosts (Lebeda et al. 2007a; Petrželová and Lebeda 2004b).

However, despite its broad host range, *B. lactucae* was found to be highly host-specific. Cross-inoculation laboratory experiments showed that pathogen populations occurring on wild Asteraceae mostly cannot serve as an inoculum source for cultivated lettuce, and inter-specific transmission was demonstrated only within the genera *Lactuca* and *Sonchus* (Lebeda and Syrovátko 1988). In another wild plant pathosystem, where the pathogen is the smut fungus *Microbotryum violaceum*, Carlsson-Granér (2006) recently showed that in a spatially fragmented metapopulation, the pathogen can alter its host species, which can increase disease spread. For cultivated lettuce, populations of *B. lactucae* on weedy-growing *L. serriola* plants, represent a very important danger to the race-specific resistance genes that originated from *L. serriola* and were introduced to lettuce (Lebeda 1984, 1989; Lebeda et al. 2002, 2007b). Our previous (Petželová and Lebeda 2004b) and recent data (Table 1) documented the relatively high frequency of *B. lactucae* on *Sonchus* species. Recently, Vieira and Barreto (2006) suggested the possibility of lettuce (*L. sativa*) infection with *B. lactucae* originating from *Sonchus* spp. However, there is still no direct evidence for natural transmission of *B. lactucae* between the genera *Lactuca* and *Sonchus*, or vice versa.

Table 1 Distribution of *Bremia lactucae* in populations of *Lactuca serriola* and other Asteraceae plants occurring in the same plant associations (1999–2006)

Species	Number of populations		Proportion (%) of infected populations within evaluated Asteraceae species
	Observed	Infected	
Arctium lappa	3	0	0
Arctium tomentosum	82	2	0.33
Carduus acanthoides	22	0	0
Carduus crispus	49	1	0.16
Cirsium arvense	144	3	0.49
Cirsium canum	6	0	0
Cirsium oleraceum	7	0	0
Cirsium vulgare	18	0	0
Lactuca serriola	768	563	91.84
Lapsana communis	9	1	0.16
Sonchus arvensis	14	4	0.65
Sonchus asper	7	3	0.49
Sonchus oleraceus	64	36	5.88
Taraxacum spp.	50	0	0
Total	1,243	613	100

Variation in phenotypic expression of *B. lactucae* infection on naturally growing *L. serriola* plants

Large variation in phenotypic expression of *B. lactucae* infection on *L. serriola* plants was observed. During the disease survey, in addition to epidemiological data (disease incidence, prevalence and severity) for each evaluated *L. serriola* population, the types of disease symptoms on leaves were also recorded. Generally, *B. lactucae* is described as a pathogen causing light green, yellow or (on older leaves) necrotic lesions visible on the upper surface of lettuce leaves. Those lesions are often surrounded by larger leaf veins, and under optimal conditions are covered with sporulation on the lower leaf surface. But it is known that there may be variation in lesion types and sporulation on infected lettuce leaves (Crute and Dixon 1981).

Disease symptoms on *L. serriola* may be divided into several groups according to the basic character of response, shape, abundance of lesions and intensity of sporulation (Fig. 2a,b). The first category of symptoms includes responses without any visible leaf necrosis (Fig. 2a). These may be characterized either by small, discrete chlorotic spots surrounded by veins that are variable in abundance on leaves according to the progress of infection, or by minute spots dispersed over the leaves with only a few conidiosporangiophores growing from each spot. Sometimes, dispersed sporulation over large parts or even the whole leaves with no obvious borders may be observed. However, there may be discrete chlorotic spots involving larger parts of leaves, with profuse sporulation on the lower leaf surface when conditions are suitable for asexual reproduction.

The second main type of macroscopic response is connected with leaf-tissue necrosis (Fig. 2b). Both

Fig. 2 Variation in expression of symptoms of *B. lactucae* on naturally infected leaves of *L. serriola*; **a** Symptoms without any visible leaf necrosis; *a1* both sides of the same leaf (*a–e* yellowing lesions well localized by main veins), *a*→*e* increasing percentage of leaf area with strongly sporulating lesions (on abaxial side) of *B. lactucae*; *f* and *g* frequent, small and localized lesions with unusual leaf discolouration, covered on abaxial side of the leaf by sporulating *B. lactucae*; *a2* abaxial side of the leaf (*h–k*: different examples of diffuse occurrence of sporulating lesions of *B. lactucae*); **b** Symptoms of *B. lactucae* infection are connected with the leaf-tissue (abaxial side) necrosis (*a*→*d*: an example of increasing percentage of necrotic lesions with reduced sporulation of *B. lactucae*)

small necrotic spots or larger ones were observed. Both may be characterized by limited to profuse pathogen sporulation.

Symptoms of *B. lactucae* infection on *L. serriola* showed considerable variability in macroscopic disease expression. In contrast, on other evaluated wild-host composites, little or no variation was found; only typical lesions surrounded by bigger veins and covered with the sporulating pathogen were observed. The main reason for this difference may be the broad genetic diversity of *L. serriola* populations, characterized by the occurrence of a large number of race-specific resistance genes and/or factors (Kuang et al. 2006; Lebeda et al. 2002; Lebeda and Petrželová 2004b, 2007; Table 6 and Fig. 4), which may differ in their phenotypic expression. Despite the high variation in symptom expression found among evaluated *L. serriola* populations, our field observations showed that infected plants within individual populations generally displayed similar disease symptoms. However, no experimental data are available to compare whether different symptoms observed on naturally infected plants are directly linked to the presence of specific R-factors in *L. serriola*, or if it is a more complex phenomenon which involves interactions among host plant and pathogen populations and the environment (Cooke et al. 2006; Drenth 2004; Frantzen 2000; Zadoks and Schein 1979). These interactions could be very variable, substantially influenced by environmental factors, such as ambient temperature (Judelson and Michelmore 1992).

Natural distribution of *B. lactucae* and disease prevalence in populations of *L. serriola*

Disease incidence, seasonal and temporal dynamics of *B. lactucae* in natural populations of *L. serriola*

Two parameters were used to assess the distribution of *B. lactucae* in natural populations of *L. serriola*. Disease incidence was expressed as the percentage of occurrence of *B. lactucae* on surveyed sites and on populations of *L. serriola*. Disease prevalence was assessed in each host population by using a visual 0–3 scale (Lebeda 2002; Petrželová and Lebeda 2004b). *Bremia lactucae* was found frequently (ranging from ca 60 to 85%) in populations of *L. serriola* at surveyed localities (Fig. 3). Disease was recorded at all developmental stages of the host plants. *Bremia lactucae* can be found throughout the growing season (April to September, in some extreme cases even up to October or November), as long as weather conditions are suitable for host plant growth. However, the highest disease incidence was recorded from June to August.

Disease prevalence in populations of *L. serriola*

Disease prevalence in infected *L. serriola* populations was surveyed, and some seasonal fluctuations within individual years were recorded. In some populations visited repeatedly during the growing season, it was possible to document the progress of infection; however, in other populations, no changes were observed. Also, the proportion of populations with different levels of disease prevalence fluctuated slightly among individual years. However, it appears that, under natural conditions, disease prevalence of *B. lactucae* infection mostly does not reach higher levels (Table 2).

Prevalence of *B. lactucae* in different types of habitats and populations of *L. serriola*

Possible influences of habitat type (Table 3) as well as of the size and density (Table 4) of host populations on the incidence and disease prevalence of *B. lactucae* were also considered (Petrželová and Lebeda 2004b). From these perspectives, there were some significant differences in the disease prevalence, which were most pronounced in urban areas (Table 3) with frequent occurrence of solitary host plants or small groups of plants, also in agricultural areas, or in habitats with moist substrates (Tables 3 and 4).

Interactions of *B. lactucae* with *G. cichoracearum* on *L. serriola*

The natural incidence of other fungal pathogens on *L. serriola* plants was also recorded. Only *Golovinomyces cichoracearum*, the causal agent of powdery mildew in Asteraceae (Braun 1995) was found to be of particular importance for incidence and prevalence of *B. lactucae* in the host populations. We focused on the co-incidence of both pathogens with results summarized in Table 5. It is evident that both pathogens are widely

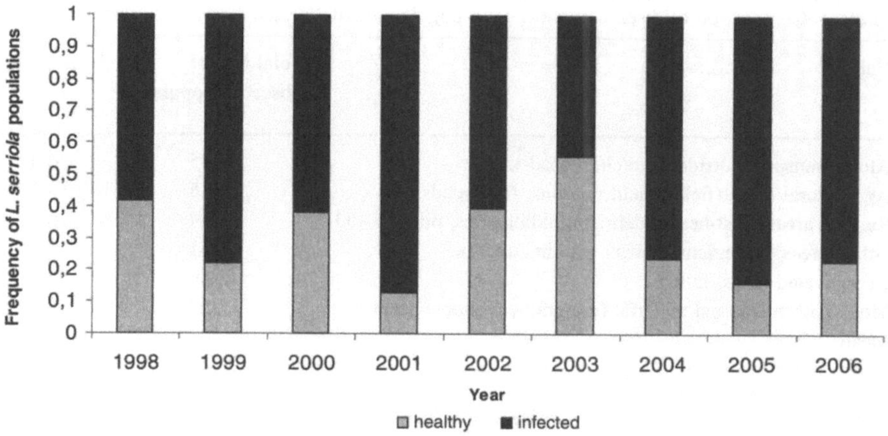

Fig. 3 Fluctuation of *Bremia lactucae* infection in natural populations of *Lactuca serriola*. Figure shows among-year fluctuations in proportion of healthy (non-infected) and *B. lactucae* infected populations of *L. serriola* within all populations surveyed in a given year

Table 2 Temporal variation in *Bremia lactucae* prevalence in natural populations of *Lactuca serriola*

Year	Number of evaluated *L. serriola* populations	Disease prevalence/% of populations			
		0	1	2	3
1998	36	41.7*	41.7*	8.3	8.3
1999	77	22.1	63.6*	10.4	3.9
2000	136	38.2	38.2*	17.7	5.9
2001	132	12.9*	59.1*	20.4*	7.6
2002	53	39.6*	35.8*	18.9	5.7
2003	59	55.9*	35.6*	3.4*	5.1
2004	101	23.8	66.3*	9.9	0.0
2005	100	16.0*	55.0	21.0*	8.0
2006	109	22.9	57.8*	16.5	2.8
Mean		30.3	50.3	11.1	5.3
P		<0.000	<0.002	<0.000	<0.200

Table shows among-year variation in proportion of *L. serriola* populations differing in disease prevalence (proportion of individuals within a population diseased with *Bremia lactucae*). For simplification we used visual 0–3 scale for expressing different levels of disease prevalence (Lebeda 2002; Petrželová and Lebeda 2004b): *0* no symptoms of *B. lactucae* infection in a surveyed *L. serriola* population; *1* low disease prevalence; *2* medium disease prevalence; *3* high disease prevalence

Mean=average of disease prevalence for the whole period (1998–2006)

P level for observed vs. expected (average) chi-square test

Methodology for recording of field observations used over the study period was continuously supplemented with some new parameters (Petrželová and Lebeda 2004b). As a result, all data are not complete for all years in Tables 2, 3 and 4.

*$P<0.05$ for differences between two percentages (marked value of prevalence and mean value/under line/) used software: StatSoft, Inc. (2001). STATISTICA Cz, Version 6. Www. StatSoft.Cz

distributed on *L. serriola* under natural conditions; however, *B. lactucae* is more frequent. Nevertheless, the percentage of populations where both pathogens were recorded together was rather high (Table 5). Some differences were found in the incidence and prevalence of these pathogens during the growing season and among individual years (Table 5).

Generally, the dynamics of a pathogen on a weed is completely different than in a crop. In a dynamic weed population, pathogen epidemiology is likely to be intimately related to the ecology of its host, influencing the host's abundance, spatial distribution, and genetic diversity. Temporal variation in these factors affects the ability of the pathogen to reproduce and spread (Cousens and Croft 2000). Other important factors for the incidence of a pathogen in weed populations include general climatic conditions and specific microclimatic conditions at individual sites.

The first pre-condition of pathogen incidence is the presence of its host. Results summarized in this paper show that *B. lactucae* occurs in a patchy fashion wherever appropriate hosts grow and conditions are suitable for pathogenesis. Populations of *B. lactucae* and *L. serriola* are a good example of a host–pathogen metapopulation structure (McDermott and McDonald 1993; Thrall et al. 2001), where individual fragmented populations are linked together by the transport of spores and gene-flow.

Long-term research has been focused on the structure and dynamics of *B. lactucae* populations on *L. serriola* (Lebeda 2002). Within the genus *Lactuca*, *L. serriola* (prickly lettuce) is the most frequent weed species in Europe, especially within the

Table 3 Comparison of disease prevalence in populations of *Lactuca serriola* occurring in different types of habitats (1998–2006)

Habitat[a]	Total No. of observed populations	Disease prevalence[b]/% of populations			
		0	1	2	3
Along transport corridors (ditches, roadsides)	385	26.0	55.5	12.5	6.0
Agricultural areas (fields, field margins, field roads)	215	15.3*	60.0	19.1	5.6
Ruderal areas, dust-heaps, debris, building sites, piles of soil	114	22.8	55.3	17.5	4.4
Urban areas (pavements, lawns, parking sites)	61	59.0*	34.5	4.9*	1.6
Uncultivated areas, fallows	86	23.2	51.2	19.8	5.8
Moist with biological material (dunghills, compost-heaps)	22	9.1*	59.1	22.7	9.1
Mean		25.9	52.6	16.1	5.4
P		<0.000	<0.135	<0.025	<0.358

StatSoft, Inc. 2001. STATISTICA Cz, Version 6. Www.StatSoft.Cz

[a] Categorization according Petrželová and Lebeda (2004b)

[b] Categorization same as used in Table 2

P level for observed vs. expected (average) Chi-square test

**P*<0.05 for differences between two percentages (marked value of prevalence and mean value/under line/)

last two decades as it has undergone a big population explosion connected to human activity (Hooftman et al. 2006; Lebeda et al. 2001b). It is a pioneer plant which colonizes disturbed areas (Feráková 1977; Lebeda et al. 2004) in various ruderal habitats, often along transport corridors (Lebeda et al. 2001b). It produces many achenes which can disperse over long distances, often generating large populations of hundreds or thousands of plants (Petrželová and Lebeda 2004b; Weaver and Downs 2003). However, our field observations showed that these habitats typically undergo a succession, where *L. serriola* is replaced by another species. For this reason, the natural distribution of *B. lactucae* on *L. serriola*

Table 4 Comparison of disease prevalence in populations of *Lactuca serriola* differing in the size and density (1998–2006)

Population size[a, c]	Total No. of observed populations	Disease prevalence[b]/% of populations			
		0	1	2	3
Individual plants[d]	89	34.8*	51.7	5.6*	7.9
Group of several dispersed plants[e]	211	29.4	53.1	12.3	5.2
Compact group of more plants[f]	364	22.8	56.9	15.9	4.4
Extensive and dense growth[g]	136	15.4	52.9	22.1*	9.6
Large areas with *L. serriola* (*Lactuca* fields)[h]	69	11.6*	60.9	18.8	8.7
Mean		22.8	55.1	14.9	7.2
P		<0.003	<0.904	<0.029	<0.588

P level for observed vs. expected (average) Chi-square test, * *P*<0.05 for differences between two percentages (marked value of prevalence and mean value /under line/; StatSoft, Inc. 2001. STATISTICA Cz, Version 6. Www.StatSoft.Cz)

[a] Categorization according to Petrželová and Lebeda (2004b)

[b] Categorization same as used in Table 2

[c] Approximate number of plants

[d] <5

[e] 5–10

[f] 11–50

[g] 51–100

[h] >100

Table 5 Temporal variation in the incidence/co-incidence of *Bremia lactucae* and *Golovinomyces cichoracearum* in populations of *Lactuca serriola*

Year	Total number of evaluated populations	Incidence of *B. lactucae* and *G. cichoracearum* (%)		
		Single infection		Co-incidence of both pathogens
		B. lactucae	*G. cichoracearum*	
1998	36	22.2	22.2	36.1
1999	77	37.7	19.5	40.3
2000	136	29.4	16.2	32.4
2001	132	48.5	5.3	38.6
2002	53	n.d.	n.d.	n.d.
2003	59	32.2	23.7	11.9
2004	101	49.5	13.9	26.7
2005	100	32.0	9.0	52.0
2006	109	45.0	11.0	32.1
In total[a]	750	38.8	13.5	34.7

n.d. not determined during the main season

[a] *L. serriola* populations (total number of evaluated populations, year 2002 excluded); % of incidence/co-incidence for the period 1998–2006

cannot be assessed only from the viewpoint of individual populations, but also at a larger geographical scale, i.e. metapopulation size.

In comparison to crops, weed populations mostly occur as mixtures of genotypes, and pathogen incidence is thus strongly dependent on the availability of susceptible host plants (Cousens and Croft 2000). Host genetic diversity has considerable influence on the occurrence of a pathogen and its variation, especially in pathosystems operating on a gene-for-gene basis (Burdon 1997). Thus, despite its persistence in the area during the studied period, it was not possible to predict the incidence of *B. lactucae* in particular sites due to the unpredictable dynamics of the host populations. Further, we observed a large discrepancy between disease incidence and disease prevalence (in the present study, expressed as the degree of infection in evaluated populations, Table 2). Although *L. serriola* populations with *B. lactucae* infection were widely distributed, the prevalence of infection in most populations was very low (Table 2). In natural populations such negative relationships between disease incidence and disease prevalence may be caused by higher levels of connectivity within the fragmented host metapopulation (Carlssson-Granér and Thrall 2002). Indeed, though *L. serriola* is distributed in a patchy fashion, it is very common in many plant associations.

The distribution of *B. lactucae* and its interaction with *L. serriola* must be also considered from the viewpoint of interactions with the physical environment. Differences in disease incidence during the main season of *B. lactucae* development were related to some extent with local temperature and rainfall in July and August. *Bremia lactucae* has a narrow optimal range of conditions for growth (Crute and Dixon 1981), preferring cool temperatures and relatively high humidity. Weather variables are considered crucial in the infection and epidemiology of *Bremia lactucae* on cultivated lettuce (Scherm and van Bruggen 1994; Su et al. 2004). A recent study (Mieslerová et al. 2007) demonstrated that in extremely dry and hot summer months (as was the case in 2003), more than 50% of populations were free of infection symptoms. In contrast, in the cool and wet Augusts of 2005 and 2006, a high frequency of infected *L. serriola* populations was observed (Mieslerová et al. 2007). The age of host plants can also influence disease severity (Petrželová and Lebeda 2004b). From our empirical data, it is evident that plants initially infected at the early stages of development (leaf rosette formation or bolting) expressed higher degrees of infection under optimal conditions than plants infected as adults. This agrees with experimental data reported by Crute and Dickinson (1976). The success of a pathogen in a host population may also be influenced by its interactions and possible competition with other pathogens attacking the same host plants (Lindow 2006). Our recent research has focused on the co-incidence of *B. lactucae* and *Golovinomyces cichoracearum* (Mieslerová et al. 2007). *Golovinomyces cichoracearum*, in comparison with *B. lactucae*, has a shorter period of incidence during the growing season, and its epidemics start later (mostly in June or July), when *B. lactucae* is already widespread among *L. serriola* plants (Lebeda 2002; Petrželová and Lebeda 2004b). However, where it did occur, *G. cichoracearum* was able to develop heavy and extensive infections within a few (3–4) weeks (Petrželová and Lebeda 2004b). The peak of incidence of both pathogens was approximately the same, i.e. August. Both pathogens can co-occur within the same population of *L. serriola* and either of them may dominate, depending strongly on different environmental optima for their growth and development, with humidity being the most pronounced parameter (Mieslerová et al. 2007). In relationship to disease severity and interactions of both pathogens, the phenomenon of induced resistance must also be seriously considered in natural plant pathosystems (Newton and Pons-Kühnemann 2007).

Host–pathogen interactions play an important role in plant populations and may have some impact on plant fitness and demography (Thrall and Burdon 2003, 2004). However, our knowledge of these ecological parameters in the *L. serriola–B. lactucae* interaction is still very limited. From our data, it is evident that there are differences in the expression of disease symptoms and disease intensity within and among individual plants inside populations and between populations, leading to some reduction in leaf surface (Fig. 2) and the number of leaves; however, we never observed the infection of reproductive (floral) parts of *L. serriola* plants. Recently, the first data about this phenomenon showed that the impact of inheriting *Bremia* resistance on reproductive plant fitness is small (Hooftman et al. 2007). Nevertheless, it was observed that infected individuals had fewer leaves at the beginning of the bolting phase than did non-infected plants, and for individual seed weight there was a significant interaction between mainlines and *Bremia* infection. However, the total seed weight per head (capitulum) was not altered by *Bremia* infection, and it was concluded that *Bremia* infection did not affect these fitness components in any consistent manner (Hooftman et al. 2007).

Variation of resistance in populations of *L. serriola*

Research was focused on the determination of resistance variation within and among Czech populations of *L. serriola* as well as on the evaluation of variation within other European populations of prickly lettuce (Lebeda and Petrželová 2004b, 2007). Samples of *L. serriola* were collected in 2001 within the framework of the EU project 'Gene-Mine' (for details see Lebeda et al. 2007a). In total, 250 individual plants from 16 Czech populations of *L. serriola* were screened following previously described methods (Lebeda and Zinkernagel 2003b) for resistance against 10 races of *B. lactucae* (NL1, NL5, NL12, NL14, NL15, NL16, BL17, BL18, BL21 and BL24) with known virulence patterns (van Ettekoven and van der Arend 1999) and were characterized by means of resistance phenotypes. Some basic results are given in Table 6 and Fig. 4.

A substantial proportion of the Czech *L. serriola* populations showed high levels of susceptibility to the *B. lactucae* isolates used (Table 6). Overall, 30% of the host individuals studied were completely susceptible, and this phenotype was widely distributed among studied populations (in 12 of 16 populations). A completely resistant phenotype was recorded in five populations, represented by 24 (9.6%) samples (Fig. 4). While most plants expressed intermediate levels of race-specific resistance, only four populations showed relatively high levels of resistance. Most plants were susceptible to at least one to three races of *B. lactucae* (among them very often BL21 and BL24).

Despite their high levels of susceptibility, Czech populations of *L. serriola* were variable in terms of their resistance to ten races of *B. lactucae*. Presence of race-specific resistance was very common. Both inter- and intra-population variation of resistance were found. In total, 45 different resistance phenotypes were recognized in the studied populations of *L. serriola*. However, 80% of the individual plants evaluated were represented by only eight resistance phenotypes; the remaining phenotypes were generally rare. When compared to other European populations of *L. serriola*, Czech populations most closely resem-

Table 6 Variation of resistance within European metapopulations of *Lactuca serriola* (Lebeda and Petrželová 2004b; Lebeda et al. 2007a)

Populations	Number of *L. serriola* populations High level[a] of				Total
	Resistance	Susceptibility	Race-specific response	Intra-population variation in race-specificity	
CZ	4	9	0	3	16
D	0	6	1	9	16
NL	0	2	6	0	8
UK	0	0	10	0	10
Totally	4	17	17	12	50

[a] Relative variation in mean resistance of populations evaluated as a relative proportion of samples with prevalence of susceptibility, resistance and race-specificity to the used set of *B. lactucae* races

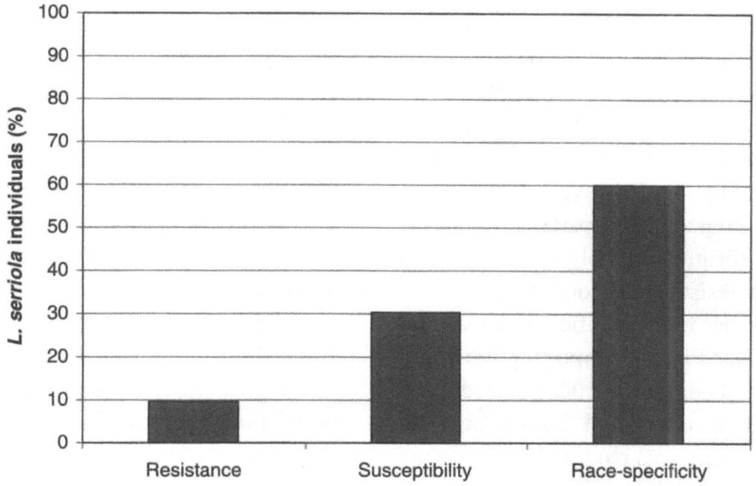

Fig. 4 Percentage of *Lactuca serriola* individuals (in total $n=250$), with different reaction patterns to ten races of *Bremia lactucae*, within the sampled Czech host populations ($n=16$; each population represented by ca 16 individual plants, from each plant tested ca 30 achenes/seedlings/). Three basic categories of reaction patterns were distinguished: *Resistance L. serriola* individual plants were resistant to all ten races of *B. lactucae* used; *Susceptibility L. serriola* individual plants were susceptible to all ten races of *B. lactucae* used; *Race-specificity* differential reaction patterns to races of *B. lactucae* used was recorded

bled resistance structures observed in their German counterparts (Table 6). However, while German populations expressed greater levels of intra-population variation, no population was completely resistant (Table 6) despite the widespread occurrence of resistant individuals among the populations. *Lactuca serriola* populations in the Netherlands also expressed a high level of intra-population variation; however, their responses showed higher levels of race-specificity (Table 6). Unlike the Czech, German and Dutch populations, disease responses of populations from the United Kingdom were much more uniform (Table 6).

During the last few decades, lettuce resistance breeding has focused on the identification and incorporation of novel sources of resistance to *B. lactucae* from wild *Lactuca* spp., especially from *L. serriola*, *L. saligna* and *L. virosa* (e.g. Beharav et al. 2006; Bonnier et al. 1992; Jeuken and Lindhout 2002; Lebeda et al. 2002, 2007b; Lebeda and Zinkernagel 2003b; Maisonneuve 2003). Mapping the distribution patterns of resistance and virulence can help us understand co-evolutionary dynamics in plant pathosystems (e.g. Carlsson-Granér 2006; Carlsson-Granér and Thrall 2002; Delmotte et al. 1999; Laine 2006; Thrall and Burdon 2003) and select appropriate resistance sources and crop-breeding strategies (Lebeda et al. 2002, 2007b). However, there is only limited information about the distribution of resistance to *B. lactucae* in natural populations of its hosts (Lebeda and Petrželová 2004b, 2007). To date, most such studies have been based on evaluation of genebank germplasm samples (Lebeda et al. 2007b) which may not represent the structure of natural host populations. Only in Europe there has been extensive field collections aimed towards obtaining large population samples of *L. serriola* (Doležalová et al. 2001; Křístková and Lebeda 1999; Lebeda et al. 2001b, 2007a).

The interaction between *L. serriola* and *B. lactucae* in Europe is an exciting model for such studies (Lebeda et al. 2001b). First, *L. serriola* is a highly invasive species and recently, due to the increasing ruderalization of the environment, it has become quite a common weed in both agricultural and natural plant ecosystems (Hooftman et al. 2006; Lebeda et al. 2001b, 2004). Furthermore, it is closely related to the cultivated lettuce, considered to be its progenitor (Lebeda et al. 2001b, 2007b). It is also used very extensively as a source of resistance against *B. lactucae* in lettuce breeding, and many of its race-specific resistance genes have been introduced into commercial lettuce cultivars (Lebeda et al. 2002, 2007b). From these perspectives, it is probably the only plant pathosystem where we can precisely study the structure, dynamics and interactions between the wild- and crop-pathosystems (Lebeda 2002; Lebeda et al. 2007c) at both the individual and population levels.

Our recent research shows that Czech populations of *L. serriola* generally display a low level of resistance to *B. lactucae*, which correlates well with the high frequency of disease occurrence recorded during field observations (Petrželová and Lebeda 2004b; Table 2 and Fig. 3). However, the frequent occurrence of differential reaction patterns to the *B. lactucae* races tested (originating only from *L. sativa*) indicates that all the resistance recorded in these *L. serriola* populations is race-specific (Lebeda and Petrželová 2004b, 2007; Fig. 4), supporting previously reported results (Lebeda et al. 2002; Lebeda and Petrželová 2001). Both inter- and intra-population variation in race-specificity was recorded in the Czech metapopulation of *L. serriola*. When the spatial distribution of individual resistance phenotypes was assessed, they were randomly distributed over the study area with no obvious aggregation of populations with more resistant and/or susceptible phenotypes. One of the probable explanations for recorded resistance patterns over the Czech metapopulation of *L. serriola* is that host migration events among locally adapted populations play an important role in shaping resistance structure.

A completely different situation was found within other European metapopulations of *L. serriola* (Lebeda et al. 2007a). Although the same resistance phenotypes could be found in different countries, other European populations generally have completely different resistance patterns with an increasing frequency of race-specificity towards the Atlantic coast. Among-population variation in resistance is largely affected by the level of patchiness and connectivity of populations within the larger metapopulation (Carlsson-Granér and Thrall 2002). In the case of data summarized in this paper, this may be an explanation for greater differentiation, both among and within *L. serriola* populations observed in continental Europe, while the decrease in variation was most pronounced for populations in the United Kingdom, where only one form of the host, *L. serriola* f. *integrifolia*, is prevalent (Lebeda et al. 2004). These populations are spatially isolated from host populations in continental Europe, both by distance per se, but also by their island location (Lebeda and Petrželová 2005). In a previous study with a limited number of plant samples, it was concluded that populations of *L. serriola* in Britain were commonly homogeneous for the *B. lactucae* resistance phenotype. There was no evidence for extensive resistance gene 'pyramiding' or population heterogeneity as defence strategies against *B. lactucae* in natural populations of *L. serriola* (Crute 1990). However, some variation in the level of field resistance was recorded (Crute 1990) which agrees with observed levels of field resistance in wild *Lactuca* spp. germplasm (Lebeda 1990).

In natural pathosystems, disease dynamics is the most important factor driving the diversity and distribution of host-resistance genotypes (Laine 2006). In the crop pathosystem, *L. sativa–B. lactucae*, geographic differences in virulence among pathogen populations from various countries and growing areas are relatively well known (e.g. Crute 1987; Lebeda and Zinkernagel 2003a) and the same should be expected in the wild pathosystem, *L. serriola–B. lactucae* (Lebeda 2002; Lebeda and Petrželová 2004a; Petrželová and Lebeda 2004c). However, for *L. serriola* we still lack sufficient data about relevant pathogen populations in much of Europe except for the Czech Republic, hindering our understanding of coevolutionary trends within this pathosystem. Thus, we can only suppose that, at the larger spatial scale, different *L. serriola* populations may have been exposed to differential selection pressures by *B. lactucae* and as a result evolved different patterns of resistance.

Temporal changes in variation of virulence in populations of *B. lactucae* occurring on *L. serriola*

Samples of *B. lactucae* from naturally infected wild populations of *L. serriola* and of cultivated lettuce (*L. sativa*) were collected and used for the virulence screening. Tests were carried out according to methods described previously (Lebeda 2002; Lebeda and Zinkernagel 2003b). Altogether, 313 isolates of *B. lactucae* from *L. serriola* were collected during the period of searching (23 in 1998, 31 in 1999, 78 in 2000, 43 in 2001, 8 in 2002, 19 in 2003, 51 in 2004 and 60 in 2005).

Virulence of isolates was examined by screening on a standard differential set (van Ettekoven and van der Arend 1999). More detailed characterization was made on 56 *L. sativa* and *L. serriola* genotypes (Lebeda and Zinkernagel 2003b) with well characterized patterns of race-specific resistance (*Dm*-genes or

R-factors). By use of both differential sets, 32 virulence factors (v-factors) were determined; however four of them (v32, v33, v41 and v42) are not included in the set of data (see Fig. 5).

Variation in virulence among natural populations of *B. lactucae* in the Czech Republic was studied from 1998 to 2005 and was analyzed at both individual and population levels. Results related to the analysis of virulence at the level of individuals have been partly (for the period 1997–2000) previously published (Lebeda and Petrželová 2004a; Petrželová and Lebeda 2004c). Variation in virulence and its temporal changes were quantified by relative frequencies of virulence factors (v-factors) in the tested isolates (Lebeda 1981, 1982; Fig. 5). With only a few exceptions (v18, v32, v33, v37, v41 and v42), most of the examined v-factors were recorded in populations of *B. lactucae* on *L. serriola* during the whole study period. Nevertheless, there were substantial differences in frequencies recorded for individual v-factors (Fig. 5). Medium to high frequencies were recorded for v-factors v5/8, v7, v11, v14–17 and v23–30. In contrast, factors v1–4, v6, v10, v12, v13, v35, v36 and v38 were detected in extremely low frequencies (Fig. 5). Frequencies of v-factors were not uneven with values varying by year. Fluctuations were most pronounced for factors v5/8, v11, v14, v16, v23 and v25–28, with v5/8 and v23 showing a rapid decrease within natural populations of *B. lactucae* (Petrželová and Lebeda 2004c). On the other hand, factors v14 and v17 increased.

Virulence patterns on the differential set recorded for individual isolates were described as virulence phenotypes (v-phenotypes). The complexity of recorded *B. lactucae* v-phenotypes varied broadly, with 6 to 33 *Lactuca* spp. differentials being infected by the tested isolates. The distribution of isolates based on the number of virulent responses showed a normal (Gaussian) distribution curve each year, with average virulence ranging from 9 to 15 virulent responses. In total, 93 different v-phenotypes were identified among 313 isolates tested in the period 1998 to 2005 (Table 7). Mean variation of v-phenotypes differed among years, ranging from 24.4% in year 2000 to 87.5% in year 2002. However, the basic spatial-distribution pattern of individual v-phenotypes was very similar among years. Each year, there was one or a few v-phenotypes prevailing in examined populations, and broad variation of other unique ones (Lebeda et al. 2007c). Some pathogen populations were also tested for the existence of intra-population variation. Different v-phenotypes were recorded even among pathogen isolates collected from *L. serriola* plants growing in close proximity in one host population (Lebeda and Petrželová 2004a).

A primary main focus for our long-term research on *Lactuca* spp.–*B. lactucae* pathosystems has been the determination of virulence variation within *B. lactucae* populations, documenting its temporal and spatial dynamics, and comparing pathogen populations in crop (*L. sativa*) and wild (*L. serriola*)

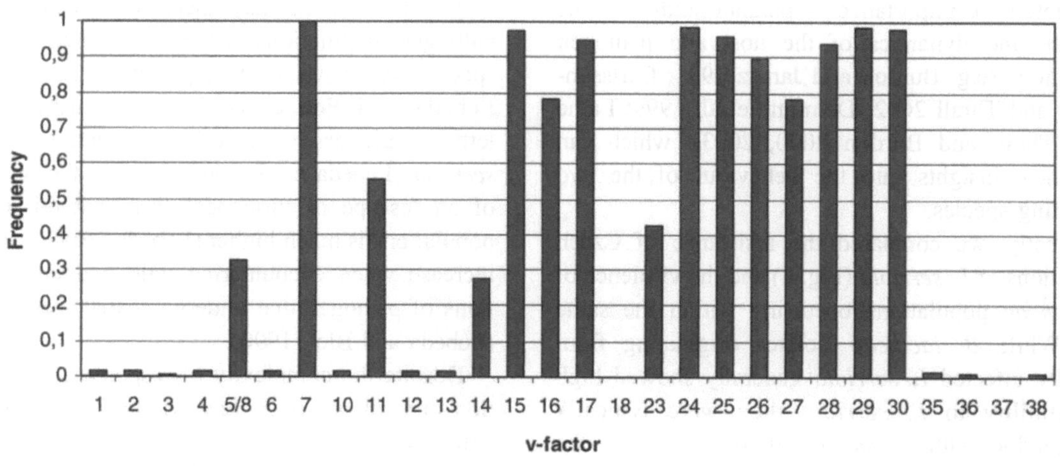

Fig. 5 Frequency of v-factors recorded in the sampled set of *Bremia lactucae* isolates (n=313) collected in the Czech pathogen populations on *Lactuca serriola* during the period 1998–2005

Table 7 Variation of virulence phenotypes in populations of *Bremia lactucae* on *Lactuca serriola* in the period 1998–2005

Year	No. of tested isolates	No. of determined v-phenotypes	Variation among studied isolates (%)
1998	23	12	52.2
1999	31	13	41.9
2000	78	19	24.4
2001	43	29	67.4
2002	8	7	87.5
2003	19	11	57.9
2004	51	16	31.4
2005	60	34	56.7
Total	313	93	29.7

pathosystems. At the population level, virulence can be thought of as the average ability of a pathogen population to overcome the diversity of resistance genes present in the corresponding host population (Thrall and Burdon 2003). From this viewpoint, *B. lactucae* populations occurring naturally on *L. serriola* showed highly complex patterns of virulence in relation to the *L. serriola* and/or *L. sativa* differentials with resistance derived from *L. serriola* (Lebeda 2002; Lebeda and Petrželová 2004a; Petrželová and Lebeda 2004c). Such complexity of pathogen isolates may have its origin in response to the heterogeneity of the host populations (Bevan et al. 1993), and it is supposed that, in gene-for-gene based pathosystems, broadly virulent isolates of pathogen are more likely to occur in highly diverse and resistant host populations (Thrall and Burdon 2003). Many papers have highlighted the importance of parallel studies of the structure and dynamics of the host and pathogen populations (e.g. Burdon and Jarosz 1991; Carlsson-Granér and Thrall 2002; Delmotte et al. 1999; Laine 2006; Thrall and Burdon 2000, 2003), which can bring new insights into the behaviour of the two interacting species.

Recently, we compared the resistance of Czech populations of *L. serriola* (Fig. 4) and the virulence of *B. lactucae* populations occurring within the same area. While *B. lactucae* isolates originating from naturally infected *L. serriola* generally showed high compatibility to *L. serriola* they were typically incompatible with *L. sativa* differentials, with the exception of those *L. sativa* genotypes carrying race-specific resistance genes derived from *L. serriola* (Lebeda 1989, 2002; Lebeda and Petrželová 2004a). Interestingly, when samples from *L. serriola* populations were screened for resistance to races of *B. lactucae* with v-phenotypes generally able to overcome resistance in *L. sativa* they typically showed relatively low levels of race-specific resistance (Lebeda and Petrželová 2004b). This raises the question of why *B. lactucae* isolates with relevant virulence are not more common in natural populations of *B. lactucae.* Only a few such isolates were found in one year (1998) in a region where lettuce is frequently cultivated (Lebeda 2002). If we suppose that populations of *B. lactucae* on *L. sativa* and *L. serriola* are fully inter-connected, isolates with such v-phenotypes should have a selection advantage in *L. serriola* populations and would appear more frequently. However, it was not true in the populations we studied, and perhaps it occurs only under certain conditions. A logical explanation is that *B. lactucae* populations on *L. sativa* and *L. serriola* are highly isolated by their host specificity, and therefore co-evolution in the wild and crop pathosystems is operating independently.

From this viewpoint, *L. serriola* appears to be a good source of resistance genes for cultivated lettuce (Lebeda et al. 2007b), and cross-inoculation experiments with isolates from *L. sativa* may reveal new sources of resistance (Beharav et al. 2006; Lebeda and Zinkernagel 2003b). However, the occasional occurrence of isolates with combined v-phenotype structure recorded both in the wild (Lebeda and Petrželová 2004a; Petrželová and Lebeda 2004c) and crop pathosystems (Lebeda et al. 2007c) indicates that genotype and gene flow between both pathosystems is possible, which may increase variation in both pathogen populations, especially when both host species are grown in close proximity (Lebeda 2002; Lebeda and Petrželová 2004a). Furthermore, in lettuce cultivars with resistance derived from *L. serriola* (Lebeda et al. 2002, 2007b), the probability of an 'escape of virulence' from natural pathogen populations is much higher (Lebeda 1984), and it may increase when encountering wild and crop populations of pathogen that undergo sexual recombination (Lebeda and Blok 1990).

Despite the complexity of responses, broad variation in virulence of *B. lactucae* to *L. serriola* differentials was found, as seen in the variable distribution of v-factors among populations. From our data, it is evident that virulence structure in *B. lactucae* populations occurring on *L. serriola* is very dynamic, undergoing both qualitative and quantitative shifts (Lebeda and Petrželová 2004a; Petrželová and Lebeda 2004c). In gene-for-gene systems, it is supposed that the genetic structure of host and pathogen populations follow each other in a dynamic interaction (Burdon 1997; Burdon et al. 1996), so a long-term decrease in particular v-factors (e.g. v5/8 or v23) in *B. lactucae* populations may be evidence of co-evolution taking place. However, increases in the frequency of other v-factors (e.g. v14 or v17) were less marked than the decreases, and may just be considered as evidence for year-to-year fluctuations over a longer time period.

On the level of individuals, no obvious changes in mean virulence were found during the study period, but there was a large variation in recorded virulence patterns (v-phenotypes) and their distribution over the pathogen metapopulation. Furthermore, great differences in the prevalence of individual v-phenotypes were observed, and many rare ones were recorded just once during our investigations. Similar population structures with uneven distribution of v-phenotypes, where just a few predominated have also been reported for other pathogens of wild plants (Bevan et al. 1993; Burdon and Jarosz 1991). It seems that populations of *B. lactucae* tend, on one hand, to higher diversity, but on the other hand to a higher prevalence of particular v-phenotypes. These v-phenotypes may have a selective advantage at a given time and place, which may be largely influenced by fluctuations in local variation in host resistance and the environmental conditions contributing to the establishment of infection in natural plant populations.

Comparison of *B. lactucae* virulence variation between wild (*L. serriola*) and crop (*L. sativa*) pathosystems

In comparison to *B. lactucae* isolates originating from *L. serriola*, isolates originating from cultivated lettuce generally displayed a highly complex response to *L. sativa* differentials and expressed a completely different virulence structure. A comparison of the frequencies of the most important v-factors (from the viewpoint of resistance breeding, see e.g. Lebeda et al. 2007b) in both pathosystems during the study period is illustrated in Fig. 6. Numerous v-factors were detected in both pathosystems; however, their frequencies differed considerably. Many of the compared v-factors were more common in the crop pathosystem (e.g. v1–4, v6, v10, v12, v13, v36, v38). More or less equal frequencies were recorded for factors v7, v11 and v16, while differences between factors v5/8, v14 and v15 were more distinct (Fig. 6). In this case, it is very interesting that all complementary race-specific resistance genes to these v-factors originate from *L. serriola* (Lebeda et al. 2002). Factor v18 was recorded only on *L. sativa*, and factor v17 was found only on *L. serriola* (Fig. 6). Finally, factor v37 was not found in either pathosystem. In the crop pathosystem temporal shifts in frequencies of some v-factors were also recorded (Lebeda et al. 2007c).

Isolates with combined virulence (for *L. sativa* and *L. serriola*) structure were recorded only occasionally. Differences were also recorded in the dynamics of occurrence of v-phenotypes in both pathosystems. Most of the v-phenotypes found on *L. sativa* were unique and did not appear repeatedly in other pathogen populations or in subsequent years (Lebeda et al. 2007c; Petrželová and Lebeda 2004a).

Large differences in virulence were recorded between populations of *Bremia lactucae* occurring on *L. sativa* and *L. serriola*. Populations in crop and wild *Lactuca–B. lactucae* pathosystems have different structures of v-factors; individual v-factors occur in different frequencies and differences were recorded in their spatial and temporal population structure and dynamics. These findings show that individual *B. lactucae* populations substantially differ in their specificity to the host species and virulence to the host genotypes, respectively. However, we observed some unexpected overlaps in virulence structure between *B. lactucae* populations on *L. sativa* and *L. serriola* (Fig. 6).

Temporal shifts in virulence were recorded in both pathosystems. In the crop pathosystem, the changes are largely influenced by increasing usage of cultivars with newly introduced race-specific resistance genes (Lebeda and Zinkernagel 2003a). Characterization of the evolutionary forces driving the wild pathosystem, *L. serriola–B. lactucae*, will require more detailed studies of resistance patterns and their changes in *L. serriola* populations (Lebeda et al. 2001b, 2007c). The virulence data for both pathosystems are very unique and comparable results are not available for any other crop and wild pathosystems.

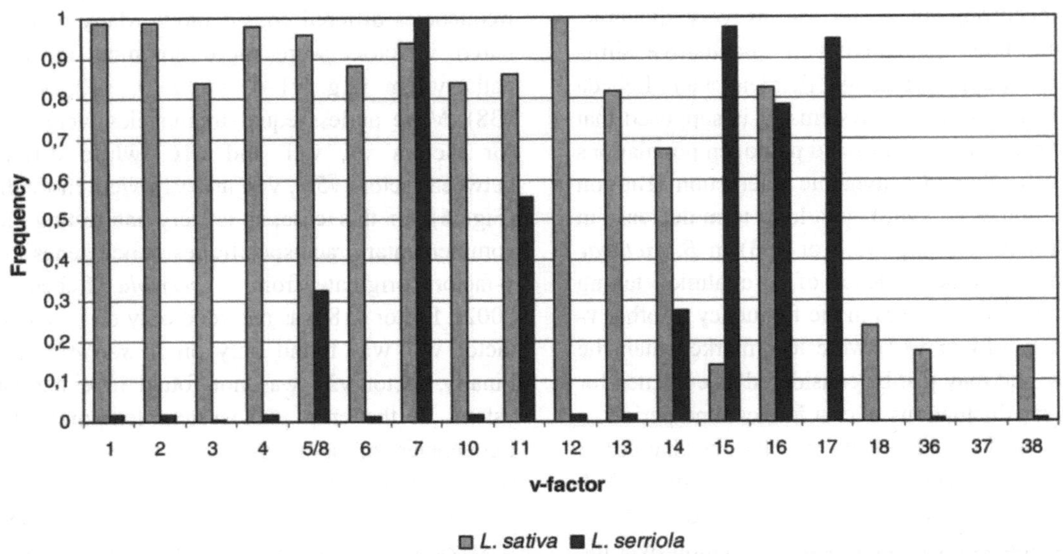

Fig. 6 Comparison of frequency of the most important v-factors recorded in the sampled set of *Bremia lactucae* isolates collected in the Czech pathogen populations on *Lactuca sativa* ($n=93$ isolates) and *Lactuca serriola* ($n=313$ isolates) during the period 1998–2005. v1, v2, v3, v4, v10, v12, v13, v14—v-factors matching *Dm* genes or R-factors in cvs of *L. sativa*. v7, v15, v16, v17, v23, v24, v25, v26, v27, v28, v29, v30—v-factors matching *Dm* genes or R-factors in *L. serriola*. v5/8, v6, v11, v18, v38—v-factors matching *Dm* genes or R-factors in cvs of *L. sativa* derived from *L. serriola*. v36, v37—v-factors matching *Dm* genes or R-factors in cvs of *L. sativa* derived from *L. saligna*. v35—v-factor matching *Dm* gene or R-factor in cvs of *L. sativa* derived from *L. virosa*. Relative frequencies of individual v-factors were expressed as the ratios between the number of isolates with given v-factor and the total number of isolates investigated for the presence of considered v-factor

Sexual reproduction and occurrence of mating types in *B. lactucae* populations on *L. serriola*

The occurrence of mating types was studied in a set of 59 *B. lactucae* isolates originating from 33 naturally infected and wild populations of *L. serriola* in the Czech Republic, including two isolates from Germany and France. The isolates were collected in the period 1997–1999. Both compatibility types were recorded; however, the majority of the isolates (96%) were determined as type B2, supporting the observation that sexual reproduction of *B. lactucae* on naturally growing *L. serriola* plants is rare (Petrželová and Lebeda 2003).

As was stated before, sexual recombination is considered to be important for generating considerable genetic variation in virulence in populations of *B. lactucae* on *L. sativa* (Crute 1992b; Lebeda and Schwinn 1994). However, our previous (Lebeda and Blok 1990) and recent results indicate that its importance for the pathogen populations on *L. serriola* is questionable (Petrželová and Lebeda 2003). From the practical viewpoint it is interesting that isolates of *B. lactucae* originating from *L. sativa* and *L. serriola* are not completely compatible when pairing together to produce oospores under laboratory conditions (Petrželová and Lebeda 2003). This is additional evidence for at least some isolation of both pathosystems (*L. sativa* versus *L. serriola*–*B. lactucae*). We may assume that the possible danger of the natural formation of new virulent races of *B. lactucae* by the crossing of pathogen isolates from crop and wild lettuce is quite low; this is also supported by the results from virulence analyses (Fig. 6).

Conclusions and future developments

From the results summarized in this paper, it is evident that the wild pathosystem, *L. serriola*–*B. lactucae*, is very complex, variable and dynamic. The frequency of pathogen incidence in host populations is very high, but disease prevalence is rather low. Other host plant species have no substantial influence on *B. lactucae* epidemiology. Pathogen incidence is most strongly influenced by ecological factors, including host habitat, density of host populations, and climatic and microclimatic conditions. Host plants are also frequently infected by powdery mildew (*Golovinomyces cichoracearum*); the frequency of co-infection by both pathogens is about 35%. We expect some competition for leaf niche, and the phenomenon of induced resistance may play a role in co-infection, but we lack clear experimental evidence to document this phenomenon.

Research showed that race-specific resistance is the dominant pattern in populations of *L. serriola*. A broad spectrum of resistance phenotypes (altogether 45) was detected in host populations occurring in the Czech Republic. In some populations, individuals with completely resistant or completely susceptible reactions were detected. Intra-population variability was rather common. Comparison of host populations from continental Europe with those from the UK showed substantial differences. The island populations were much more homogeneous with regard to variation in resistance, and reactions were always race-specific in inoculation studies.

From the viewpoint of virulence, the pathogen population is enormously variable, and most of the known v-factors were detected. However, there are substantial differences in the frequency of individual v-factors. In the pathogen population, v-factors that match R-factors originating from *L. serriola* prevail, but v-factors matching race-specific *Dm* genes from cultivated lettuce are very rare. At the individual level, we recorded many v-phenotypes, but only a few were common in *B. lactucae* populations. Comparative studies of the virulence structure of pathogen populations in wild and crop pathosystems clearly demonstrated completely different compositions of v-factors and v-phenotypes. It seems that there is no direct epidemiological linkage between both pathosystems. However, some unexpected overlaps in virulence structure were recorded, and these differences may be evidence for the existence of potential migration or gene flow between the pathosystems. This phenomenon, although rare, may be quite important and could influence the stability of resistance derived from *L. serriola* in cultivated lettuce.

The data obtained about this pathosystem demonstrate that *B. lactucae* belongs to a group of pathogens with high evolutionary potential (*sensu* McDonald and Linde 2002), and the wild host population is extremely variable from the viewpoint of resistance. The evolutionary forces operating in and between both pathosystems are not well known. Future research on these pathosystems can contribute to our understanding of this exciting area of plant pathology and oomycete population biology. We propose that these pathosystems are quite suitable as model systems for the study of structure and variability from a spatial and temporal viewpoint, and that interactions between the wild and crop pathosystems are especially relevant for crop improvement and agricultural production.

Acknowledgements Critical reading and valuable remarks by Dr. M. P. Widrlechner (USDA-ARS, Iowa State University, Plant Introduction Station, Ames, Iowa, USA) and Dr. P. H. Thrall (CSIRO Plant Industry, Canberra, Australia) are gratefully acknowledged. We are grateful to Dr. Milena Kršková for help with statistical analyses of the data. This research was supported by grants MSM 6198959215 and QH71229 (NAZV) and by the National Programme of Genepool Conservation of Microorganisms and Small Animals of Economic Importance of the Czech Republic.

References

Achar, P. N. (1996). First report of downy mildew disease in lettuce caused by *Bremia lactucae* in Natal, Southern Africa. *Plant Disease*, D-1996-0229-01N, (on-line).

Beharav, A., Lewinsohn, D., Lebeda, A., & Nevo, E. (2006). New wild *Lactuca* genetic resources with resistance against *Bremia lactucae*. *Genetic Resources and Crop Evolution*, *53*, 467–474.

Bevan, J. R., Clarke, D. D., & Crute, I. R. (1993). Resistance to *Erysiphe fischeri* in two populations of *Senecio vulgaris*. *Plant Pathology*, *42*, 636–646.

Bonnier, F. J. M., Reinink, K., & Groenwold, R. (1992). New sources of major gene resistance in *Lactuca* to *Bremia lactucae*. *Euphytica*, *61*, 203–211.

Braun, U. (1995). *The powdery mildews (Erysiphaceae) of Europe*. Jena: Gustav Fischer Verlag.

Burdon, J. J. (1987). *Diseases and plant population biology*. Cambridge: Cambridge University Press.

Burdon, J. J. (1997). The evolution of gene-for-gene interactions in natural pathosystems. In I. R. Crute, E. B. Holub, & J. J. Burdon (Eds.), *The gene-for-gene relationships in plant–parasite interactions* (pp. 245–262). Wallingford, UK: CAB International.

Burdon, J. J., & Jarosz, A. M. (1991). Host–pathogen interactions in natural populations of *Linum marginale* and *Melampsora lini*: I. Patterns of resistance and racial variation in a large host population. *Evolution*, *45*, 205–217.

Burdon, J. J., Thrall, P. H., & Ericson, L. (2006). The current and future dynamics of disease in plant communities. *Annual Review of Phytopathology*, *44*, 19–39.

Burdon, J. J., Wennström, A., Elmqvist, T., & Kirby, G. C. (1996). The role of race specific resistance in natural plant populations. *Oikos*, *76*, 411–416.

Carlsson-Granér, U. (2006). Disease dynamics, host specificity and pathogen persistence in isolated host populations. *Oikos, 112*, 174–184.

Carlsson-Granér, U., & Thrall, P. H. (2002). The spatial distribution of plant populations, disease dynamics and evolution of resistance. *Oikos, 97*, 97–110.

Cooke, B. M., Jones, D. G., & Kaye, B. (Eds.) (2006). *The Epidemiology of plant diseases*. Dordrecht: Springer.

Cousens, R., & Croft, A. M. (2000). Weed populations and pathogens. *Weed Research, 40*, 63–82.

Crute, I. R. (1987). The geographical distribution and frequency of virulence determinants in *Bremia lactucae*: Relationships between genetic control and host selection. In M. S. Wolfe & C. E. Caten (Eds.), *Populations of plant pathogens: Their dynamics and genetics* (pp. 193–212). Oxford, UK: Blackwell Scientific Publications.

Crute, I. R. (1990). Resistance to *Bremia lactucae* (downy mildew) in British populations of *Lactuca serriola* (prickly lettuce). In J. J. Burdon & S. R. Leather (Eds.), *Pests, pathogens and plant communities* (pp. 203–217). Oxford: Blackwell Scientific Publications.

Crute, I. R. (1992a). From breeding to cloning (and back again?): a case study with lettuce downy mildew. *Annual Review of Phytopathology, 30*, 485–506.

Crute, I. R. (1992b). Downy mildew of lettuce. In H. S. Chaube, J. Kumar, A. N. Mukhopadhyay, & U. S. Singh (Eds.), *Plant diseases of international importance. vol. ii. diseases of vegetable and oil seed crops* (pp. 165–185). Englewood Cliffs, New Jersey: Prentice Hall.

Crute, I. R., & Dickinson, C. H. (1976). The behaviour of *Bremia lactucae* on cultivars of *Lactuca sativa* and other composites. *Annals of Applied Biology, 82*, 433–450.

Crute, I. R., & Dixon, G. R. (1981). Downy mildew diseases caused by the genus *Bremia* Regel. In D. M. Spencer (Ed.), *The downy mildews* (pp. 421–460). London, New York, San Francisco: Academic Press.

Crute, I. R., & Johnson, A. G. (1976). The genetic relationship between races of *Bremia lactucae* and cultivars of *Lactuca sativa*. *Annals of Applied Biology, 83*, 125–137.

Datnoff, L. E., Nagata, R. T., & Raid, R. N. (1994). Pathotyping of *Bremia lactucae* in Florida. *Plant Disease, 78*, 584–857.

Delmotte, F., Bucheli, E., & Shykoff, J. A. (1999). Host and parasite population structure in a natural plant–pathogen system. *Heredity, 82*, 300–308.

Dinoor, A., & Eshed, N. (1984). The role and importance of pathogens in natural plant communities. *Annual Review of Phytopathology, 22*, 443–466.

Doležalová, I., Lebeda, A., & Křístková, E. (2001). Prickly lettuce (*Lactuca serriola* L.) germplasm collecting and distribution study in Slovenia and Sweden. *Plant Genetic Resources Newsletter, 128*, 41–44.

Drenth, A. (2004). Fungal epidemics—does spatial structure matter. *New Phytologist, 163*, 4–7.

Farrara, B., Illott, T. W., & Michelmore, R. W. (1987). Genetic analysis of factors for resistance to downy mildew (*Bremia lactucae*) in lettuce (*Lactuca sativa*). *Plant Pathology, 36*, 499–514.

Feráková, V. (1977). *The genus Lactuca in Europe*. Czechoslovakia, Bratislava: Komenský University Press.

Frantzen, J. (2000). Resistance in populations. In A. J. Slusarenko, R. R. S. Fraser, & L. C. van Loon (Eds.), *Mechanisms of resistance to plant diseases* (pp. 161–187). Dordrecht: Kluwer.

Gilbert, G. S. (2002). Evolutionary ecology of plant diseases in natural ecosystems. *Annual Review of Phytopathology, 40*, 13–43.

Hammond-Kosack, K. E., & Jones, J. D. G. (1997). Plant disease resistance genes. *Annual Review of Plant Physiology and Plant Molecular Biology, 48*, 575–607.

Hooftman, D. A. P., Nieuwenhuis, B. P. S., Posthuma, K. I., Oostermeijer, J. G. B., & den Nijs, H. J. C. M. (2007). Introgression potential of downy mildew resistance from lettuce to *Lactuca serriola* and its relevance for plant fitness. *Basic and Applied Ecology, 8*, 135–146.

Hooftman, D. A. P., Oostermeijer, J. G. B., & den Nijs, H. J. C. M. (2006). Invasive behaviour of *Lactuca serriola* (Asteraceae) in the Netherlands: Spatial distribution and ecological amplitude. *Basic and Applied Ecology, 7*, 507–519.

Jeuken, M., & Lindhout, P. (2002). *Lactuca saligna*, a non-host for lettuce downy mildew (*Bremia lactucae*), harbors a new race-specific *Dm* gene and three QTLs for resistance. *Theoretical and Applied Genetics, 105*, 384–391.

Judelson, H. S., & Michelmore, R. W. (1992). Temperature and genotype interactions in the expression of host resistance in lettuce downy mildew. *Physiological and Molecular Plant Pathology, 40*, 233–245.

Koike, S. T., & Ochoa, O. E. (2007). Downy mildew caused by *Bremia lactucae* on strawflower (*Helichrysum bracteatum*) in California. *Plant Disease, 91*, 326.

Křístková, E., & Lebeda, A. (1999). Collection of *Lactuca* spp. genetic resources in the Czech Republic. In A. Lebeda & E. Křístková (Eds.), *Eucarpia leafy vegetables '99, Proceedings of the Eucarpia Meeting on Leafy Vegetables Genetics and Breeding* (pp. 109–116). Olomouc, Czech Republic: Palacký University in Olomouc.

Kuang, H., Ochoa, O. E., Nevo, E., & Michelmore, R. W. (2006). The disease resistance gene *Dm3* is infrequent in natural populations of *Lactuca serriola* due to deletions and frequent gene conversions at the *RGC2*locus. *The Plant Journal, 47*, 38–48.

Laine, A.-L. (2006). Evolution of host resistance: looking for coevolutionary hotspots at small spatial scales. *Proceedings of the Royal Society, 273*, 267–273.

Lebeda, A. (1979). Identification of races of *Bremia lactucae* in Czechoslovakia. *Journal of Phytopathology, 94*, 208–217.

Lebeda, A. (1981). Population genetics of lettuce downy mildew (*Bremia lactucae*). *Journal of Phytopathology, 101*, 228–239.

Lebeda, A. (1982). Population genetic aspects in the study of phytopathogenic fungi. *Acta Phytopathologica Academiae Scientarum Hungarica, 17*, 215–219.

Lebeda, A. (1984). Response of differential cultivars of *Lactuca sativa* to *Bremia lactucae* isolates from *Lactuca serriola*. *Transactions of the British Mycological Society, 83*, 491–494.

Lebeda, A. (1986). Specificity of interactions between wild *Lactuca* species and *Bremia lactucae* isolates from *Lactuca serriola*. *Journal of Phytopathology, 117*, 54–64.

Lebeda, A. (1989). Response of lettuce cultivars carrying the resistance gene *Dm*11 to isolates of *Bremia lactucae* from *Lactuca serriola*. *Plant Breeding, 102*, 311–316.

Lebeda, A. (1990). The location of sources of field resistance to *Bremia lactucae* in wild *Lactuca* species. *Plant Breeding, 105*, 75–77.

Lebeda, A. (2002). Occurrence and variation in virulence of *Bremia lactucae* in natural populations of *Lactuca serriola*. In P. T. N. Spencer-Phillips, U. Gisi, & A. Lebeda (Eds.), *Advances in downy mildew research* (pp. 179–183). The Netherlands, Dordrecht: Kluwer Academic Publishers.

Lebeda, A., & Blok, I. (1990). Sexual compatibility types of *Bremia lactucae* isolates originating from *Lactuca serriola*. *Netherlands Journal of Plant Pathology, 96*, 51–54.

Lebeda, A., & Boukema, I. W. (1991). Further investigation of the specificity of interactions between wild *Lactuca* spp. and *Bremia lactucae* isolates from *Lactuca serriola*. *Journal of Phytopathology, 133*, 57–64.

Lebeda, A., Doležalová, I., Feráková, V., & Astley, D. (2004). Geographical distribution of wild *Lactuca* species (Asteraceae, Lactuceae). *The Botanical Review, 70*, 328–356.

Lebeda, A., Doležalová, I., Křístková, E., Dehmer, K. J., Astley, D., van de Wiel, C. C. M., et al. (2007a). Acquisition and ecological characterization of *Lactuca serriola* L. germplasm collected in the Czech Republic, Germany, the Netherlands and United Kingdom. *Genetic Resources and Crop Evolution, 54*, 555–562.

Lebeda, A., Doležalová, I., Křístková, E., & Mieslerová, B. (2001b). Biodiversity and ecogeography of wild *Lactuca* spp. in some European countries. *Genetic Resources and Crop Evolution, 48*, 153–164.

Lebeda, A., & Petrželová, I. (2001). Occurrence and characterization of race-specific resistance to *Bremia lactucae* in wild *Lactuca* spp. In *Eucarpia section genetic resources—"Broad variation and precise characterization — limitation for the future* (p. 93)". Poznań, May 16.–20. 2001—Book of abstracts.

Lebeda, A., & Petrželová, I. (2004a). Variation and distribution of virulence phenotypes of *Bremia lactucae* in natural populations of *Lactuca serriola*. *Plant Pathology, 53*, 316–324.

Lebeda, A., & Petrželová, I. (2004b). Occurrence of race-specific resistance to *Bremia lactucae* in *Lactuca serriola* germplasm originating from four European countries. In J. Vollmann, H. Grausgruber, & P. Ruckenbauer (Eds.), *Genetic variation for plant breeding* (pp. 113–116). Vienna, Austria: EUCARPIA & BOKU—University of Natural Resources and Applied Life Sciences.

Lebeda, A., & Petrželová, I. (2005). Comparison of resistance to *Bremia lactucae* in populations of *Lactuca serriola* occurring in Central Europe (Czech Republic) and the British Isles (England, U.K.). In S. Bullitta (Ed.), *XVII Eucarpia Genetic Resources Section Meeting; "Plant Genetic Resources of Geographical and "other" islands (Conservation, evaluation and use for plant breeding)* (p. 14)". Castelsardo (Italy), 30 March–2 April 2005; Book of Abstracts, CNR-ISPAAM, sezione Sassari, Italy.

Lebeda, A., & Petrželová, I. (2007). Race-specific resistance to *Bremia lactucae* in European populations of *Lactuca serriola*. In: *EUCARPIA Leafy Vegetables 2007, Conference Abstracts* (p. 11). 18–20 April 2007; University of Warwick, Warwick HRI, UK, 2007; Oral Presentations.

Lebeda, A., Petrželová, I., & Maryška, Z. (2007c). Comparative analysis of variation and dynamics of virulence of *Bremia lactucae* populations on *Lactuca sativa* and *Lactuca serriola*. In B. MacDonald, P. Brunner, P. Ceresini & Lebeda, A. (Eds.), *Population and evolutionary biology of fungal symbionts* (pp. 50–51.). Book of Abstracts, International Meeting, April 29–May 4, 2007, Ascona, Switzerland; ETHZ—Swiss Federal Institute of Technology of Zurich and Palacký University in Olomouc; Zurich and Olomouc. JOLA, v.o.s., Kostelec na Hané, Czech Republic.

Lebeda, A., Pink, D. A. C., & Astley, D. (2002). Aspects of the interactions between wild *Lactuca* spp. and related genera and lettuce downy mildew (*Bremia lactucae*). In P. T. N. Spencer-Phillips, U. Gisi, & A. Lebeda (Eds.), *Advances in downy mildew research* (pp. 85–117). The Netherlands, Dordrecht: Kluwer Academic Publishers.

Lebeda, A., Pink, D. A. C., & Mieslerova, B. (2001a). Host–parasite specificity and defense variability in the *Lactuca* spp.-*Bremia lactucae* pathosystem. *Journal of Plant Pathology, 83*, 25–35.

Lebeda, A., Ryder, E. J., Grube, R., Doležalová, I., & Křístková, E. (2007b). Lettuce (Asteraceae; *Lactuca* spp.), Chapter 9. In R. Singh (Ed.), *Genetic resources, chromosome engineering, and crop improvement series, volume 3—vegetable crops* (pp. 377–472). Boca Raton, FL, USA: CRC Press.

Lebeda, A., & Schwinn, F. J. (1994). The downy mildews—an overview of recent research progress. *Journal of Plant Diseases and Protection, 101*, 225–254.

Lebeda, A., & Syrovátko, P. (1988). Specificity of *Bremia lactucae* isolates from *Lactuca sativa* and some *Asteraceae* plants. *Acta Phytopathologica et Entomologica Hungarica, 23*, 39–48.

Lebeda, A., & Zinkernagel, V. (2003a). Evolution and distribution of virulence in the German population of *Bremia lactucae*. *Plant Pathology, 52*, 41–51.

Lebeda, A., & Zinkernagel, V. (2003b). Characterization of new highly virulent German isolates of *Bremia lactucae* and efficiency of resistance in wild *Lactuca* spp. germplasm. *Journal of Phytopathology, 151*, 274–282.

Lindow, S. (2006). Phyllosphere microbiology: a perspective. In M. J. Bailey, A. K. Lilley, T. M. Timms-Wilson, & P. T. N. Spencer-Phillips (Eds.), *Microbial ecology of aerial plant surfaces* (pp. 1–20). Wallingford, UK: CABI International.

Maisonneuve, B. (2003). *Lactuca virosa*, a source of disease resistance genes for lettuce breeding: results and difficulties for gene introgression. In T. J. L. van Hintum, A. Lebeda, D. A. C. Pink, & J. W. Schut (Eds.), *Eucarpia leafy vegetables '03* (pp. 31–35). The Netherlands: CGN Wageningen.

Marlatt, R. B. (1974). Biology, morphology, taxonomy and disease relations of the fungus *Bremia*, Florida. *Agricultural Experimental Station Technical Bulletin, 764*, 1–25.

McDermott, J. M., & McDonald, B. A. (1993). Gene flow in plant pathosystems. *Annual Review of Phytopathology, 31*, 353–373.

McDonald, B. A., & Linde, C. (2002). Pathogen population genetics, evolutionary potential, and durable resistance. *Annual Review of Phytopathology, 40*, 349–379.

Michelmore, R. W. (1981). Sexual and asexual reproduction sporulation in the downy mildews. In D. M. Spencer (Ed.), *The downy mildews* (pp. 165–181). London: Academic.

Mieslerová, B., Petrželová, I., Lebeda, A., & Česneková, E. (2007). Occurrence of lettuce downy mildew and powdery mildew in natural populations of prickly lettuce. In A. Lebeda & P. T. N. Spencer-Phillips (Eds.), *Advances in downy mildew research, vol. 3. Proceedings of The 2nd International Downy Mildews Symposium* (pp. 59–64). v.o.s., Kostelec na Hané (Czech Republic): Palacký University in Olomouc and JOLA.

Newton, A., & Pons-Kühnemann, J. (2007). Induced resistance in natural ecosystems and pathogen population biology: exploiting interactions. In D. Walters, A. Newton, & G. Lyon (Eds.), *Induced resistance for plant defence. a sustainable approach to crop protection* (pp. 133–142). Oxford, UK: Blackwell.

Petrželová, I., & Lebeda, A. (2003). Distribution of compatibility types and occurrence of sexual reproduction in natural populations of *Bremia lactucae* on wild *Lactuca serriola* plants. *Acta Phytopathologica et Entomologica Hungarica, 38*, 43–52.

Petrželová, I., & Lebeda, A. (2004a). Comparison of virulence of *Bremia lactucae* isolates originating from *Lactuca sativa* and *Lactuca serriola*. *Acta fytotechnica et zootechnica, 7*, 248–250.

Petrželová, I., & Lebeda, A. (2004b). Occurrence of *Bremia lactucae* in natural populations of *Lactuca serriola*. *Journal of Phytopathology, 152*, 391–398.

Petrželová, I., & Lebeda, A. (2004c). Temporal and spatial variation in virulence of natural populations of *Bremia lactucae* occurring on *Lactuca serriola*. In P. T. N. Spencer-Phillips & M. Jeger (Eds.), *Advances in downy mildew research, vol. 2* (pp. 141–163). The Netherlands, Dordrecht: Kluwer.

Scherm, H., & van Bruggen, A. H. C. (1994). Weather variables associated with infection of lettuce by downy mildew (*Bremia lactucae*) in coastal California. *Phytopathology, 84*, 860–865.

Sharaf, K., Lewinsohn, D., Nevo, E., & Beharav, A. (2007). Virulence patterns of *Bremia lactucae* in Israel. *Phytoparasitica, 35*, 100–108.

Su, H., van Bruggen, A. H. C., Subbarao, K. V., & Scherm, H. (2004). Sporulation of *Bremia lactucae* affected by temperature, relative humidity, and wind in controlled conditions. *Phytopathology, 94*, 396–401.

Thrall, P. H., & Burdon, J. J. (2000). Effect of resistance variation in a natural plant host–pathogen metapopulation on disease dynamics. *Plant Pathology, 49*, 767–773.

Thrall, P. H., & Burdon, J. J. (2003). Evolution of virulence in a plant host–pathogen metapopulation. *Science, 299*, 1735–1737.

Thrall, P. H., & Burdon, J. J. (2004). Host–pathogen life history interactions affect biological control success. *Weed Technology, 18*(Suppl. S), 1735–1737.

Thrall, P. H., Burdon, J. J., & Young, A. (2001). Variation in resistance and virulence among demes of a plant host–pathogen metapopulation. *Journal of Ecology, 89*, 736–748.

Trimboli, D. S., & Crute, I. R. (1983). The specific virulence characteristics of *Bremia lactucae* (lettuce downy mildew) in Australia. *Australasian Plant Pathology, 12*, 58–60.

Van Ettekoven, K., & Van der Arend, A. J. M. (1999). Identification and denomination of "new" races of *Bremia lactucae*. In A. Lebeda & E. Křístková (Eds.), *Eucarpia leafy vegetables '99* (pp. 171–175). Olomouc, Czech Republic: Palacký University.

Vieira, B. S., & Barreto, R. W. (2006). First record of *Bremia lactucae* infecting *Sonchus oleraceus* and *Sonchus asper* in Brazil and its infectivity to lettuce. *Journal of Phytopathology, 154*, 84–87.

Weaver, S. E., & Downs, M. P. (2003). The biology of Canadian weeds. 122. *Lactuca serriola* L. *Canadian Journal of Plant Science, 83*, 619–628.

Zadoks, J. C., & Schein, R. D. (1979). *Epidemiology and plant disease management*. New York: Oxford University Press.

Eur J Plant Pathol (2008) 122:147–155
DOI 10.1007/s10658-008-9346-6

Development of detection systems for the sporangia of *Peronospora destructor*

Roy Kennedy · Alison J. Wakeham

Received: 6 November 2007 / Accepted: 9 June 2008

Abstract A monoclonal antibody that recognises components of the wall of sporangia of *Peronospora destructor* was raised. Tests using spores of higher fungi and other species of mildew demonstrated the specificity of the monoclonal. The antibody was used to develop lateral flow devices for sporangia of *P. destructor*. A competitive lateral flow format was developed which could detect onion downy mildew sporangia. Five-microliter gold anti-mouse IgM solution pre-mixed with 10 µl of *P. destructor* monoclonal antibody (EMA 242) proved the optimal concentration for detection of sporangia of *P. destructor* when applied to sample pads of lateral flow devices. Limits of approximately 500 sporangia of *P. destructor* could be detected by the absence of a test line on the lateral flow device within test samples. Using a scanning densitometer improved the sensitivity of detection. Further development and validation of the test is required if it is to be used for risk assessments of onion downy mildew in the field.

Keywords Onion downy mildew · Monoclonal antibody · *Peronospora destructor* · Lateral flow assay · Detection · PTA ELISA · Immunofluorescence

R. Kennedy (✉) · A. J. Wakeham
Warwick HRI, University of Warwick, Wellesbourne, Warwick, Warwickshire CV35 9EF, UK
e-mail: Roy.Kennedy@warwick.ac.uk

Introduction

Foliar diseases of onion crops (onion downy mildew and *Botrytis* leaf blight) can cause heavy yield losses in bulb and salad onion crops. Onion downy mildew (*Peronospora destructor*) is the most serious disease in bulb and salad onions in the UK (Gilles et al. 2004; Clarkson et al. 2000). Actual yield losses in bulb onions of 60 to 75% have been recorded (Cook 1932; Cruickshank 1958). These losses mainly result from severe infections in bulb onion crops causing early defoliation, reduced bulb sizes and poor storage quality of bulbs (Rondomanski 1967). In salad onions, yield losses can be as high as 100% with whole crops being discarded, as downy mildew symptoms on the plant make them unmarketable. Fungicidal control of onion downy mildew is difficult and fungicides are only effective if they are applied before or immediately after disease first appears in the crop (Kennedy 1998). The environmental requirements for infection and sporulation by *P. destructor* have been reported (Yarwood 1937, 1943; Hildebrand and Sutton 1982). Mathematical models describing climatic effects on sporulation and infection have been described (Gilles et al. 2004; Battilani et al. 1996; Jesperson and Sutton 1987). However, despite the rapid development of onion downy mildew and the requirements for reductions in fungicide usage by consumers, the practical use of these systems in risk assessment has been limited to date.

New approaches in forecasting diseases of onion crops based on estimation of spore numbers in air samples have been reported (Carisse et al. 2005; Berger 1970). Detection and quantification of air-borne spore numbers can be used to predict disease accurately before it is visible in the crop. Peaks of airborne spores are always detected prior to crops becoming infected. It has been reported that one or two peaks in sporangial concentration in the air of the potato blight pathogen *Phytophthora infestans* preceded the first observed symptoms of the disease in the field (Bugiani et al. 1998; Phillon 2003). In these studies the information on spore number had to be collected manually using a microscope which was slow and time consuming. Tests which can be conducted in the field are necessary if information on air-borne inoculum concentration is to be of more practical value. However, there are few reported systems for detecting and differentiating airborne spores (Wakeham et al. 2004). Molecular techniques exist for detection of spores using Hirst types spore samplers (Williams et al. 2001). Day et al. (2002) reported the development of cell flow cytometric differentiation of air-borne sporangia of *P. infestans* using 'in field' systems. Both techniques required either laboratory processing of results or the development and use of sensitive equipment not fully demonstrated under field conditions. The development and use of detection systems for estimating air-borne spore numbers would be a further development in risk assessment for onion downy mildew. This study reports on the development of an immunomonitoring system for conidia of *P. destructor*.

Materials and methods

Production of *P. destructor* immunogen for antibody production

The isolate (PD HL00) of *P. destructor* used in the study was as reported by Gilles et al. (2004). Leaf surface wax material of ten onion sets (*Allium cepa* cv. White Lisbon) was removed by gentle agitation with sheep's wool (found to remove leaf wax without leaf damage) prior to inoculation with *P. destructor*. Twenty-five 20-μl droplets of *P. destructor* (1×10^4 conidia ml^{-1} H_2O) were applied to each sheep's wool-treated leaf. To induce infection, inoculated plants were incubated in high humidity for three days after which plants were removed and placed in a temperature-controlled glasshouse (18°C) for a further 2 weeks. Inoculated plants were returned to a high humidity environment for a period of 48 h to induce sporulation by *P. destructor* on infected leaves.

Collection of *P. destructor* spores from leaf surfaces

A hand-held Burkard surface cyclone sampler (Burkard Manufacturing Co., Rickmansworth, Herts, UK) was used to collect sporangia of *P. destructor* from the surface of the infected leaf material. The 2 ml Eppendorf containing the collected spores was removed and 1 ml of chilled sterile distilled water (SDW) was added. The collected *P. destructor* sporangia were suspended in water and 0.5 ml volume of chilled SDW was added. The sporangial suspension was filtered through a stainless steel membrane (47 μm pore size) to remove any large contaminating material. The liquid phase was collected and bacterial and other small leaf contaminants removed by filtering using a polyester membrane (10 μm pore size). The filtrate was collected and resuspended in 1 ml phosphate buffered saline solution, pH 7.0 (PBS). Bright field microscopy was used to determine the presence of *P. destructor* sporangia which were adjusted to a concentration of 3.5×10^4 conidia ml^{-1}.

Immunization of mice with *P. destructor* sporangia

The spore suspension was agitated, using a Gallenkamp spinmix, continuously for a period of 5 min after 3 h at 0–4°C. A microfuge (MSE Microcentaur) was used at 13 rpm for 5 min to separate particulate spore material from the soluble spore fraction of the sample. The soluble fraction of the sample was retained and concentrated at first by freeze-drying (Modulyo 4 k, Edwards) and then rehydrating to a final volume of 100 μl PBS. Two Balb C female mice (coded 7996, 7997) were immunised (by intraperitoneal injection) each with 50 μl of the concentrated soluble *P. destructor* sporangial preparation mixed with an equal volume of Titermax adjuvant. All further immunisations were as described above. Tail bleeds were taken seven days after the second immunisation procedure and a PTA-ELISA (described below) was carried out to determine whether the mice had produced an immune response to *P. destructor*.

The mice received a final pre-fusion boost of the *P. destructor* soluble sporangial immunogen mixed with adjuvant (100 μl). The spleen of mouse 7996 was removed 4 days later and the fusion was carried out according to those methods reported by Dewey (1992). Hybrids were fed on days 3, 6 and 10 and cell culture supernatants screened by PTA ELISA and immunofluorescence 14 days after cell fusion for the presence of antibodies which recognised sporangial components of *P. destructor*.

Monoclonal antibody screening

Plate trapped antigen ELISA (PTA ELISA)

One hundred μl of *P. destructor* soluble sporangial washings in PBS were aliquoted in to each of 96-well Polysorp microtitre well strips (Nunc, Roskilde, Denmark; Cat. No.469957). The strips were incubated overnight in an enclosed chamber at 18°C. Unbound material was then removed and the microtitre wells were washed once with 200 μl PBS. The microtitre wells were blocked with 200 μl of 1% Casein buffer (1% (w/v) casein PBS) and incubated at 37°C for 45 min. Residual blocking buffer was removed and wells were washed four times for 1 min each with 200 μl PBS, 0.05 % Tween 20 and 0.1 % Casein (PBSTw C). Each well received 100 μl of fusion hybridoma tissue culture supernatant mixed with PBSTw C. Following incubation in a Wellwarm shaker incubator (30°C) for a period of 45 min as above, wells were washed three times for 1 min each with 200 μl PBSTw. A DAKO duet amplification system was used according to manufacturer's instructions (DAKO Ltd, Cambridge, UK) to amplify the signal generated by bound tissue culture supernatant antibodies. Wells were washed as described above and 100 μl of 3,3′,5,5′-tetramethylbenzidene substrate (Sigma, UK) was added to each well. The reaction was stopped by adding 25 μl of a 20% 1 M H_2S0_4 solution to each well. Absorbance at 450 nm was determined with a Biohit BP800 ELISA plate reader (Alpha Laboratories, Hampshire, UK).

Immunofluorescence

Twenty microliters of a 1×10^3 spores ml^{-1} *P. destructor* conidial spore suspension was aliquoted to individual multiwell glass slides (Cel-Line/Series Scientific Corp, USA; Cat No. 10-3404). Following air drying any unbound spore material was removed with a PBSTwC wash. Material remaining bound to the multiwell glass slides was incubated with 20 μl of hybridoma tissue culture supernatant antibodies (TCS) mixed with PBSTwC for a period of 30 min at room temperature. A counterstain of Evans blue and Eriochrome black was incorporated within the TCS antibody suspension to quench *P. destructor* spore autoflourescence (Kennedy et al. 1999). Each multiwell received a wash as described above and following air drying was incubated with anti-mouse antibodies which had been conjugated to fluorescein isothiocyanate dye. A counter-stain was included to ensure quenching of sporangial autoflourescence. Incubation was carried out at room temperature in darkness to prevent photobleaching of the conjugated antibody. The processed multiwells received a final wash of PBSTwC and after air drying were mounted and viewed by episcopic fluorescence microscopy for the presence of antibody/ fluorescein-tagged sporangia of *P. destructor*. Hybridoma antibody tissue culture supernatants, identified as positive to *P. destructor* sporangial material using either PTA ELISA and/or IF, were selected and twice cloned to monoclonal Ab status.

Selection of specific *P. destructor* monoclonals

To determine specificity, the selected *P. destructor* monoclonal cell lines were determined by PTA-ELISA and IF against a range of fungal species. Tests were carried out on spores and mycelium taken from pure cultures of *Bremia lactucae*, *Peronospora parasitica*, *Paecilomyces variotii*, *Botrytis cinerea*, *B. squamosa*, *Stemphyllium* sp., *Aureobasidium pullulans*, *Phoma betae*, *Ascochyta rabei*, *Fusarium culmorum*, *Penicillium roquefortii*, *Pyrenophora teres* and sporangia of *P. destructor*. Stains used in these tests are as designated in Kennedy et al. (2000). With the exception of *P. destructor*, *P. parastica*, *B. lactucae* and *Ascochyta* (all of which were grown directly on plant material) the fungal species used in the reactivity tests were grown on a synthetic medium covered with a sterile Supor membrane filter prior to inoculation. Fourteen days after inoculation (of cultures grown on agar) 5 ml of PBS (pH 7.5) solution was applied to the culture surface. Surface washings were taken by gently stroking the culture surface with a glass spreader. All collected spore concentrations were

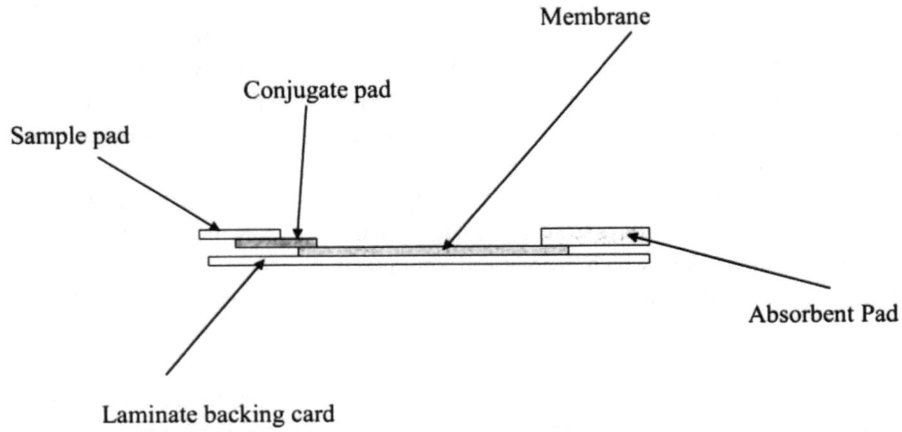

Fig. 1 Lateral flow cross-section (5 mm strip)

adjusted to a final concentration of 1×10^5 spores ml^{-1} PBS. The spore solutions were individually aliquoted into each microtitre well (100 µl per well) of a Polysorp microtitre strip. The wells were covered and incubated overnight at 4°C. Unbound material was removed and the microtitre wells were washed once with 200 µl PBS. An ELISA was carried out as previously described.

Development of a competitive lateral flow assay format for the detection of conidia of *P. destructor*

A competitive lateral flow format (Fig. 1), comprising a Millipore 135 HiFlow™ cellulose ester membrane direct cast onto 2 ml Mylar backing (Millipore Corp, USA.), an absorbent pad (Schleicer and Schuell, Germany) and a sample pad (Millipore Corp., USA) was constructed for the detection of *P. destructor* sporangia. Control lines of an anti-mouse serum were sprayed directly onto the membrane surface using a flat bed air jet dispenser (Biodot Ltd, West Sussex, UK). A collected soluble fraction of a *P. destructor* sporangial sample, prepared as described earlier, was adjusted to a protein concentration of 500 µg ml^{-1} , 250 µg ml^{-1} and 125 µg ml^{-1} in PBS and applied as a test line again using a flat bed air jet dispenser. Membranes were air-dried at 35°C for a period of 4 h. The test and control line-labelled lateral flows were cut in to 5 mm strips and each strip housed within a plastic case (Schleicer & Schuell, Germany).

Antibody conjugation

A British Biocell gold anti-mouse IgM solution was pre-mixed (conjugated) with a selected hybridoma cell line (coded EMA 242) before drop application to a 5 mm sample pad and air-drying. Variable concentrations of gold conjugated antibody EMA 242 were applied to different sample pads to investigate the antibody conjugate concentration which gave optional test line formation on the lateral flow device (*lfd*). Sample conjugate Ab pads were attached to each *lfd* strip as shown in Fig. 1. A test antigen of a 60 µl sporangial suspension (3×10^3 *P. destructor* sporangia) was then applied to the sample pad of an *lfd* strip. The competitive lateral flow devices (*clfd*) were viewed 5 min post-sample application. For each test, a spore-free suspension was applied to a *clfd* as a negative control. The variable antibody dilutions of MAb used in these tests are shown in Table 1.

Visual detection threshold of a competitive lateral flow device employing two membrane types for *P. destructor* sporangia

Studies were carried out using a *clfd* format for the detection of known concentrations of *P. destructor* spores. Two different membrane types were examined: a Millipore 135 HiFlow™ cellulose ester membrane direct cast on to 2 ml Mylar backing and a Millipore

Table 1 Antibody dilutions in the samples pad and corresponding *P. destructor* test line concentration

P. destructor MAb type and dilution factor EMA 242	*P. destructor* protein concentrations µg ml^{-1} at test line		
1 in 160	500	250	125
1 in 320	500	250	125
1 in 640	500	250	125

240 HiFlow™ cellulose ester cast membrane. The *lfd* devices were prepared as described above and a test line of 250 μg ml^{-1} *P. destructor* soluble antigen in PBS was applied. The membranes were air-dried at 35°C, cut into 5 mm strips and each strip housed within a plastic case as previously described.

A known sporangial concentration of *P. destructor* sporangia (60 μl) was mixed with EMA 242 gold conjugate (5 μl) to produce a final antibody dilution of either 1:150, 1:400 or 1:600. The mixture was applied to the sample pad of each *clfd* and results viewed 5 min post-sample application. For each membrane type a 'spore free suspension' was mixed with MAb EMA 242 gold conjugate to act as a negative control.

Semi-quantitative tests with lateral flow format using Millipore 135 HiFlow™ cellulose ester membrane

A 5 μl British Biocell gold anti-mouse IgM solution was pre-mixed with 10 μl EMA 242 and then applied drop-wise to lateral flow sample pads at a test volume of 15 μl, each dried as previously described. This was chosen as the optimal concentration for detecting approximately 500 sporangia of *P. destructor*. Sporangia of *P. destructor* in sample buffer were applied drop-wise (70 μl) to the sample pads of the pre-prepared lateral flows. Sporangial concentrations ranged from 240 to 960 sporangia applied. The lateral flow devices were viewed 20 min post-sample application for the formation of a test and control line and test line optical density values were generated using a BioDot lateral flow reader (BioDot, Chichester). A negative control of lateral flow running buffer alone (0 downy mildew conidia) was also included within these tests.

Results

Monoclonal antibody screening

Plate trapped antigen ELISA (PTA ELISA)

Eleven hybridoma cell lines were identified (using PTA ELISA) as producing antibodies which recognised components associated with the sporangial material of *P. destructor*. A preliminary screen against a range of plant fungal pathogens identified three tissue culture supernatants for expansion to monoclonal status. These were selected, cloned to monoclonal antibody status and coded EMA 240, 242, and 243. Monoclonal antibody cell line EMA 240 was not used in reactivity tests as it was observed to react with other downy mildew species when tested by ELISA (data not shown). Monoclonal antibodies EMA 242 and 243 reacted to their homologous antigen (*P. destructor*), and demonstrated a high level of specificity when tested against other fungal species (Fig. 2).

Immunofluorescence

Only eight of the eleven cell lines were identified as producing antibodies which recognised components directly associated with the conidia of *P. destructor* when visualised by immunofluorescence. Of these, six were excluded following preliminary reactivity studies (data not shown). Those selected for further testing by IF were EMA 242 and 243 (Table 2). EMA 240 did not react with material directly associated with *P. destructor*; however, an area of diffuse speckling was noted surrounding the spore. In immunofluorescence studies both EMA 242 and 243 reacted with the spore wall of *P. destructor* and retained a high level of specificity when tested against other fungal species. This indicated a high degree of similarity between these cell lines.

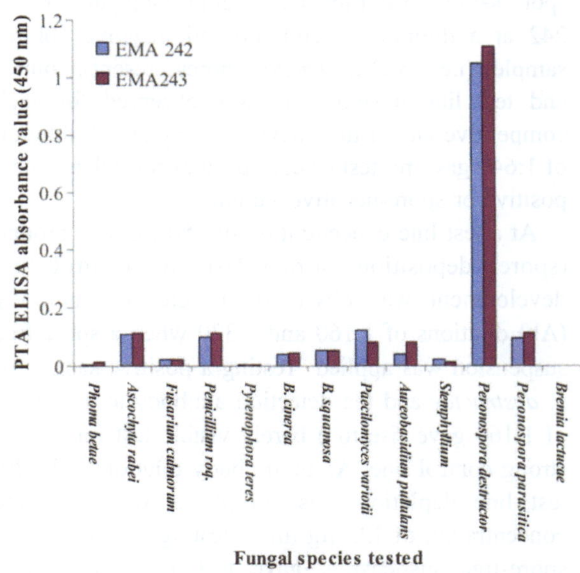

Fig. 2 Reactivity of monoclonal antibodies EMA 242 and 243 to a range of airborne fungal species as tested by PTA ELISA (each value represents the mean of two replications, SD 0.0892)

Table 2 Reactivity of monoclonal antibodies EMA 242 and 243 to a range of airborne fungal species as tested by immunofluorescence

Fungal species tested	Isolate Code	EMA 242		EMA 243	
		Mycelium	Spores	Mycelium	Spores
Phoma betae	WPhbl	☒	☒	☒	☒
Aschochyta rabei	cbs765.01	Not tested	☒	Not tested	☒
Fusarium culmorum	C2751	☒	Not Tested	☒	Not Tested
Penicillium roqueforti	C2709	Not tested	☒	Not tested	☒
Botrytis cinerea	Cbs121.39	☒	☒	☒	☒
Botrytis squamosa	Cbs105.23	☒	☒	☒	☒
Paecilomyces variottii	C2745	Not tested	☒	Not tested	☒
Aureobasidium pullulans	C1718	Not tested	☒	Not tested	☒
Stemphyllium Peronospora	Wsa1	☒	☒	☒	☒
Peronospora destructor	WPd05	Not tested	☑	Not tested	☑
Peronospora parasitica	WHpBo717	Not tested	☒	Not tested	☒
Bremia lactucae	WB103	Not tested	☒	Not tested	☒

☒ No fluorescence observed

☑ fluorescence observed denoting reactivity

cbs The Centraalbureau voor Schimmelcultures

W Warwick HRI Culture collection, *c* Rothamsted culture collection

Assessment of competitive lateral flow assay format for the detection of *P. destructor*

At a test line application of 500 μg ml^{-1} spore protein deposition, test line inhibition (i.e. no test line development) was observed when a *P. destructor* spore sample was mixed with gold conjugated EMA 242 at a dilution >1:160. For all negative control samples (i.e. no *P. destructor* spores present), control and test line development was observed for each competitive lateral flow device. Using an Ab dilution of 1:640 gave no test or control lines for either spore-positive or spore-negative samples.

At a test line concentration of 250 μg ml^{-1} protein (spore) deposition, strong test and control line development was observed at detection antibody (Ab) dilutions of 1:160 and 1:320 when a spore-free suspension was applied. Testing a positive sample of *P. destructor* and the detection antibody at a dilution of 1:160 gave rise to a barely visible test line, but a strong control line. At an antibody dilution of 1:320, test line depletion was complete. At a test line concentration of 125 μg ml^{-1}, testing a *P. destructor* spore-free suspension, gave test and control line development when a detection antibody dilution of 1:160 was used. Using a positive *P. destructor* spore sample gave rise to a clear control line but no test line development (a positive test for the competitive lateral flow format). At all other antibody dilutions, control lines were barely visible and no test line development was noted for any of the samples tested.

Visual detection threshold of a competitive lateral flow device employing two membrane types for *P. destructor* sporangia

Using a Millipore HiFlow™ 135 Membrane and antibody EMA 242 gold conjugate at a dilution of 1:150, test line formation was observed for all spore samples tested. This denoted that the detection sensitivity of the test was poor and unable to detect 2,000 sporangia of *P. destructor*. However, by diluting the activity of the antibody conjugate to 1:400, test sensitivity was improved (Table 3a, Fig. 3a). At an antibody conjugate dilution of 1:600, the test became void with no test line formation for any of the samples tested.

Using a Millipore HiFlow™ 240 membrane competitive lateral flow device and EMA 242 conjugated to gold spheres at a dilution of 1:150, test line formation was again observed for all spore samples tested. As previously noted, by diluting the activity of the antibody to 1:400, the sensitivity of the test was improved (Table 3b). For this membrane, an

Table 3 Detection of sporangia of *P. destructor* with varying antibody dilutions using (a) Millipore HiFlow™ Membrane 135 (b) Millipore HiFlow™ Membrane 240

(a)

No *P. destructor* spores in sample

EMA 242 Ab dilution	0	62	125	250	500	1000	2000
1 in 150	✓	✓	✓	✓	✓	✓	✓
1 in 400	✓	✓	✓	✓	✗/✓	✗/✓	✗/✓
1 in 600	✗	✗	✗	✗	✗	✗	✗
(b)							
EMA 242 Ab dilution	0	62	125	250	500	1000	2000
1 in 150	✓	✓	✓	✓	✓	✓	✓
1 in 400	✓	✓	✓	✓	✓	✗	✗
1 in 600	✓	✓	✓	✓	✗	✗	✗

✓ Clear test line (*P. destructor* not detected in sample

✗ no test line development (*P. destructor* presence detected by *clfd*)

✗/✓ weak test line development

antibody dilution of 1:600 was required to achieve a detection assay where sporangia in excess of 250 could be detected (Fig. 3b, Table 3b).

Semiair-quantitative lateral flow prototype tests with onion downy mildew

The results of using increasing amounts of *P. destructor* sporangia on the lateral flow device are shown in Table 4. When a negative sample (0 sporangia) was applied to a lateral flow device, strong test line development was observed. As spore concentrations increased, the test line colour formation decreased. When a *P. destructor* sporangial concentration of 960 was applied to a lateral flow device, no test line development was observed. Using a Bio-dot lateral flow reader, an optical density value of 2.2 was observed with control suspensions (0 sporangia). However, when 960 sporangia of *P. destructor* was added to the device, the optical density of the line decreased to 0.3.

Discussion

In this study detection tests for *P. destructor* sporangia were developed although these have not yet been used in the field. Detection of the presence of *P. destructor* sporangia could be important in onion downy mildew control regimes. Control of plant pathogens could be improved if inoculum could be detected quickly in the field directly by the grower. Airborne inoculum plays a vital role in the development of epidemics caused by *Botrytis* leaf blight on onion crops (Carisse et al. 2003, 2005). In this work, a linear relationship was found between number of lesions on plants and airborne *Botrytis* conidial concentrations. Airborne conidial concentrations of 25 to 35 conidia m^{-3} of air were associated with 2.5 lesions per leaf. When detection of *Botrytis* inoculum was used as a control criterion under field conditions, it led to a reduction in fungicide usage of 75 and 56% in 2002 and 2003. A similar relationship between spore number and disease intensity has been reported for *Cercospora apii* on celery (Berger 1969). In both these studies, microscopes were used to determine spore numbers from air samples.

One of the objectives of the work reported in this paper was to construct rapid tests for sporangia of *P. destructor*. If rapid tests were suitable for use in the field they could be used potentially to detect sporangia of *P. destructor* using air samplers as a means of forecasting the onset of disease development. To date no strains of onion downy mildew have been reported in the UK. However it is likely that the antibodies used to construct lateral flow tests developed in this study would react equally with all populations of *P. destructor* found in the field. By using techniques outlined in this paper, early detection of *P. destructor* in samples could be made possible. The lateral flow device would, however, need to be tested with portable air samplers in the

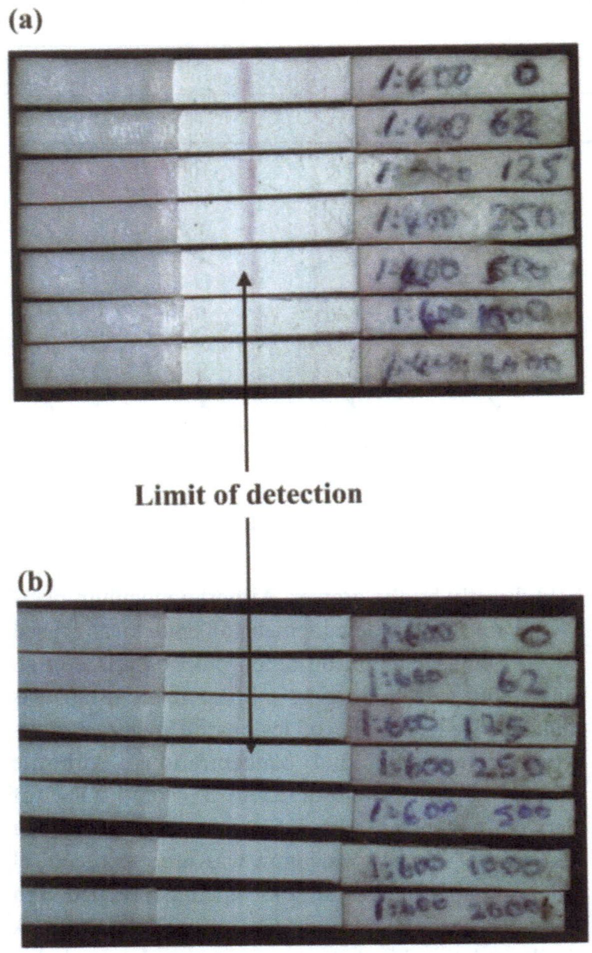

Fig. 3 Development of a competitive lateral flow for onion downy mildew conidia **a** competitive lateral flow test employing detection antibody concentration of 1:400, **b** competitive lateral flow test employing detection antibody concentration of 1:600

field to determine the optimal trapping format for *P. destructor* sporangia. Trapping formats for sporangia would also need to be integrated with numbers of sporangia found above infected crops in the absence of onion downy mildew symptoms on plants. However, by integrating the device with the output from a scanning densitometer, the sensitivity of the device can be improved to enable it to detect the presence of sporangia of *P. destructor* at lower concentrations. Scanning densitometers are becoming more portable and their use might enable low numbers of *P. destructor* sporangia to be detected using a lateral flow device. Symptoms of onion downy mildew within crops are difficult to detect at low levels, especially in large cropping areas. Additionally, the symptomology of the disease on young plants is poorly understood and observed. By detecting the presence of *P. destructor* sporangia, it would be possible to determine action thresholds for onion crops at different stages in their development. The lateral flow device, if used to detect sporangia of *P. destructor* in the field, would require validation in different onion-producing areas.

Using advanced monitoring techniques, the optimal criteria for applying fungicide applications to an onion crop could be investigated. Disease development might also be detected in the absence of visible symptoms. This is a critical point in considerations of disease control, since if early applications of fungicide can be targeted to when *P. destructor* sporangia are present, improved control could be achieved. Rapid diagnostic tests, similar to those reported in this study, exist for identifying *Phytophthora* spp. (Lane et al. 2007). However the test kits reported reacted to a range of *Phytophthora* spp. which are commonly found in soils. These lateral flow devices and those reported by Thornton et al. (2004) were used to identify infected plant tissues in soil samples. However, in studies reported in this paper, lateral flow devices reacted selectively to *P. destructor* sporangia (no infected plant material present). If the device were to be used in conjunction with air samples, it would pose fewer problems in comparison to using lateral flow devices for detecting infected soil or plant tissues.

Detecting *P. destructor* sporangia would be particularly useful early in the season as a method of preventing disease transfer between over-wintered salad onion crops and bulb onions grown as sets or as seeded crops. The use of weekly estimates of inoculum in air samples has also been reported (Kennedy and Wakeham 2006) for other diseases, notably *Pyrenopeziza brassicae* (light leaf spot of horticultural and arable brassicas). Tests which can be

Table 4 Optical density values of the test line at varying *P. destructor* sporangial numbers

Onion downy mildew sporangial number	Optical density value
0	2.2
240	1.8
480	0.7
960	0.3

conducted in the field are necessary if information on air-borne inoculum concentration is to be of more practical value. Results of the trials reported in this paper demonstrate the development of lateral flow devices that can detect plant pathogenic inoculum. Potential exists for linking these estimates of *P. destructor* inoculum to mathematical models describing the environmental factors which affect onion downy mildew sporulation (Gilles et al. 2004). Use of this approach might improve the efficiency of both the inoculum detection system and disease forecasts. However this would require investigation in future work.

References

Battilani, P., Rossi, V., Racca, P., & Giosuè, S. (1996). ONIMIL, a forecaster for primary infection of downy mildew of onion. *European Mediterranean Plant Protection Organisation Bulletin*, *26*, 567–576.

Berger, R. D. (1970). Forecasting Helminthosporium turcicum attacks in Florida sweetcorn. *Phytopathology*, *60*, 1285.

Bugiani, R., Govoni, P., & Cobelli, L. (1998). First large scale application of IPI model for potato late blight prediction in the Po valley. In H. Schepers, & E. Bouma (Eds.), *Proceedings of the Workshop on the European Network for the Development of an Integrated Control Strategy of Potato Late Blight* (pp. 188–199). Carlow, Ireland: PAV.

Carisse, O., McCartney, H. A., Gagnon, J. A., & Brodeur, L. (2005). Quantification of air-borne inoculum as an aid in the management of leaf blight of onion caused by *Botrytis squamosa*. *Plant Disease*, *89*, 726–733 doi:10.1094/PD-89-0726.

Carisse, O., Rolland, D., Lefebvre, A., & Talbot, B. (2003). Using aerobiology data to manage onion blight caused by *Botrytis squamosa*. *Proceedings of the 8th International Congress of Plant Pathology*, *1*, 5.

Clarkson, J. P., Kennedy, R., & Phelps, K. (2000). The effect of temperature and water potential on the production of conidia of sclerotia of *Botrytis squamosa*. *Plant Pathology*, *49*, 119–128 doi:10.1046/j.1365-3059.2000.00417.x.

Cook, H. T. (1932). Studies on the downy mildew of onion and the causal organism, *Peronospora destructor* (Berk.) Caspary. New York agricultural experimental. *Station, Ithaca*, *143*, 1–40.

Cruickshank, I. A. M. (1958). Environment and sporulation of phytopathogenic fungi. IV. The effect of light on the formation of conidia of *Peronospora tabacina* Adam. *Australian Journal of Biological Sciences*, *16*, 87–98.

Day, J. P., Kell, G., & Griffiths, G. W. (2002). Differentiation of *Phytophthora infestans* from other air-borne biological particles by flow cytometry. *Applied and Environmental Microbiology*, *68*, 37–45 doi:10.1128/AEM.68.1.37-45.2002.

Dewey, F. M. (1992). Detection of plant invading fungi by monoclonal antibodies. In J. M. Duncan, & L. Torrance (Eds.), *Techniques for the rapid detection of plant pathogens* (pp. 47–62). Oxford: Blackwell.

Gilles, T., Clarkson, J. P., Phelps, K., & Kennedy, R. (2004). Development of MILIONCAST, an improved model for predicting downy mildew sporulation on onions. *Plant Disease*, *88*, 695–702 doi:10.1094/PDIS.2004.88.7.695.

Hildebrand, P. D., & Sutton, J. C. (1982). Weather variables in relation to an epidemic of onion downy mildew. *Phytopathology*, *72*, 219–224.

Jesperson, G. D., & Sutton, J. C. (1987). Evaluation of a forecaster for downy mildew of onion (*Allium cepa* L.). *Crop Protection (Guildford, Surrey)*, *6*, 95–103 doi:10.1016/0261-2194(87)90106-2.

Kennedy, R. (1998). Bulb onions: Evaluation of fungicides for control of downy mildew (*Peronospora destructor*). Horticultural Development Council, Annual Report (Year 1) for project FV 189, pp. 10.

Kennedy, R., & Wakeham, A. J. (2006). Impact of fungicide resistance on light leaf spot control in vegetable brassicas in Scotland. *Aspects of Applied Biology*, *78*, 51–58.

Kennedy, R., Wakeham, A. J., Byrne, K. G., Meyer, U. M., & Dewey, F. M. (2000). A new method to monitor airborne inoculum of the fungal plant pathogens *Mycosphaerella brassicicola* and *Botrytis cinerea*. *Applied and Environmental Microbiology*, *66*, 297–307 doi:10.1128/AEM.66.7.2996-3003.2000.

Kennedy, R., Wakeham, A. J., & Cullington, J. E. (1999). Production and immunodetection of ascospores of *Mycosphaerella brassicicola*: The ringspot pathogen of vegetable crucifers. *Plant Pathology*, *48*, 297–307 doi:10.1046/j.1365-3059.1999.00341.x.

Lane, C. R., Hobden, E., Walker, L., Barton, V. C., Inman, A. J., Hughes, K. J. D., et al. (2007). Evaluation of a rapid diagnostic field test kit for identification of *Phytophthora* spp. including *P. ramorum* and *P. kernoviae* at the point of inspection. *Plant Pathology*, *56*, 828–835 doi:10.1111/j.1365-3059.2007.01615.x.

Phillon, V. (2003). Timing of sprays against potato late blight, based on daytime average airborne spore concentration. *Proceedings of the 8th International Congress of Plant Pathology*, *1*, 6.

Rondomanski, W. (1967). Studies on the epidemiology of onion downy mildew, *Peronospora destructor* (Berk.) Fries. Technical Report for 1962–67. Research Institute for Vegetable Crops, Skierniewice, Poland, 23.

Thornton, C. R., Groenhof, A. C., Forrest, R., & Lamotte, R. (2004). A one step immunochromatographic lateral flow device specific to *Rhizoctonia solani* and certain related species and its use to detect and quantify *R. solani* in soil. *Phytopathology*, *94*, 280–288 doi:10.1094/PHYTO.2004.94.3.280.

Wakeham, A. J., Kennedy, R., & McCartney, H. A. (2004). Using ELISA to monitor the collection and retention of a range of common airborne spore types in air-samples. *Journal of Aerosol Science*, *35*, 835–850 doi:10.1016/j.jaerosci.2004.01.005.

Williams, R. H., Ward, E., & McCartney, H. A. (2001). Methods of integrating air-sampling and DNA analysis for the detection of airborne fungal spores. *Applied and Environmental Microbiology*, *67*, 2453–2459 doi:10.1128/AEM.67.6.2453-2459.2001.

Yarwood, C. E. (1937). Relation of light to the diurnal periodicity of sporulation of certain downy mildews. *Journal of Agricultural Research*, *54*, 365–373, 8.

Yarwood, C. E. (1943). Onion downy mildew. *Hilgardia*, *14*, 595–691.

Eur J Plant Pathol (2008) 122:157–167
DOI 10.1007/s10658-008-9290-5

Fungicide modes of action and resistance in downy mildews

Ulrich Gisi · Helge Sierotzki

Received: 24 September 2007 / Accepted: 11 February 2008 / Published online: 9 April 2008

Abstract Among oomycetes, *Plasmopara viticola* on grape and *Phytophthora infestans* on potato are agronomically the most important pathogens requiring control measures to avoid crop losses. Several chemical classes of fungicides are available with different properties in systemicity, specificity, duration of activity and risk of resistance. The major site-specific fungicides are the Quinone outside inhibitors (QoIs; e.g. azoxystrobin), phenylamides (e.g. mefenoxam), carboxylic acid amides (CAAs; e.g. dimethomorph, mandipropamid) and cyano-acetamide oximes (cymoxanil). In addition, multi-site fungicides such as mancozeb, folpet, chlorothalonil and copper formulations are important for disease control especially in mixtures or in alternation with site-specific fungicides. QoIs inhibit mitochondrial respiration, phenylamides the polymerization of r-RNA, whereas the mode of action of the other two site-specific classes is unknown but not multi-site. The use of site-specific fungicides has in many cases selected for resistant pathogen populations. QoIs are known to follow maternal, largely monogenic inheritance of resistance; they bear a high resistance risk for many but not all oomycetes. For phenylamides, inheritance of resistance is based on nuclear, probably monogenic mechanisms involving one or two semi-dominant genes; resistance risk is high for all oomycetes. The molecular mechanism of resistance to QoIs is mostly based on the G143A mutation in the cytochrome b gene; for phenylamides it is largely unknown. Resistance risk for CAA fungicides is considered as low to moderate depending on the pathogen species. Resistance to CAAs is controlled by two nuclear, recessive genes; the molecular mechanism is unknown. For QoIs and CAAs, resistance in field populations of *P. viticola* may gradually decline when applications are stopped.

Keywords CAA fungicides · Cytochrome b gene · G143A substitution · Inheritance of resistance · Monogenic resistance · Phenylamide fungicides · *Phytophthora infestans* · *Plasmopara viticola* · QoI fungicides · Recessive resistance · Segregation of resistance

U. Gisi (✉) · H. Sierotzki
Research Biology, Syngenta Crop Protection,
4332 Stein, Switzerland
e-mail: ulrich.gisi@syngenta.com

U. Gisi
Institute of Botany, University of Basel,
4056 Basel, Switzerland

Introduction

In spite of cultural practices and breeding for resistant cultivars, downy mildews are among the most devastating plant diseases. To avoid yield losses, disease control is required mainly by using chemical products. The economic importance of single downy mildew diseases (within Peronosporales and Sclerosporales) can be ranked according to the size of crop area treated with chemicals. In 2006, the sales value of the global fungicide market for the control of diseases caused by oomycetes was about US$ 1.2 billion. By far the biggest segment of downy mildews worldwide is represented by *Plasmopara viticola* on grape (54%), followed by *Pseudoperonospora cubensis* on cucurbits (12%), *Bremia lactucae* on lettuce (8%), *Peronospora* spp. on leek and onion (6%), on tobacco (4%), on field crops such as peas, brassicas and sugar beet (each 3%) and on soybeans and corn (each 2%), *Pseudoperonospora humuli* on hops and *Plasmopara halstedii* on sunflower (each 1%), and the systemic pathogens *Peronosclerospora* and *Sclerophthora* spp. in corn (1%) (Syngenta internal data). Not included in this list are pathogens of the Pythiales, especially *Phytophthora* spp., of which *Phytophthora infestans* on potato and tomato is the most predominant segment (about the same as the *P. viticola* segment). *Phytophthora infestans* is included in this review, because most fungicides controlling downy mildews are also active against *Phytophthora* spp., and the modes of action and mechanisms of resistance have often been studied in *Phytopthora* spp., which are easier to handle under laboratory and glasshouse conditions than the biotrophic downy mildews. Chemical control is the most effective measure currently used to protect crops from downy mildews. Surprisingly, the rather 'old' multi-site fungicides including dithiocarbamates (e.g. mancozeb), phthalimides (folpet), chloronitriles (chlorothalonil) and copper formulations account still for about 50% of the downy mildew fungicide market. Among the single-site fungicides, four chemical classes dominate the market: the Quinone outside inhibitors (QoIs; 'strobilurins', mainly azoxystrobin, famoxadone, fenamidone), the phenylamides (PAs, mainly mefenoxam), the carboxylic acid amides (CAAs; mainly dimethomorph, iprovalicarb, benthiavalicarb, mandipropamid) and the cyanoacetamid-oximes (cymoxanil). Smaller market shares are taken by phosphonates (mainly fosetyl-Al), dinitroanilines (fluazinam), carbamates (propamocarb) and plant defence inducers such as the benzothiadiazoles (BTH, acibenzolar-S-methyl/Bion).

For many decades, multi-site contact fungicides were the only compounds available for the control of downy mildews. Within the last 30 years, chemical control of downy mildews has undergone dramatic changes with the detection and introduction of single-site fungicides such as cymoxanil (1976), fosetyl-Al (1977), phenylamides (1977–1983), propamocarb (1978), CAAs (1992–2005) and QoIs (1996–2000; Gisi 2002). Generally, single site-fungicides act against a very specific step in the metabolism of pathogens and have only few side effects on other processes or non-target organisms. Most single-site fungicides penetrate into the leaf and are protected against wash-off by rain; some are also systemic and move into untreated parts of the plant. In contrast to multi-site fungicides, most single-site inhibitors bear a high intrinsic risk of causing the evolution of resistant pathogen sub-populations. This development is a common phenomenon in agricultural practice and is based on the selection of resistant individuals by the use of fungicides. However, a robust disease control programme will result also in a successful resistance management, because the probability of resistant survivors is smaller if the initial inoculum density is low. Therefore, all agronomic measures reducing disease pressure will also contribute to a reduced fungicide resistance risk. Effective resistance management and successful disease control are supported by the use of effective fungicide rates, alternation and mixtures of fungicides, appropriate spray intervals and an early onset of applications in the disease cycle.

QoI fungicides

QoI fungicides are inhibitors of mitochondrial respiration; they inhibit the electron transport at cytochrome b (complex III) by binding to the Qo site, the ubiquinol oxidizing pocket, which is located at the positive, outerside of mitochondrial membranes. In the Qo pocket, the amino acid glutamic acid (Glu) at position 272 of the ef protein loop is responsible for binding to an oxygen moiety in the toxophore of the fungicide molecule (O–H–N bridge; Gisi et al. 2002). The cytochrome b *(cyt b)* gene is the molecular target for QoI fungicides; it is located in the mitochondrial genome. Long before the introduction of agricultural QoI fungicides, resistance to QoI molecules (e.g. myxothiazol) was described as being based on several mutations in the *cyt b* gene in a range of genera such as yeast (*Saccharomyces*), bacteria (*Rhodobacter*), protozoa (*Paramecium*), sea urchin, algae (*Chlamydomonas*) and mice (Di Rago et al. 1989; Geier et al. 1992; Degli-Esposti et al. 1993; Brasseur et al. 1996). However, it was not known which mutation would appear in plant pathogens.

Sierotzki et al. (2000a, b) detected the G143A substitution (exchange of glycine by alanine at position 143) for the first time in QoI-resistant isolates of *Blumeria* (*Erysiphe*) *graminis* f.sp. *tritici* and *Mycosphaerella fijiensis*. This substitution is based on a single nucleotide polymorphism in the triplet at position 143 from GGT to GCT in the *cyt b* gene. It was described in the following years in resistant isolates of many important plant pathogen species such as *P. viticola, P. cubensis, Venturia inaequalis* and *Mycosphaerella graminicola* (Heaney et al. 2000; Steinfeld et al. 2002; Gisi et al. 2002). It is associated with high levels of resistance (high resistance factors RF) or 'complete' resistance which leads to a complete loss of disease control if QoIs are used as solo products. A second mutation, F129L (exchange of phenylalanine by leucine at position 129) was discovered in resistant isolates of a few pathogen species such as *P. viticola* (Sierotzki et al. 2005), *Pythium aphanidermatum* (Gisi et al. 2002) and *Pyrenophora teres* (Sierotzki et al. 2007), resulting in a 'partial', less pronounced resistance leading to reduced disease control. However, in *P. infestans, Bremia lactucae, Peronospora* spp. and in all rust genera (e.g. *Puccinia, Uromyces, Phakopsora, Hemileia*), no resistant isolates (and no mutations) were detected until now. For rusts, the lack of resistance (based on G143A) has been elucidated recently: an intron is present in *cyt b* between positions 143 and 144 which has to be spliced for correct transcription and translation. The splice site recognition is based on a GGT triplet (De La Salle et al. 1982). If mutated from GGT to GCT, splicing will not occur resulting in a non-functional cytochrome b which is lethal (Grasso et al. 2006).

QoI resistance in *P. viticola* populations in Europe was first detected in 2000 (Heaney et al. 2000) and evolved quickly with a rapid increase of resistant isolates reaching a frequency in 2003 of 70% to 80% in France and about 30% in northwest Spain (Galicia; Fig. 1a,b). In the following years, the frequency of resistance remained more or less stable in the two countries (except for France with a further increase to about 90% in 2006). In the north of Italy and in Switzerland, frequencies reached high levels in certain areas, whereas in Portugal, Germany and Austria, they are still low (Sierotzki et al. 2008). Since the collected leaves always represented a bulk population, the measured frequency of resistance

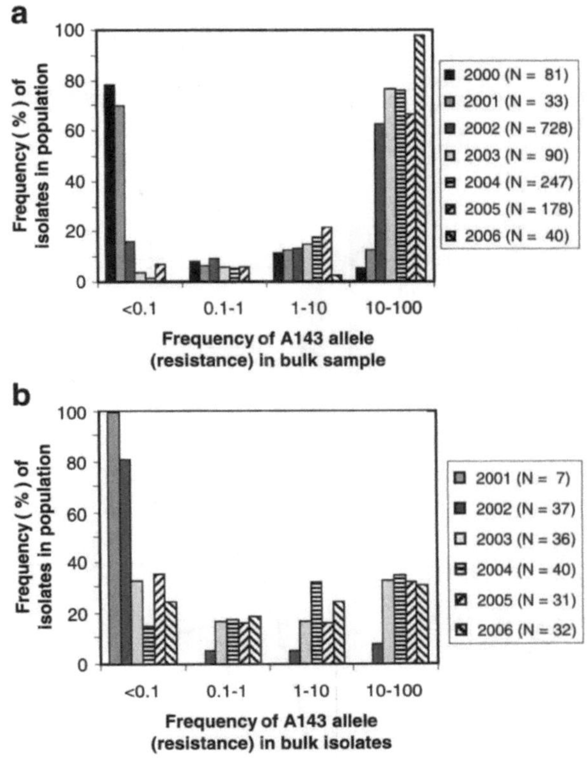

Fig. 1 Frequency of bulk isolates carrying the A143 allele (resistance to QoIs) *in Plasmopara viticola* populations collected in 2000 to 2006 in France (**a**) and in Spain (**b**; after Sierotzki et al. 2008)

(A143 allele in Q-PCR tests) is representative for the entire population at a specific vineyard. However, if single sporangiophore isolates are picked from these samples, they are always either completely sensitive (100% G143 allele) or completely resistant (100% A143 allele); heteroplasmic stages were never detected.

At a specific trial site in Brazil (Holambra), QoI treatments were carried out during several years until November 2000, when they were stopped for three years (2001–2003), re-started again in 2004, stopped in 2005 and started again in 2006. A decline of resistance was observed when QoI applications were stopped and a rapid increase when QoIs were used again (Fig. 2a,b; Sierotzki et al. 2008). This fluctuation of resistance resulting from the use strategy of QoIs might be based on a reduced fitness of QoI-resistant isolates, as it was described also by Heaney et al. (2000) and Genet et al. (2006). Similar declines of

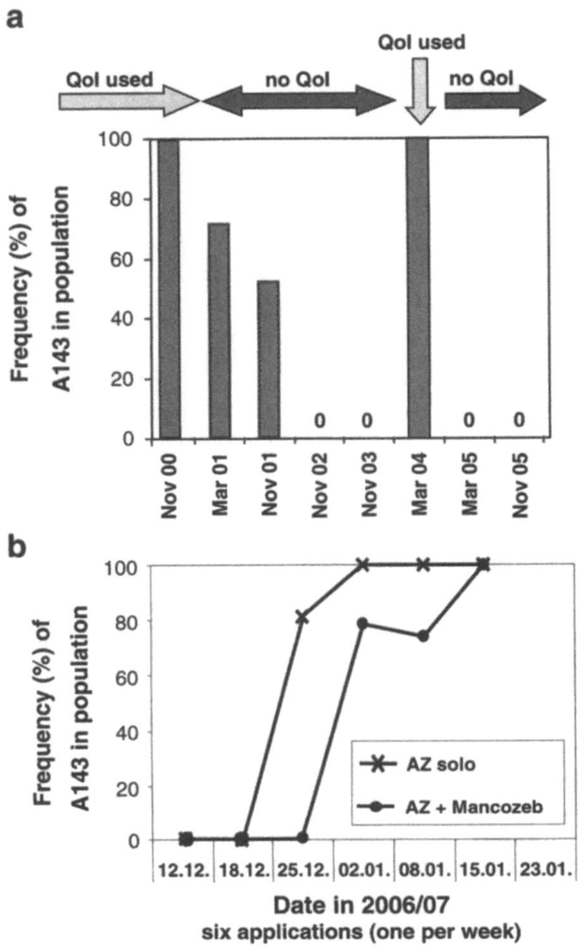

Fig. 2 Change in frequency of QoI-resistant isolates (carrying the A143 allele) in *Plasmopara viticola* populations as a result of different QoI usages at a trial site location (Holambra) in Brazil. **a** season-long applications (six treatments),stop and restart of azoxystrobin applications between 2000 and 2005; **b** six applications (one per week) of azoxystrobin (AZ) solo or in mixture with mancozeb in the 2006/2007 season (after Sierotzki et al. 2008)

resistance after QoI treatments were stopped have also been observed in oospore populations of *P. viticola* in some vineyards in Italy (Toffolatti et al. 2006) and in *P. cubensis* populations in glasshouses in Japan (Ishii 2003 personal communication).

In order to investigate the segregation pattern of QoI resistance, a sensitive P2 and a resistant P1 mating type single sporangiophore isolate (Scherer and Gisi 2006) were crossed by co-inoculating a 1:1 sporangial mixture onto grape leaves. After 14 days of incubation, plenty of oospores were produced in the leaves which were further incubated in dry conditions in the dark for another 8 weeks. Then, the rotted leaves with oospores were ground to powder, mixed with perlite and moistened with water for inducing oospore germination (Gisi et al. 2007b) as originally described for *P. infestans* (Rubin and Cohen 2006). Young grape leaves were incubated on top of the oospore/perlite mixture for 1 to 3 weeks until first sporangiophores appeared which were picked and propagated for producing F1 progeny isolates. Based on the mitochondrial origin of QoI resistance, a maternal inheritance of resistance (0: 1 or 1: 0, depending whether resistance is in the male or female parent) was expected. Surprisingly, a segregation of resistance r:s=8:23 (or ~1:3) was observed; all resistant offspring carried the A143, the sensitive offspring the G143 allele (Fig. 3; Blum and Gisi 2008). Possible reasons for the unexpected segregation might be mitochondrial leakage, irregularities in the mating process or involvement of a recessive nuclear gene with epistatic regulation of the mitochondrial gene. The first possibility can be ruled out because no heteroplasmic stages were detected with Q-PCR; the second hypothesis is based on 'femaleness' of isolates (both parents may have the potential for oospore production) and was described for *P. infestans* (Judelson 1997). Thus, many basic features of the biology of *P. viticola* are still not well understood. Indeed, if inheritance of QoI resistance does not follow a 0:1 (in a single cross) or 1:1 pattern (in populations, assuming P1: P2 ratio is about 1:1), evolution of resistance is not easy to predict, not even under 'controlled' conditions. The G143A mutation was obviously very rare in unselected populations of *P. viticola* prior to the use of QoI fungicides, but was quickly selected through the continuous use of these fungicides. It is an open question as to whether the G143A mutation (A143 allele) might be lost when QoI applications are stopped, as quickly as it appeared through selection. In addition, *P. viticola* is a pathogen with a high rate of sexual recombination and high genetic diversity resulting in many different genotypes every season (Scherer and Gisi 2006) but with a low migration rate resulting in local epidemics. Thus, resistance evolution might be driven mainly by 'local' processes (micro-climate, fungicide use strategies, disease pressure on different varieties).

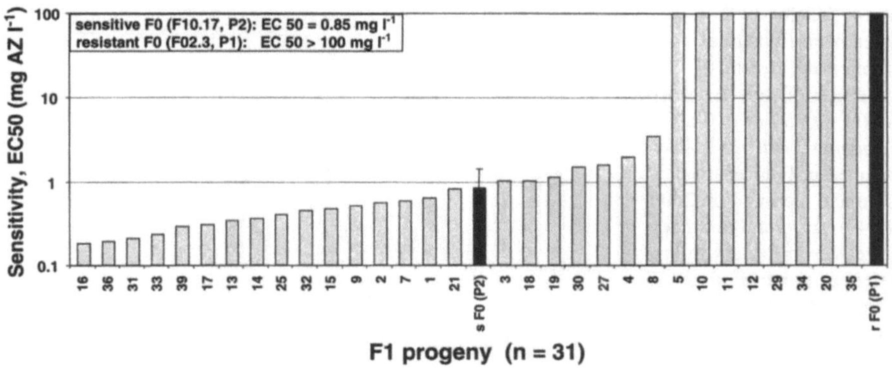

Fig. 3 Sensitivity to azoxystrobin (AZ; EC 50 mg l^{-1}) of F1 progeny isolates derived from a cross between an AZ- sensitive and AZ-resistant parent (F0, black columns) in *Plasmopara viticola* (after Blum and Gisi 2008)

Phenylamide fungicides

Phenylamide fungicides such as mefenoxam (metalaxyl-*M*), metalaxyl and benalaxyl inhibit ribosomal RNA synthesis, specifically RNA polymerization (polymerase complex I; Davidse 1995). The molecular target gene is unknown and no sequence data and no mutations are available, although phenylamides have been in use for more than 25 years. Resistance to phenylamides developed rather quickly after their introduction in many pathogen species of the oomycetes such as *P. viticola, P. cubensis, Peronospora tabacina, B. lactucae* and *P. infestans* (Table 1). Although resistance is widely spread nowadays, the frequency in populations rarely reaches 100%; it fluctuates not only from year to year but also within the season. In French vineyards, resistance in *P. viticola* populations in 1987 to 1998 varied from 15 to 75% (Gisi 2002). In 2004, the proportion of sensitive, intermediate and resistant isolates was 35%, 45%, and 20%, respectively (samples from France; Gisi et al. 2007b) and 15%, 40% and 45%, respectively, in 2006 (samples from France, Italy, Spain, Germany; Fig. 4a). Resistance increased during the season (Fig. 4a) and was higher in mefenoxam-treated than untreated fields (Fig. 4b). The decrease of sensitive and the increase of intermediate and resistant isolates during the season was observed earlier over a period of four consecutive years (Fig. 4c) and seems to follow a seasonal pattern every year. Similar observations were also made for *P. infestans* populations (Gisi and Cohen 1996). Whether sexual recombination,

Table 1 Published cases of resistance to phenylamides in field populations of important pathogens in Peronosporales (first publication, in chronological order)

Pathogen species	Host plant	References
Pseudoperonospora cubensis	cucurbits	Reuveni et al. 1980
Phytophthora infestans	potato	Davidse et al. 1981
Plasmopara viticola	grape	Staub and Sozzi 1981
Peronospora tabacina	tobacco	Bruck et al. 1982
Phytophthora cinnamomi	avocado	Darvas and Becker 1984
Phytophthora parasitica	tobacco	Shew 1985
Bremia lactucae	lettuce	Crute 1987
Pythium spp.	turf, carrot	Sanders and Soika 1988; White et al. 1988
Plasmopara halstedii	sunflower	Albourie et al. 1998
Phytophthora erythroseptica	potato	Lambert and Salas 1994
Peronospora viciae	pea	Falloon et al. 2000
Pythium spp.	ornamentals	Moorman and Kim 2004
Peronospora destructor	onion	Wright 2004

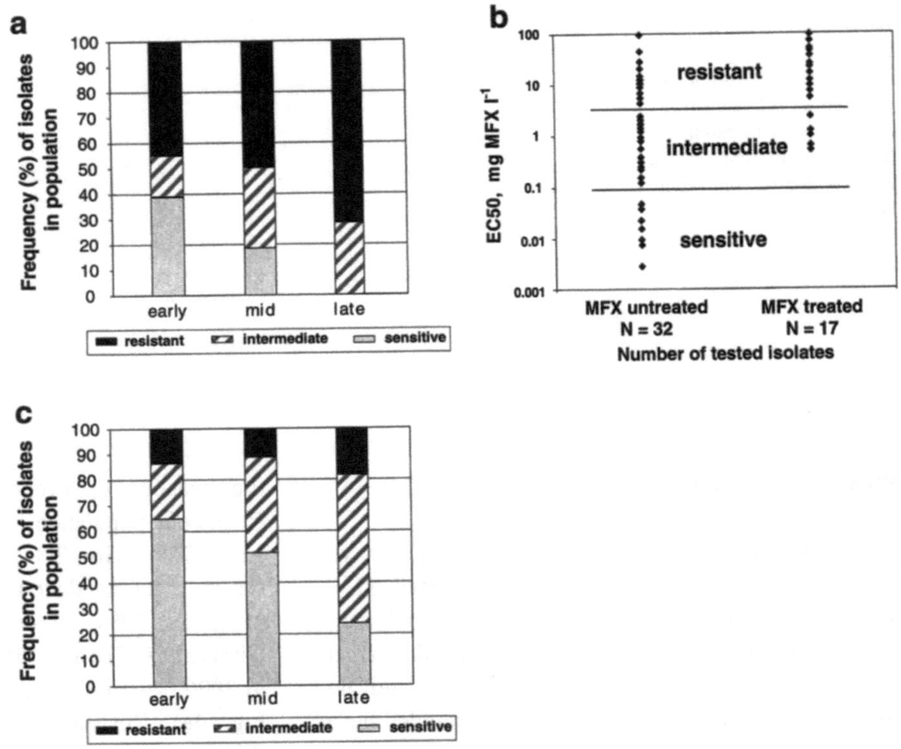

Fig. 4 Sensitivity to mefenoxam (sensitive, intermediate, resistant; proportion of isolates in (**a**) and (**c**), EC 50 in (**b**)) of *Plasmopara viticola* isolates collected in France, Italy, Spain and Germany in 2006 (*N*=49) early, mid and late in the season (**a**) and from mefenoxam-untreated or treated fields (**b**), and in France (Armagnac area) in four consecutive years (1997 to 2000, represented as the average of the four years) early (*N*=132), mid (*N*=85) and late (*N*=110) in the season

pathogenic fitness and/or over-seasoning capacity of resistant isolates play a major role for these dynamics is still to be elucidated.

Inheritance of phenylamide resistance in the F1 progeny was studied in crosses of *P. viticola* (Fig. 5a) and *P. infestans* (Fig. 5b) by co-inoculation of mefenoxam-sensitive and resistant parents according to the method described for QoI fungicides (Fig. 3). The segregation pattern of phenylamide resistance in *P. infestans* corresponded to the expected Mendelian mechanisms for inheritance based on one semi-dominant gene: all F1 progeny isolates were intermediates (Fig. 5b, Knapova et al. 2002). However, in *P. viticola* the segregation pattern was somewhat unexpected (Fig. 5a) with a proportion of r:i:s=21:8:2 (Blum and Gisi 2008). In this experiment, the sensitive parent may be considered as 'truly sensitive' or as intermediate (depending on categorization of EC50 value in the bioassay). The segregation of resistance in the F1, based on one semi-dominant gene in the resistant parent should result either in an entirely intermediate progeny (if the other parent is considered as sensitive) or in r:i=1:1 (if the other parent is considered as intermediate). The observed segregation pattern does not fit to either of the expected segregation. If resistance is based on one dominant gene, the expected segregation would differ even more from the observed pattern. Since for *P. viticola*, no other data on resistance segregation in the F1 are available in the literature, it remains unclear whether the genetic background of mefenoxam resistance is different in *P. viticola* compared to *P. infestans*, or whether sexual recombination and oospore production follows some yet unknown modifications in *P. viticola* as was speculated to occur for the inheritance of QoI resistance (see above). Similarly, a distorted segregation pattern was also described in *P. infestans* when oospores were produced *in planta* as compared to in agar cultures (Van der Lee et al. 2004; Rubin and Cohen 2006). In addition, the segregation pattern of phenylamide resistance in *P. viticola* was described for F2 progeny by Gisi et al. (2007b): a proportion of s:i:r=1:2.7:2 was observed which was considered to be based on one semi-dominant gene affected by minor genes as described for *P. infestans* (Judelson and Roberts 1999). In

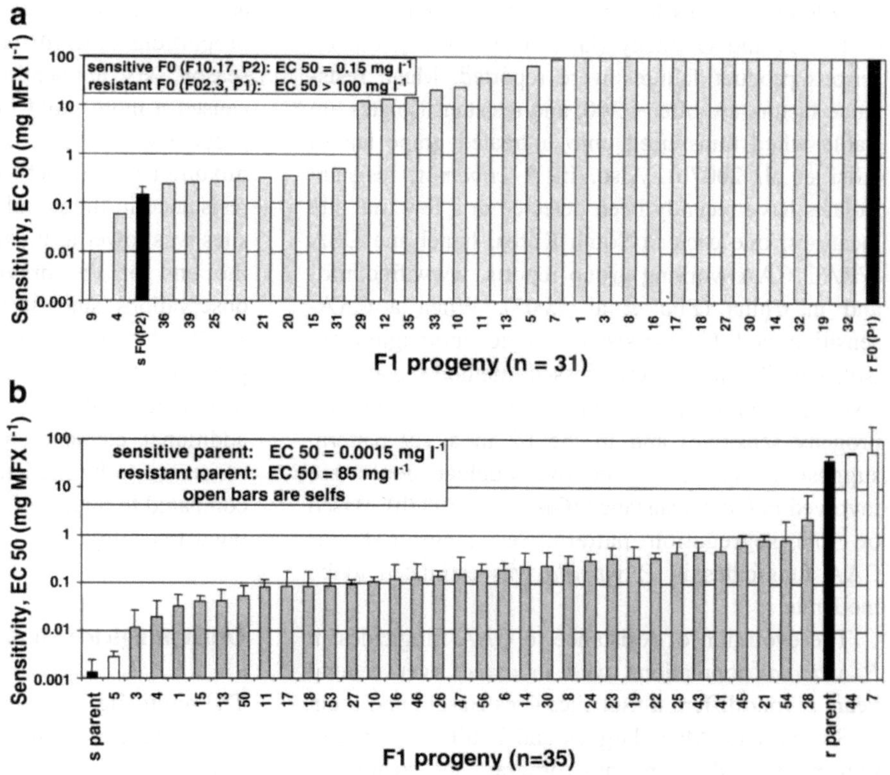

Fig. 5 Sensitivity to mefenoxam (MFX; EC 50 mg l^{-1}) of F1 progeny isolates derived from a cross between an MFX-sensitive and MFX-resistant parent (F0, black columns) of **a** *Plasmopara viticola* (after Blum and Gisi 2008) and of **b** *Phytophthora infestans* (after Knapova et al. 2002)

summary, the risk of resistance for phenylamide fungicides is considered as high, the evolution as fast with a certain stabilization effect over time, and a re-appearance of sensitivity in unselected populations after sexual recombination.

CAA fungicides

The biochemical mode of action of CAA fungicides (including dimethomorph, flumorph, iprovalicarb, benthiavalicarb, mandipropamid) is still speculative; potential targets are phospholipid biosynthesis (Griffiths et al. 2003) and cell wall deposition (Jende et al. 2002; Cohen and Gisi 2007; Gisi et al. 2007a). Although CAAs may interfere with cell membranes, it is doubtful whether the observed effects on phosphocholinetransferase, the last step in the Kennedy pathway of lecithin biosynthesis (Griffiths et al. 2003) can be considered as primary effects caused by CAA fungicides. Similarly, the observed changes in cell wall architecture and deposition during germination of cystospores (Jende et al. 2002) may be a secondary effect, because some of the key enzymes of cell wall biosynthesis such as glucanases and synthases of β-1,3 glucans and cellulose may not be inhibited directly (Mehl and Buchenauer 2002; Gisi et al. 2007a). Most likely, the target site for CAA fungicides may be membrane-bound at the interface between plasmalemma and cell wall (Syngenta internal data). So far, the target gene(s) have not been identified and no mutations conferring resistance are known, although CAA-resistant field isolates of *P. viticola* are available.

In spite of an intensive monitoring programme, no resistant isolates have been detected in *P. infestans* populations (Cohen et al. 2007; FRAC CAA working group reports, www.frac.info), although CAA fungicides (dimethomorph) have been used commercially for more than 10 years. Also, enforced selection experiments and mutagenesis did not yield isolates in *P. infestans* with stable resistance to CAAs (Bagirova et al. 2001; Stein and Kirk 2004; Yuan et al. 2006; Cohen et al. 2007; Rubin et al. 2008). Therefore, resistance risk for CAA fungicides in *P. infestans* can be considered as low. The entire genus *Pythium* is insensitive to CAA fungicides; therefore, there are no resistance issues for CAAs in this genus. However, CAA-resistant isolates have been detected in *P.*

viticola populations for several years in some regions of France and Germany (Gisi et al. 2007b), but no serious product failures were reported. Mean resistance factors are often >300, and resistant isolates are stable when transferred onto untreated grape leaves (Gisi et al. 2007b). Also, in *P. cubensis*, resistant isolates have recently been detected in a few trial site locations, one each in South Korea, Israel and USA (FRAC CAA working group reports, www.frac.info) and in China (Zhu et al. 2007). When a CAA-sensitive and CAA-resistant single sporangiophore isolate of *P. viticola* were crossed (method see above), resistance segregated in the F1 in a 0:1 (entire F1 progeny sensitive) and in the F2 in a 1:9 pattern, suggesting that two recessive nuclear genes are involved in CAA resistance (Gisi et al. 2007b). Based on this segregation pattern, resistance risk for CAA fungicides in *P. viticola* was estimated to be moderate.

The evolution of resistance to CAA fungicides in *P. viticola* populations was followed over several years at four different trial sites (Les Barges and Stein in Switzerland, Marsillargues and Estillac in France) in response to different spray programmes (treatments stopped or applications as solo product or in mixture with multi-site fungicides; Fig. 6). At locations with fully resistant initial populations (resulting from season-long selection with recommended rates during several years), resistance clearly declined (in some cases to zero) within 2 years after mandipropamid applications were stopped. At locations with low proportions of CAA resistance at the beginning of the season, six applications (full recommended rates) caused a more rapid increase of resistance when the product was used solo as compared to when used in mixture (with Folpet). Interestingly, the proportion of resistance in the populations at the Stein and Estillac sites was lower at the beginning of the season than at the end of the previous season (Fig. 6). These observations suggest that CAA-resistant isolates of *P. viticola* may be less fit in the absence of selection pressure than sensitive isolates. The decline of resistance in the absence of CAA treatments is an additional element supporting the classification of resistance to be moderate for CAAs in *P. viticola* (as compared to high for QoI and phenylamide fungicides; www.frac.info).

Other fungicide classes against oomycetes

The multi-site inhibitors such as copper formulations, dithiocarbamates like mancozeb, phthalimides like folpet and chloroisophthalonitriles like chlorothalonil are non-systemic, preventive fungicides forming a protectant barrier at the surface of the plant against pathogens, and inhibit pathogen development prior to penetration into the tissue. They interact mostly unspecifically with many biochemical steps in the pathogen metabolism, such as the formation of

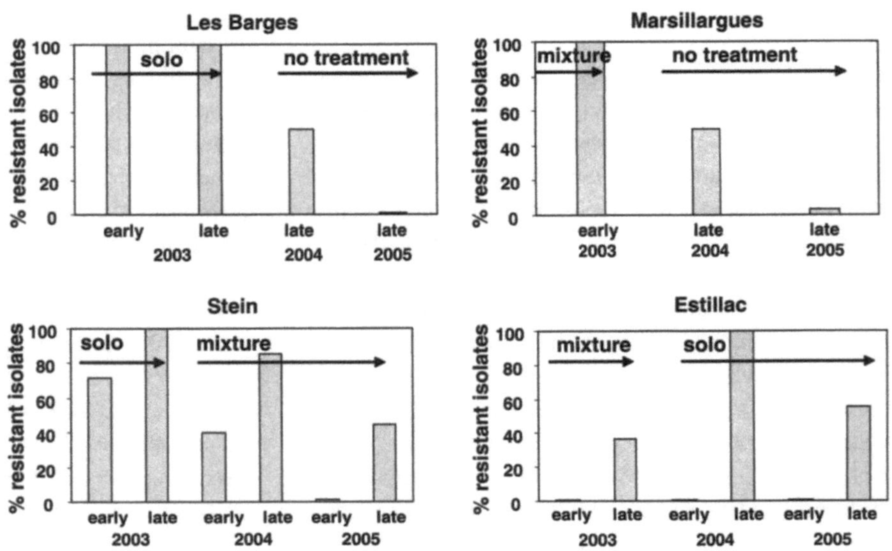

Fig. 6 Evolution of resistance to CAA fungicides at four trial sites during the season and over several years as a result of season-long application of mandipropamid (six to eight applications) as a solo product or in mixtures with multi-site fungicides and after treatments were stopped

complexes with enzymes possessing sulphydryl-groups (Gisi 2002). As a consequence, the enzymes are inactivated leading to a general disruption of metabolism and cell integrity. Based on the multi-site mode of action, resistance to such inhibitors has never developed and is unlikely to evolve. Multi-site fungicides are important elements in spray programmes (about 50% of the total oomycete fungicide market) either as stand alone products or in mixtures with single-site fungicides to improve their activity and to delay resistance evolution.

Based on its short persistence, the systemic cyanoacetamide oxime fungicide cymoxanil is used against oomycetes always in mixtures with multi-site fungicides. The biochemical mode of action of cymoxanil is unknown. Reduced sensitive (or resistant) isolates have been reported in field populations of *P. viticola* (but not of *P. infestans*) in several vineyards of Italy and France (Gullino et al. 1997; Genet and Vincent 1999). Depending on the proportion of resistance in populations, the curative activity of cymoxanil can be significantly reduced. The dinitroaniline fungicide fluazinam is an uncoupler of phosphorylation from electron transport by disrupting the proton gradient; as a consequence, ATP production is blocked (Gisi 2002). Resistance in field populations of oomycetes has never been reported. The carbamate fungicide propamocarb is reported to affect the permeability of cell membranes; as a consequence, leakage of cell components has been observed, but the precise biochemical mode of action is not well understood. Field isolates resistant to propamocarb have been detected in *Pythium* species (Moorman and Kim 2004). Within the chemical class of phosphonates, fosetyl-Al and its breakdown product phosphorous acid (H_3PO_3) are readily taken up by plant tissue and translocated systemically within the phloem (symplastically). Fosetyl-Al may have an indirect effect against downy mildews by stimulating the plant defence reactions; but also a direct antifungal activity has been reported (change in phosphorylated sugar content and cell wall composition; Gisi 2002). Nevertheless, the primary site of action is not known, and resistant isolates in field populations have never been detected. For 'true' plant defence inducers such as acibenzolar-*S*-methyl (Bion), it is generally assumed that pathogens cannot develop resistance very easily, because several mechanisms would have to be overcome (Gisi 2002).

Conclusions

Resistance evolution depends on the specific action of the fungicide (biochemical and molecular mechanisms), the agronomic usage of the fungicide and the pathogen biology. These parameters have been used by FRAC for resistance risk assessment for all chemical classes ('FRAC grid' in Monograph 1, FRAC classification lists, www.frac.info). In *P. viticola*, fungicide resistance has emerged quite quickly (within a few years) after product introduction for almost every chemical class of single-site fungicides (phenylamides, cymoxanil, QoIs, CAAs), whereas in *P. infestans* it emerged only for phenylamides (Kuck and Russell 2006). Therefore, resistance risk assessment has to be carried out carefully and for each single pair of fungicide and pathogen species. In order to delay resistance evolution and ensure robust disease control for as long as possible, all available chemical classes of fungicides such as QoIs, phenylamides, CAAs, cyanoacetamide-oximes, phosphonates and multi-site fungicides should be integrated in a spray programme, either in sequence and/or in mixtures. Fungicide applications should start before or at the onset of the epidemics; the recommended rates and spray intervals have to be strictly followed and adapted to the local disease and weather conditions.

Acknowledgements We acknowledge the excellent experimental contributions of Maya Waldner, Regula Frey, Dominique Edel, Noemy Kraus and Mathias Blum.

References

Albourie, J. M., Tourville, J., & Tourviellie De Labrouhe, D. (1998). Resistance to metalaxyl in isolates of the sunflower pathogen *Plasmopara halstedii*. *European Journal of Plant Pathology, 194*, 235–242.

Bagirova, S. F., Li, A. Z., Dolgova, A. V., Elansky, S. N., Shaw, D. S., & Dyakov, T. (2001). Mutants of *Phytophthora infestans* resistant to dimethomorph fungicide. *Journal of Russian Phytopathological Society, 2*, 19–24.

Blum, M., & Gisi, U. (2008). Inheritance of resistance in *Plasmopara viticola*. In H.W. Dehne, U. Gisi, K.H. Kuck, P.E. Russell & H. Lyr (Eds.), *Modern Fungicides and Antifungal Compounds V.* BCPC, Alton (in press).

Brasseur, G., Saribas, A. S., & Daldal, F. (1996). A compilation of mutations located in the cytochrome b subunit of the bacterial and mitochondrial bc1 complex. *Biochimica and Biophysica Acta, 1275*, 61–69.

Bruck, R., Gooding Jr., G. V., & Main, C. E. (1982). Evidence for resistance to metalaxyl in isolates of *Peronospora hyoscyami*. *Plant Disease, 66*, 44–45.

Cohen, Y., & Gisi, U. (2007). Different activity of carboxylic acid amide fungicides against various development stages of *Phytophthora infestans*. *Phytopathology, 97*, 1274–1283.

Cohen, Y., Rubin, E., Hadad, T., Gotlieb, D., Sierotzki, H., & Gisi, U. (2007). Sensitivity of *Phytophthora infestans* to mandipropamid and the effect of enforced selection pressure in the field. *Plant Pathology, 56*, 836–842.

Crute, I. R. (1987). The occurrence, characteristics, distribution, genetics and control of a metalaxy-resistant genotype of *Bremia lactucae* in the United Kingdom. *Plant Disease, 71*, 763–767.

Darvas, J. M., & Becker, O. (1984). Failure to control *Phytophthora cinnamomi* and *Pythium splendens* with metalaxyl after its prolonged use. *Citrus and Subtropical Fruit Journal, 603*, 9–11.

Davidse, L. C. (1995). Phenylamide fungicides: biochemical action and resistance. In H. Lyr (Ed.), *Modern Selective Fungicides* (2nd ed., pp. 347–354). Jena: Gustav Fischer.

Davidse, L. C., Looijeu, D., Turkenstein, L. J., & Van Der Wal, D. (1981). Occurrence of metalaxyl resistant strains of potato blight in Dutch potato fields. *The Netherlands Journal of Plant Pathology, 87*, 65–68.

Degli-Esposti, M., De Vries, S., Crimi, M., Ghelli, A., Patarnello, T., & Meyer, A. (1993). Mitochondrial cytochrome b: evolution and structure of the protein. *Biochimica and Biophysica Acta, 1143*, 243–271.

De La Salle, H., Jacq, C., & Slonimski, P. P. (1982). Critical sequences with mitochondrial introns: pleiotropic mRNA maturase and *cis*-dominant signals of the box intron controlling reductase and oxidase. *Cell, 28*, 721–732.

Di Rago, J. P., Coppée, J. Y., & Colson, A. M. (1989). Molecular basis for resistance to myxothiazol, mucidin (strobilurin A) and stigmatellin. *Journal of Biological Chemistry, 264*, 14543–14548.

Falloon, R. E., Follas, G. B., Butler, R. C., & Goulder, D. S. (2000). Resistance in *Peronospora viciae* to phenylamide fungicides: reduced efficacy of seed treatments of pea (*Pisum sativum*) and assessment of alternatives. *Crop Protection, 19*, 313–325.

Geier, B. M., Schägger, H., Brandt, U., Colson, A. M., & Von Jagow, G. (1992). Point mutation in cytochrome b of yeast ubihydroquinone-cytochrome-c-oxidoreductase causing myxothiazol resistance and facilitated dissociation of the iron-sulfur subunit. *European Journal of Biochemistry, 208*, 375–380.

Genet, J. L., Jaworska, G., & Deparis, F. (2006). Effect of dose rate and mixtures of fungicides on selection for QoI resistance in populations of *Plasmopara viticola*. *Pest Management Science, 62*, 188–194.

Genet, J. L., & Vincent, O. (1999). Sensitivity of European *Plasmopara viticola* populations to cymoxanil. *Pesticide Science, 55*, 129–136.

Gisi, U. (2002). Chemical control of downy mildews. In P. T. N. Spencer-Phillips, U. Gisi, & A. Lebeda (Eds.), *Advances in downy mildew research* (pp. 119–159). Dordrecht: Kluwer Academic.

Gisi, U., & Cohen, Y. (1996). Resistance to phenylamide fungicides: A case study with *Phytophthora infestans* involving mating type and race structure. *Annual Review of Phytopathology, 34*, 549–572.

Gisi, U., Lamberth, C., Mehl, A., & Seitz, T. (2007a). Carboxylic Acid Amide (CAA) Fungicides. In W. Krämer, & U. Schirmer (Eds.) *Modern crop protection compounds* (pp. 651–674). Weinheim: Wiley.

Gisi, U., Sierotzki, H., Cook, A., & McCaffery, A. (2002). Mechanisms influencing the evolution of resistance to Qo inhibitor fungicides. *Pest Management Science, 58*, 859–867.

Gisi, U., Waldner, M., Kraus, N., Dubuis, P. H., & Sierotzki, H. (2007b). Inheritance of resistance to carboxylic acid amide (CAA) fungicides in *Plasmopara viticola*. *Plant Pathology, 56*, 199–208.

Griffiths, R. G., Dancer, J., O'Neill, E., & Harwood, J. L. (2003). A mandelamide pesticide alters lipid metabolism in *Phytophthora infestans*. *New Phytologist, 158*, 345–353.

Grasso, V., Palermo, S., Sierotzki, H., Garibaldi, A., & Gisi, U. (2006). Cytochrome b gene structure and consequences for resistance to Qo inhibitor fungicides in plant pathogens. *Pest Management Science, 62*, 465–472.

Gullino, M. L., Mescalchin, E., & Mezzalama, M. (1997). Sensitivity to cymoxanil in populations of *Plasmopara viticola* in northern Italy. *Plant Pathology, 46*, 729–736.

Heaney, S. P., Hall, A. A., Davies, S. A., & Olaya, G. (2000). Resistance to fungicides in the QoI-STAR cross-resistance group: current perspectives. *Brighton Crop Protection Conference, 2*, 755–762.

Jende, G., Steiner, U., & Dehne, H. W. (2002). Microscopical characterization of fungicidal effects on infection structures and cell wall formation of *Phytophthora infestans*. In H. W. Dehne, U. Gisi, K. H. Kuck, P. E. Russell, & H. Lyr (Eds.), *Modern Fungicides and Antifungal Compounds III* (pp. 83–90). Bonn: AgroConcept.

Judelson, H. S. (1997). Expression and inheritance of sexual preference and selfing potential in *Phytophthora infestans*. *Fungal Genetics and Biology, 21*, 188–197.

Judelson, H. S., & Roberts, S. (1999). Multiple loci determining insensitivity to phenylamide fungicides in *Phytophthora infestans*. *Phytopathology, 89*, 754–760.

Knapova, G., Schlenzig, A., & Gisi, U. (2002). Crosses between isolates of *Phytophthora infestans* from potato and tomato and characterization of F1 and F2 progeny for phenotypic and molecular markers. *Plant Pathology, 51*, 698–709.

Kuck, K. H., & Russell, P. E. (2006). FRAC: Combined resistance risk assessment. In R. J. Bryson, F. J. Burnett, V. Foster, B. A. Fraaije & R. Kennedy (Eds.), *Fungicide Resistance: are we winning the battle but losing the war?*, AAB Conference Proceedings, Aspects of Applied Biology 78, 3–10, Edinburgh, UK.

Lambert, D. H., & Salas, B. (1994). Metalaxyl insensitivity of *Phytophthora erythroseptica* isolates causing pink rot of potato in Maine. *Plant Disease, 78*, 1010.

Mehl, A., & Buchenauer, H. (2002). Investigations of the biochemical mode of action of iprovalicarb. In H. W. Dehne, U. Gisi, K. H. Kuck, P. E. Russell, & H. Lyr (Eds.), *Modern Fungicides and Antifungal Compounds III* (pp. 75–82). Bonn: AgroConcept.

Moorman, G. W., & Kim, S. H. (2004). Species of *Pythium* from greenhouses in Pennsylvania exhibit resistance to propamocarb and mefenoxam. *Plant Disease, 88*, 630–632.

Reuveni, M., Eyal, H., & Cohen, Y. (1980). Development of resistance to metalaxyl in *Pseudoperonospora cubensis*. *Plant Disease*, *64*, 1108–1109.

Rubin, E., & Cohen, Y. (2006). An improved method for infecting tomato leaves or seedlings with oospores of *Phytophthora infestans* used to investigate F1 progeny. *Plant Disease*, *90*, 741–749.

Rubin, E., Gotlieb, D., Gisi, U., & Cohen, Y (2008). Mutagenesis of *Phytophthora infestans* for resistance against carboxylic acid amide (CAA) and phenylamide fungicides. *Plant Disease*, *92* (in press).

Sanders, P. L., & Soika, M. D. (1988). Metalaxyl resistance frequency in overwintering populations of *Pythium aphanidermatum* from metalaxyl control failure sites. *Phytopathology*, *78*, 1510.

Scherer, E., & Gisi, U. (2006). Characterization of genotype and mating type in European isolates of *Plasmopara viticola*. *Journal of Phytopathology*, *154*, 489–495.

Shew, H. D. (1985). Response of *Phytophthora parasitica* var. *nicotianae* to metalaxyl exposure. *Plant Disease*, *69*, 559–562.

Sierotzki, H., Frey, R., Wullschleger, J., Palermo, S., Karlin, S., Godwin, J., et al. (2007). Cytochrome b gene sequence and structure of *Pyrenophora teres* and *P. tritici-repentis* and implications for QoI resistance. *Pest Management Science*, *63*, 225–233.

Sierotzki, H., Kraus, N., Assemat, P., Stanger, C., Cleere, S., Windass, J., et al. (2005). Evolution of resistance to QoI fungicides in *Plasmopara viticola* populations in Europe. In H. W. Dehne, U. Gisi, K. H. Kuck, P. E. Russell, & H. Lyr (Eds.), *Modern Fungicides and Antifungal Compounds IV* (pp. 73–80). Alton: BCPC.

Sierotzki, H., Kraus, N., Fernandes, N., & Gisi, U. (2008). Dynamics of QoI resistance in *Plasmopara viticola*. In H. W. Dehne, U. Gisi, K. H. Kuck, P. E. Russell & H. Lyr (Eds.), *Modern Fungicides and Antifungal Compounds V*, BCPC, Alton, (in press).

Sierotzki, H., Parisi, S., Steinfeld, U., Tenzer, I., Poirey, S., & Gisi, U. (2000a). Mode of resistance to respiration inhibitors at the cytochrome bc1 enzyme complex of *Mycosphaerella fijiensis* field isolates. *Pest Management Science*, *56*, 833–841.

Sierotzki, H., Wullschleger, J., & Gisi, U. (2000b). Point-mutation in cytochrome b gene conferring resistance to strobilurin fungicides in *Erysiphe graminis* f.sp. *tritici* field isolates. *Pesticide Biochemistry and Physiology*, *68*, 107–112.

Staub, T., & Sozzi, D. (1981). Résistance au métalaxyl en pratique et les conséquences pour son utilisation. *Phytiatrie–Phytopharmacie*, *30*, 283–291.

Stein, J. M., & Kirk, W. W. (2004). The generation and quantification of resistance to dimethomorph in *Phytophthora infestans*. *Plant Disease*, *88*, 930–934.

Steinfeld, U., Sierotzki, H., Parisi, S., & Gisi, U. (2002). Comparison of resistance michanisms to strobilurin fungicides in *Venturia inaequalis*. In H. W. Dehne, U. Gisi, K. H. Kuck, P. E. Russell, & H. Lyr (Eds.), *Modern Fungicides and Antifungal Compounds III* (pp. 167–176). Bonn: AgroConcept.

Toffolatti, S., Serrati, L., Sierotzki, H., Gisi, U., & Vercesi, A. (2006). Assessment of QoI resistance in *Plasmopara viticola* oospores. *Pest Management Science*, *63*, 194–201.

Van der Lee, T. H., Testa, A., Robold, A., van't Klooster, J., & Govers, F. (2004). High-density genetic linkage maps of *Phytophthora infestans* reveal trisomic progeny and chromosomal rearrangements. *Genetics*, *167*, 1643–1661.

White, J. G., Stanghellini, M. E., & Ayoubi, L. M. (1988). Variation in the sensitivity to metalaxyl of *Pythium* spp. isolated from carrot and other sources. *Annals of Applied Biology*, *113*, 269–278.

Wright, P. J. (2004). Resistance of *Peronospora destructor* (onion downy mildew) to metalaxyl fungicides. *Proceedings of New Zealand Plant Protection Conference*, *57*, 347.

Yuan, S. K., Liu, X. L., Si, N. G., Dong, J., Gu, B. G., & Jiang, H. (2006). Sensitivity of *Phytophthora infestans* to flumorph: In vitro determination of baseline sensitivity and the risk of resistance. *Plant Pathology*, *55*, 258–263.

Zhu, S. S., Liu, X. L., Wang, Y., Wu, X. H., Liu, P. F., Li, J. Q., et al. (2007). Resistance of *Pseudoperonospora cubensis* to flumorph on cucumber in plastic houses. *Plant Pathology*, *56*, 967–975.

Eur J Plant Pathol (2008) 122:169–183
DOI 10.1007/s10658-008-9327-9

Activity of carboxylic acid amide (CAA) fungicides against *Bremia lactucae*

Yigal Cohen · Avia (Evgenia) Rubin · Dror Gotlieb

Received: 31 August 2007 / Accepted: 5 May 2008

Abstract Four carboxylic acid amide (CAA) fungicides, mandipropamid (MPD), dimethomorph (DMM), benthiavalicarb (BENT) and iprovalicarb (IPRO) were examined for their effects on various developmental stages of *Bremia lactucae,* the causal agent of downy mildew in lettuce, in vitro and in planta. Spore germination in vitro or on leaf surfaces was inhibited by all CAA fungicides (technical or formulated). MPD was more effective in suppressing germination than DMM or BENT, whereas IPRO was least effective. CAA induced no disruption of F-actin microfilament organisation in germinating spores of *B. lactucae*. CAA applied to germinating spores in vitro prevented further extension of the germ tubes. When applied to germinated spores on the leaf surface they prevented penetration. Preventive application of CAA to intact plants inhibited infection. MPD was more effective in suppressing infection than DMM or BENT, whereas IPRO was least effective. Curative application was effective at ≤3 h post-inoculation (hpi) but not at ≥18 hpi. CAA (except IPRO) applied to upper leaf surfaces inhibited spore germination on the lower surface and hence reduced infection. CAA suppressed sporulation of *B. lactucae* on floating leaf discs and when sprayed onto infected plants two days before onset of sporulation. BENT and DMM were more effective in suppressing sporulation than MPD or IPRO. Epidemics of downy mildew in shade-house grown lettuce were suppressed by CAA. A single spray applied to five-leaf plants before transplanting controlled the disease for 50 days. The results suggest that CAA are effective inhibitors of spore germination and therefore should be used as preventive agents against downy mildew of lettuce caused by *B. lactucae.*

Y. Cohen (✉) · A. (E.) Rubin · D. Gotlieb
The Mina and Everard Goodman Faculty of Life Sciences, Bar-Ilan University,
Ramat Gan 52900, Israel
e-mail: ycohen@mail.biu.ac.il

Keywords Benthiavalicarb · Cinnamic acid amides · Dimethmorph · Disease control · Downy mildew · Iprovalicarb · Lettuce · Mandelic acid amides · Mandipropamid · Oomycete · Valinamid carbamates

Introduction

Mandipropamid (MPD) is a new mandelic acid amide fungicide (Lamberth et al. 2006) which together with dimethomorph (DMM) and flumorph (cinnamic acid amides), iprovalicarb (IPRO) and benthiavalicarb (BENT; valinamid carbamates), belongs to the carboxyl acid amide (CAA) fungicides (Anon. 2006). CAA fungicides are effective against oomycete foliar plant pathogens.

Field studies indicate that MPD is highly effective against late blight in potato and tomato, downy mildew in grapes and several downy mildews in vegetable crops (Harp et al. 2006; Huggenberger et al. 2005;

Huggenberger and Kuhn 2006). A recent report (Harp et al. 2007) has shown effective control of downy mildew in lettuce with Revus (a.i. MPD). The fungicide was shown (Hermann et al. 2005) to quickly bind to the wax layer of the leaf surface thus providing strong rain-fastness and long-lasting efficacy. A small amount of MPD is taken up by the leaf tissue, providing curative and translaminar activities against disease. All four CAA fungicides belong to one cross-resistance group, as field isolates of *Plasmopara viticola* exhibit resistance to all (Gisi et al. 2006).

A recent study (Cohen and Gisi 2007) provided a comprehensive analysis of the effects of three CAA fungicides, MPD, DMM and IPRO, on all stages in the asexual life-cycle of *Phytophthora infestans*. The most sensitive stage to CAA was shown to be germination of cystospores and sporangia (direct germination) and the most active CAA was MPD. Nano-molar doses of MPD were sufficient to block spore germination in vitro and in vivo. Mycelium growth and sporulation were less sensitive to CAA. Of the three CAA fungicides tested, MPD was most effective in suppressing late blight epidemics in shade-house grown potatoes (Cohen and Gisi 2007).

Benthiavalicarb-isopropyl, another CAA fungicide, was not studied in the previous research (Cohen and Gisi 2007). It was reported (Miyake et al. 2003) to be strongly inhibitory to germination of sporangia and cystospores, mycelial growth and sporulation of various oomycetes. The compound has not only strong preventive activity, but also curative and penetrant activity, with excellent residual effects and rainfastness.

Several attempts were made to reveal the mode of action of CAA. Morphological studies (Albert et al. 1988, 1991; Bagirova et al. 2001; Cohen et al. 1995; Cohen and Gisi 1996; Dereviagina et al. 1999; Hermann et al. 2005; Huggenberger et al. 2005; Jende et al. 1999, 2001; Matheron and Porchas 2000; Miyake et al. 2003; Reuveni 2003; Stenzel et al. 1998) indicated that DMM, IPRO and BENT, as well as the experimental CAA XR-539 (Young et al. 2005), inhibit cell wall deposition/assembly in cystospores of *P. infestans*. Biochemical studies (Griffiths et al. 2003) with the mandelamide SX 623509 in mycelium of *P. infestans* suggested alterations in phospholipid biosynthesis, with an inhibition of phosphatidylcholine (lecithin) biosynthesis as a main target. Unpublished data (Cohen and Gisi) indeed indicate that lecithin (phosphatidylcholine) compromised the inhibitory activity of CAA on cystospore germination. Recently, Zhu et al. (2007) compared a flumorph-sensitive and a flumorph-resistant *Phytophthora melonis* and suggested that the primary site of action of flumorph is the disruption of F-actin organisation.

The aim of the present study was to examine the effects of CAA on the development of *Bremia lactucae*, the causal agent of downy mildew in lettuce, in vitro and in planta. Here, four CAA fungicides were tested, including benthiavalicarb (BENT) which was not tested with *P. infestans*.

Materials and methods

Fungicides

Four CAA (carboxylic acid amide) fungicides were used: mandipropamid (MPD; Syngenta, mw=412), dimethomorph (DMM; BASF, mw=266), iprovalicarb (IPRO; Bayer, mw=320) and benthiavalicarb (BENT; Kumiai Chemicals, mw=339). Technical grade fungicides were dissolved in DMSO (10 mg ml^{-1}) and diluted in double distilled water (DDW) to the desired concentrations. Formulated fungicides used were: mandipropamid 250SC; dimethomorph (Forum) 50WP and iprovalicarb 50WG. Benthiavalicarb (10SC) in Agsolex-8 (*N*-octylpyrrolidone) was a gift from Makhteshim, Beer Sheba, Israel. All concentrations are represented in units of active ingredient (a.i.).

Pathogen

All studies, unless stated otherwise used isolate ISR-60 of *B. lactucae* Regel (Sharaf et al. 2007) carrying 13 virulence factors (0, 1, 2, 3, 4, 5/8, 6, 7, 10, 11, 13, 15, 16, 17) obtained from A. Beharav, The Institute of Evolution, Haifa University, Israel. Some studies used isolates BL-18, BL-21, BL-24 and BL-25 (a gift from A. Lebeda, Olomouc University, Czech Republic). Isolates were maintained by repeated inoculation of lettuce cotyledons in Petri dishes in a growth chamber (15°C, 12 h light/day).

Plants

The susceptible lettuce (*Lactuca sativa*) cv. Noga (cup type; Hazera Genetics, Mivhor, Israel) was used. For growth chamber studies, plants were grown from seeds in 175 ml pots containing 40 g peat/vermiculite mixture (1/1, *v/v*) to give 20 plants per pot. Plants grown in the greenhouse (18–32°C) were used 1 week after sowing, when two cotyledon leaves had developed. In other experiments, plants were grown either in Speedling (Hishtil, Petah-Tiqwa, Israel) trays (25 ml cells), one plant per cell and used when they had five to six true leaves, or in 0.5-l pots and used at the five to six true leaf stage.

Application of compounds

Compounds were diluted in water to a series of concentrations and applied to lettuce plants by spraying onto the upper leaf surfaces to initial run-off. Depending on the experiment, a compound was applied either before or after inoculation. In other experiments, compounds were applied as 10 µl droplets to detached cotyledon leaves, true leaves, or leaf discs. In the field, compounds were applied by spraying to initial run-off with the aid of a backpack sprayer.

Inoculation

Spores of *B. lactucae* were collected from freshly sporulating lettuce leaves into ice-cold DDW, their density adjusted to 1×10^4 spores per milliliter and then sprayed onto the upper leaf surfaces of the test plants to initial run-off with the aid of a glass atomiser. Plants were thereafter placed in a dew chamber (18°C, in the dark) for 20 h and then transferred to a growth chamber at 20°C (12 h light/day, 100 µE m^{-2} s^{-1}). At 5 days post-inoculation (dpi) plants were placed in Perspex boxes (100% relative humidity) for two days to induce sporulation of the pathogen.

Disease assessment

The number of sporulating plants was determined at the cotyledon stage with the aid of a magnifying lens (×10) at 7 dpi, unless stated otherwise.

Germination of spores in vitro

Spores were mixed (1:1) with CAA and applied to depressions in glass slides (20 µl per depression, *n*=3). Slides were kept in moist Petri dishes at 13°C for 20 h in the dark. Germination of 100 spores was recorded per depression with the aid of a dissecting microscope at ×160.

F-Actin distribution

Spores were mixed (1:1) with water (as control), 0.01 mg l^{-1} MPD or 0.1 mg l^{-1} DMM and applied to depressions in glass slides (20 µl per depression, *n*=9). Slides were kept in moist Petri dishes at 13°C for 18 h in the dark to allow for spore germination. Spores were then fixed for 10 min with formaldehyde (3.7%, 10 µl per depression), collected from the depressions, washed twice, and treated for 10 min with Alexa Fluor® 488 phalloidin (Invitrogen, Molecular Probes, Eugene, OR, USA). F-Actin distribution was visualised with a confocal microscope (Karl Zeiss, LSM-510 META), and images captured with a Zeiss AxioCam camera and analysed using Zeiss AxioVision software.

Germination of spores on leaves

Tests were performed with either cotyledon leaves or the first formed true leaf detached from 7- or 10 day-old plants, respectively. Cotyledons or leaves (*n*=4) were placed on moistened filter paper in 9 cm Petri dishes and each inoculated on the upper surface with a 20 µl droplet of spore suspension (containing 500 spores) mixed 1:1 with CAA. Dishes were incubated at 13–15°C for 20 h in the dark. A 10 µl droplet of 0.02% calcofluor (Polyscience Inc., Warrington, PA, USA) was added to each cotyledon and leaf, and germination was visualised with the aid of an UV epifluorescence microscope (Olympus AX70) equipped with an excitation filter of 390–420 nm and an emission filter of 425–450 nm. *Bremia lactucae* structures fluoresced blue; 100 spores were recorded per sample.

Temporary exposure of spores to CAA

The effect of temporary exposure of spores to CAA was assessed by incubating spores in 0, 0.5 or 1 mg l^{-1} CAA on ice for 1 h, then washing with water and allowing them to either germinate for 6 h at 13°C in the dark in vitro or to infect lettuce leaves.

Pathogen development in planta

To detect pathogen structures inside the tissue, leaves were cleared by boiling in 90% ethanol for

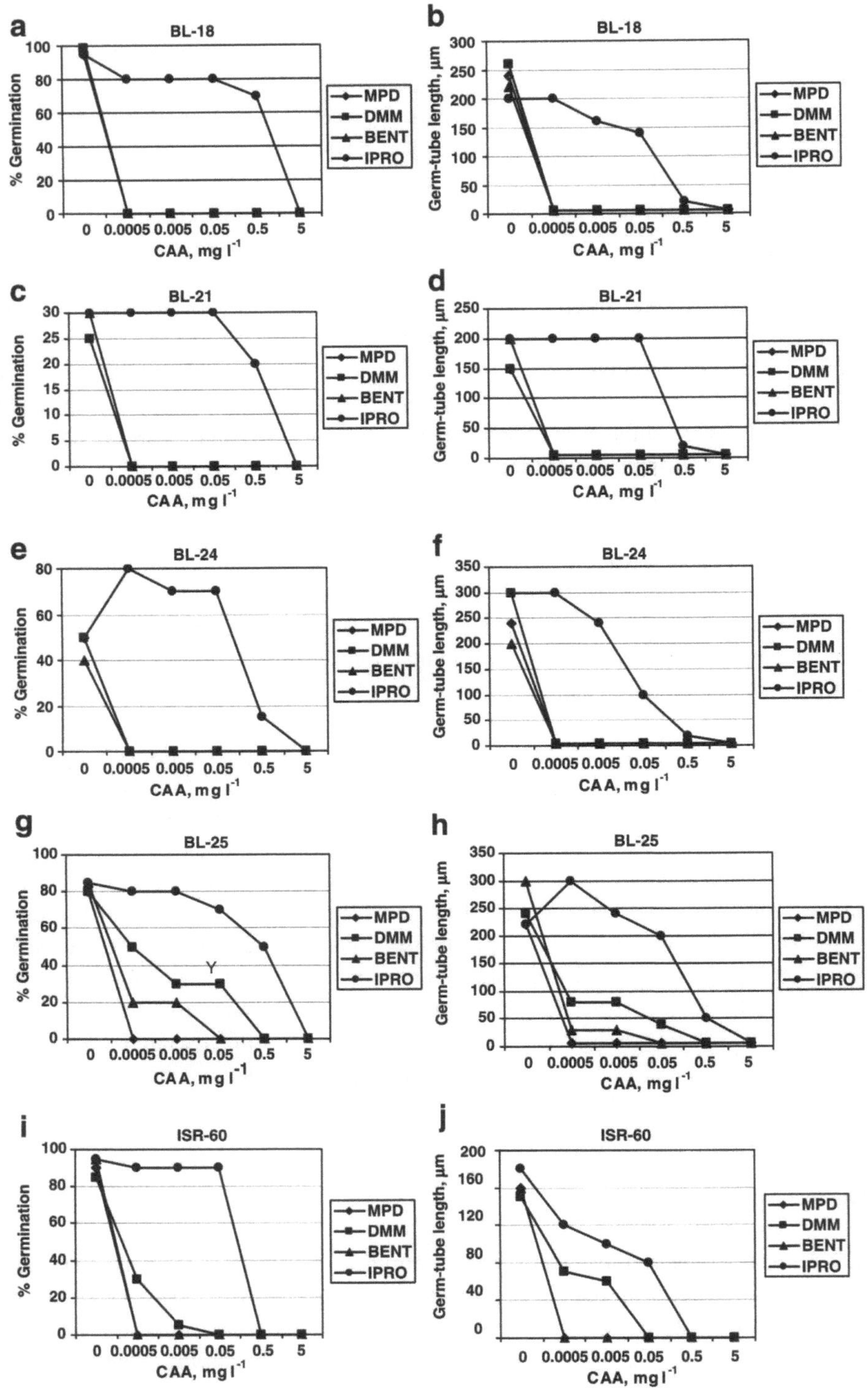

Fig. 1 a–j Percentage spore germination (means and standard deviation of the mean, n=100; germ-tube length, >5 µm) and germ-tube length of four European (BL-18 to BL-25) and one Israeli (ISR-60) isolates of *B. lactucae* in vitro in the presence of technical CAA compounds (*MPD* mandipropamid, *DMM* dimethomorph, *IPRO* iprovalicarb, *BENT* benthiavalicarb). **k** Epifluorescence micrographs of calcofluor-stained *B. lactucae* spore germination in vitro in the presence of 0.0005 mg l^{-1} of four technical CAA compounds. Scale bar=100 µm. **l** Germination of *B. lactucae* spores in the presence of water and CAA: Only very small germ-tubes are produced in the latter (*arrows*). Scale bar=20 µm

20 min, and then placed in 0.05% aniline blue in 70 mM potassium phosphate buffer (pH 8.9) at 4°C for 24 h. Stained leaves were placed on glass slides, a drop of 0.02% calcofluor applied to the surface and then examined by epifluorescence microscopy, as above (Cohen et al. 1989, 1990). Spores and germ tubes on the leaf surface fluoresced blue; β-1–3 glucans in walls of intercellular hyphae

Fig. 1 Continued

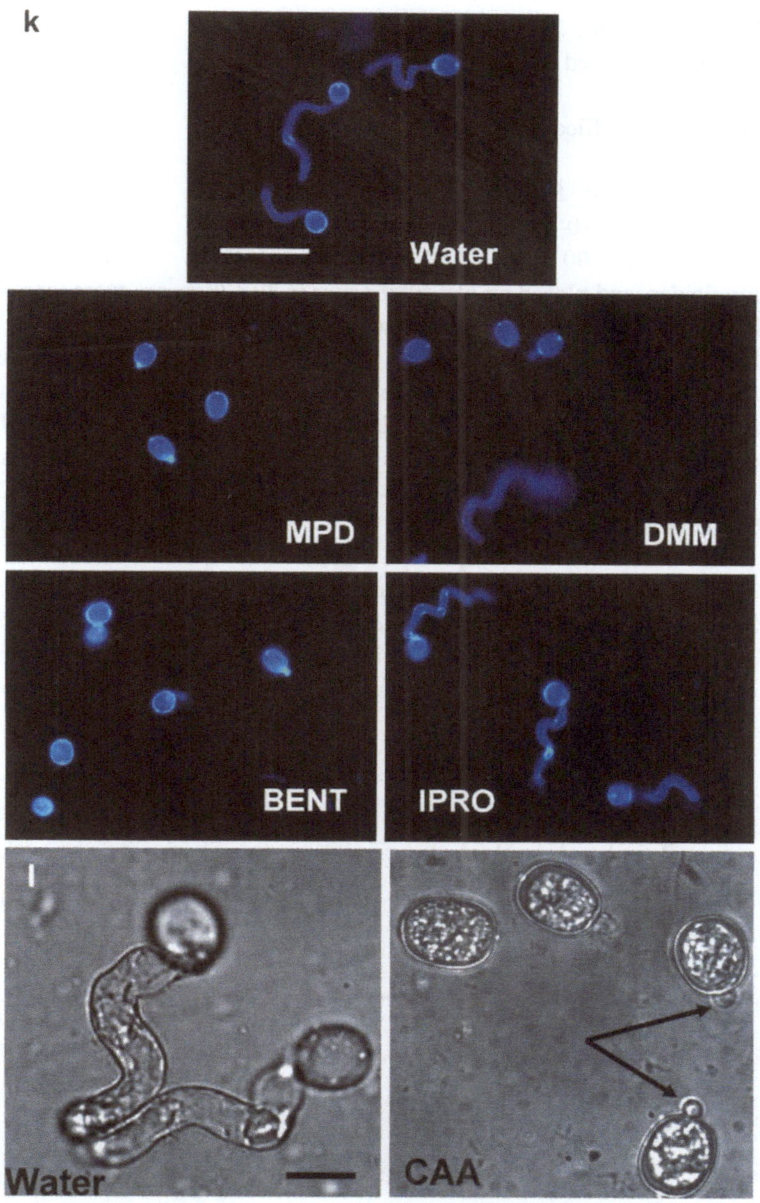

and callose deposited around haustorial necks fluoresced yellow.

Sporulation on leaves, plants and leaf discs

Leaves of plants with one true leaf were detached at 7 dpi and floated on technical CAA in 5 cm Petri dishes at 18°C for 48 h (12 h light/day). Intensity of sporulation was estimated with the aid of a magnifying lens (×10), and spore counts for leaves floating on 100 mg l^{-1} were made using a cytometer. Plants at the cotyledon stage were sprayed with technical CAA at 4 dpi, the number of sporulating plants counted at 7 dpi and spore counts measured at 11 dpi. Leaf discs (1 cm diam) from infected 10-leaf plants at 7 dpi were floated on 1 ml of either water or formulated CAA at 20°C for 20 h in the dark in six-well titer plates, and the number of sporophores per disc was determined by fluorescence microscopy after staining with calcofluor. Similar discs were incubated at 18°C with 12 h light per day for 2 days, and the number of spores produced assessed as above.

Translaminar efficacy

Leaves formed fifth from the stem base were detached from 10-leaf plants, sprayed on upper leaf surface with 100 mg l^{-1} of four technical CAA fungicides, and placed on wet filter paper in 20×20×3 cm plates. After 3 h, leaves were inverted and inoculated on their lower surface with spores of *B. lactucae*, and spore germination determined at 20 hpi as before. The number of sporulating lesions was counted at 10 dpi.

Shade-house experiments

Two experiments were conducted during 2006–2007 with whole plants of *L. sativa* cv. Noga to evaluate the efficacy of CAA in controlling *B. lactucae*. Plants were raised from seeds in Speedling trays in the greenhouse, and transplanted when they had five true leaves into polystyrene containers (1.2×0.6×0.2 m) filled with peat and vermiculite (1:1, *v*/*v*), to give 10 plants per container. Containers were located in shade-houses in the field at Bar-Ilan University Farm. Shade-houses were covered with white plastic nets (50 holes per square inch, mesh) to avoid aphid and viral infections. In 2006, plants were sprayed twice with three concentrations of formulated MPD, with the first spray at 4 weeks after planting, when plants had reached the 10–12 leaf stage. The second spray was applied after a further 8 days. In 2007, plants were sprayed once with 500 mg l^{-1} of each of three formulated CAA fungicides, before transplanting into the shade-house. BENT was not tested outdoors because no formulated solo product was available. In both experiments plants were inoculated with a spore suspension of *B. lactucae* (1×10^3 ml^{-1}) in the evening on the same day as the treatment. After inoculation, plants were covered with plastic sheets

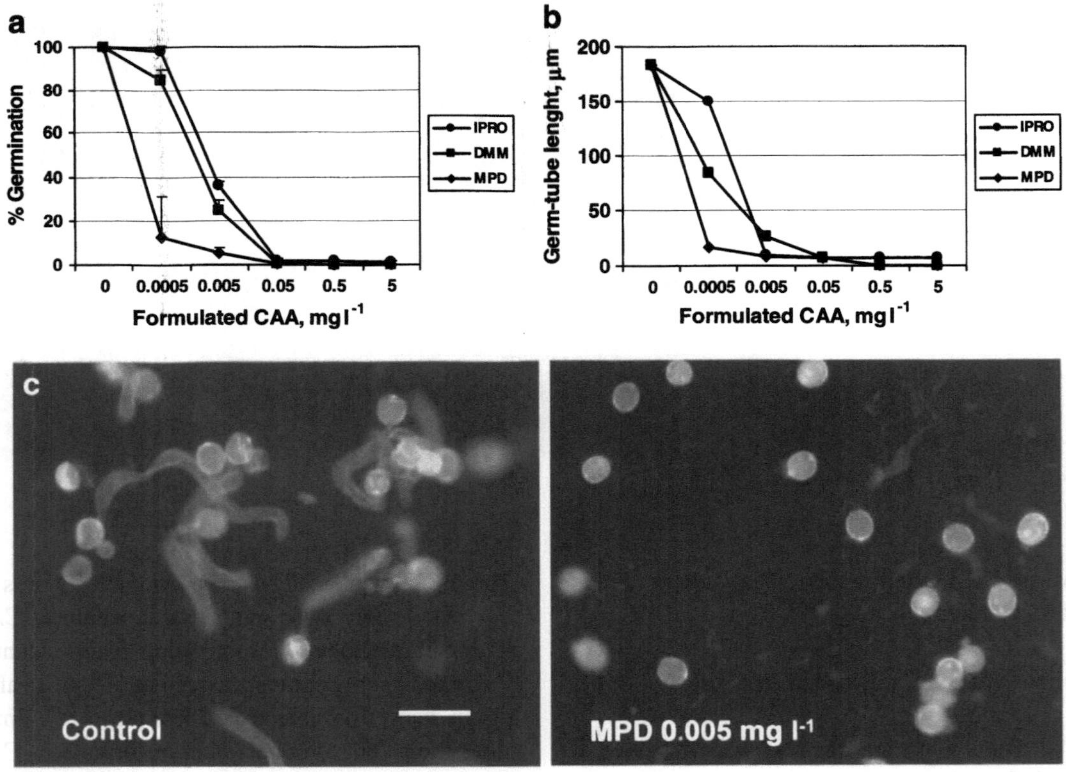

Fig. 2 Percent spore germination (means and standard deviation of the mean, n=100) (**a**) and germ-tube length (**b**) of *B. lactucae* isolate ISR-60 on leaf surfaces treated with three formulated CAA compounds (see Fig. 1 for abbreviations). **c** Epifluorescence micrographs of calcofluor-stained *B. lactucae* spore germination on leaf surfaces treated with water (control) and technical MPD. Scale bar=50 µm

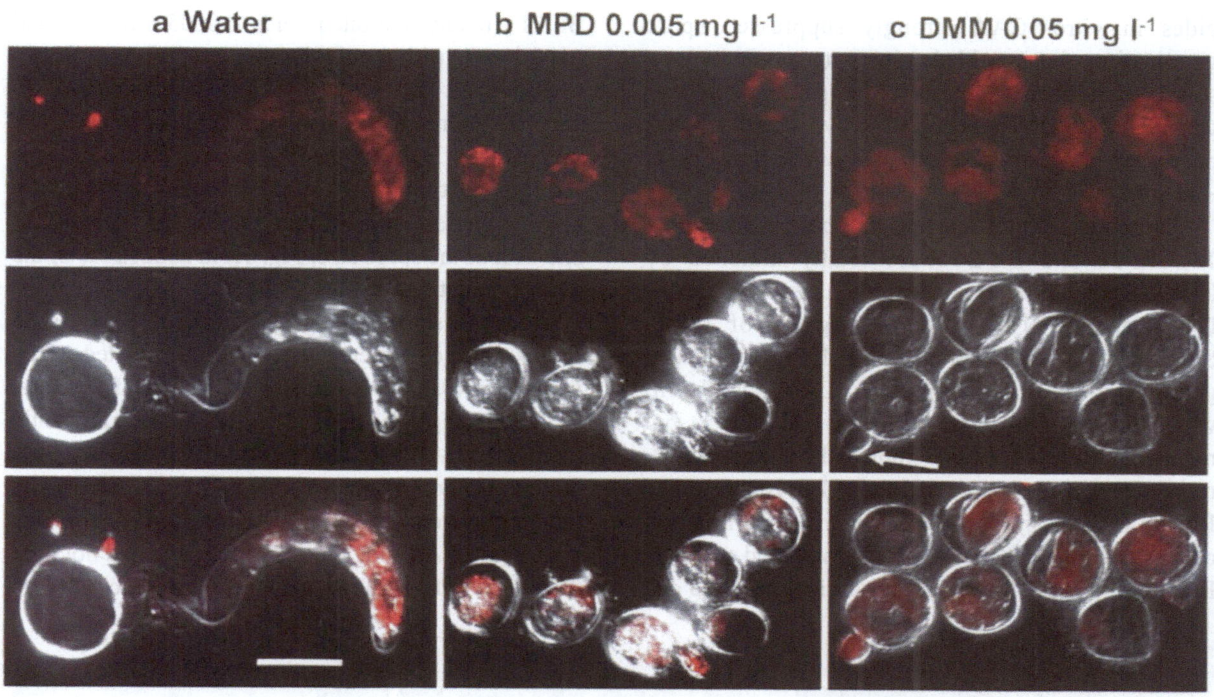

Fig. 3 Laser scanning confocal micrographs of phalloidin-stained *B. lactucae* spore germination in vitro showing no effect of CAA on F-actin organization in the germinating spores (*arrow* small germ-tube). Scale bar=20 μm

for the night to ensure infection. In the first experiments, CAA was applied with a hand sprayer, or a manual backpack sprayer, at a rate of about 20–30 ml per plant. Disease severity was recorded at 42 or 50 dpi by counting the number of downy mildew lesions developing on 10 plants in a container.

Fig. 4 Epifluorescence micrographs of calcofuor-stained germinating spores of *B. lactucae* after adding 0.005 mg l^{-1} MPD 3 h after start of germination in vitro. **a** Germ-tube at 3 h after start of germination (time of treatment). Control (**b**) and treated (**c**) germ-tubes at 48 h after start of germination. Scale bar=20 μm

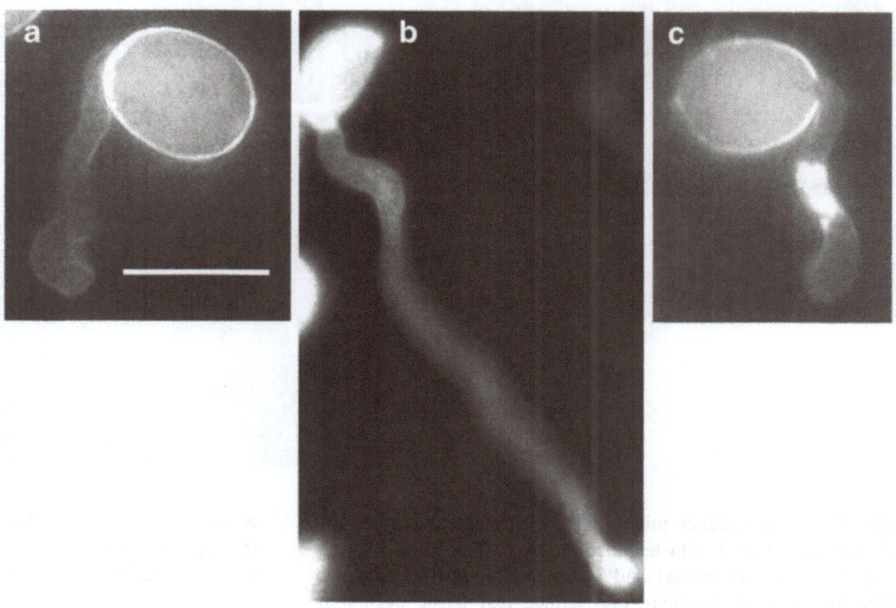

Results

Spore germination

Four European and one Israeli isolate of *B. lactucae* were tested for sensitivity to technical CAA fungicides in vitro. CAA strongly suppressed spore germination, but enabled spores to produce very small germ-tubes (Fig. 1l). True germination was considered to have occurred only when a spore produced a germ-tube of ≥5 μm (spore diam ~20 μm).

Figure 1a–f shows that the European isolates BL-18, BL-21 and BL-24 were totally inhibited by 0.0005 mg l^{-1} (lowest concentration tested) of MPD, DMM and BENT, and by 5 mg l^{-1} of IPRO. Strangely, low concentrations of IPRO stimulated germination of BL-24 (Fig. 1e). BL-25 (Fig. 1g, h) was similarly sensitive to MPD and IPRO, but required ×100 and ×1,000 more BENT and DMM, respectively, to be suppressed fully. The Israeli isolate ISR-60 (Fig. 1i, j) was more sensitive to IPRO, and controlled fully with 0.5 mg l^{-1}, compared with the European isolates. It was highly sensitive to MPD and BENT (totally inhibited with 0.0005 mg l^{-1}) and ×100 less sensitive to DMM.

Figure 1k shows the in vitro germination of *B. lactucae* spores in water and in four technical CAA fungicides at 0.0005 mg l^{-1} (for numerical data see Fig. 1a–j). At inhibitory concentrations, all fungicides allowed the formation of very short germ-tubes (about one fifth of the spore size, Fig. 1l) in about 20% of the spore population.

Figure 2 shows the germination of *B. lactucae* spores in the presence of three formulated CAA fungicides (0.0005–5 mg l^{-1}) on plant leaf surfaces. At 0.0005 mg l^{-1}, MPD (1.2 nM) was significantly more effective than DMM or IPRO, causing 90% inhibition. Formulated IPRO in planta was much more effective than technical IPRO in vitro (compare with Fig. 1). All fungicides strongly affected not only the number of germinating spores (Fig. 2a, c) but also

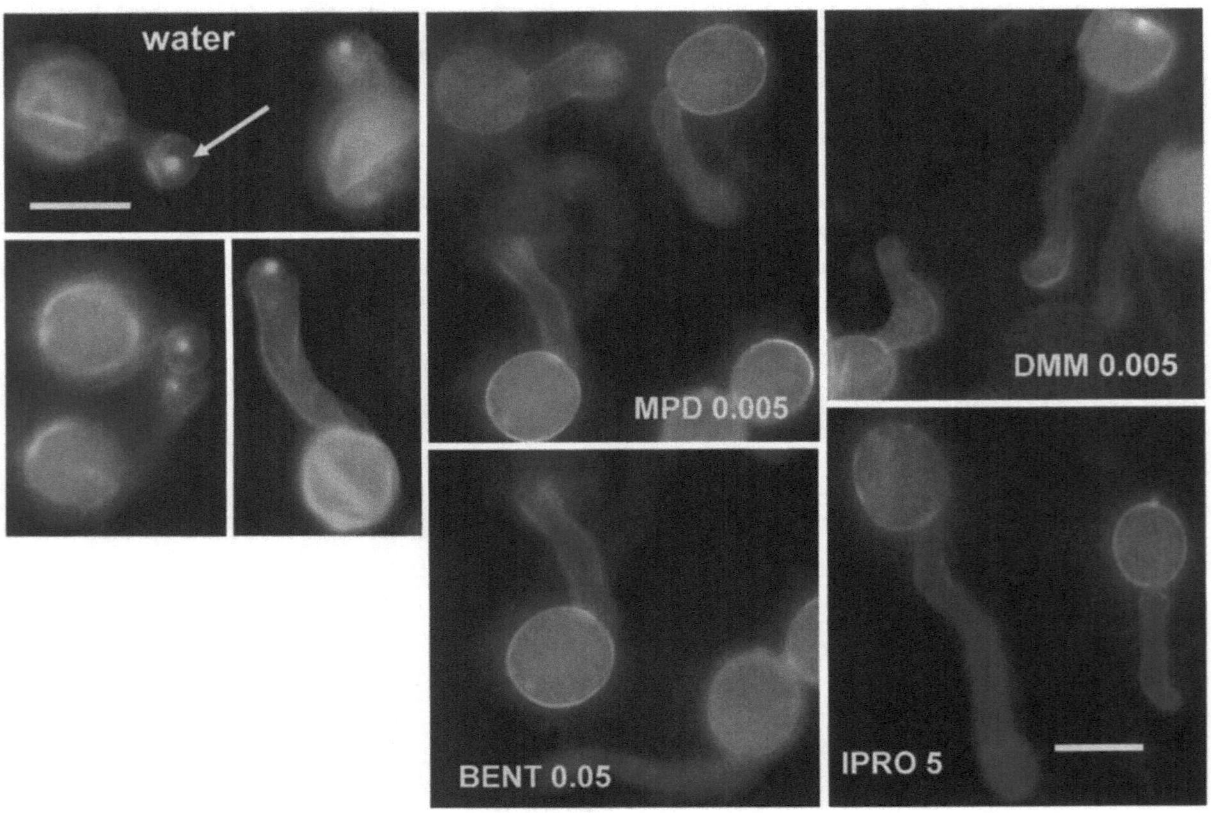

Fig. 5 Epifluorescence micrographs of germinating spores of *B. lactucae* treated with technical CAA (0.005–5 mg l^{-1}; see Fig. 1 for abbreviations) and water at 3 h post-inoculation. Penetration of *B. lactucae* into lettuce leaf tissue occurs in water-treated leaf tissue but not in CAA-treated tissue. Arrow indicates point of penetration (yellow fluorescence). Aniline blue staining followed by calcofluor staining; scale bar=20 μm

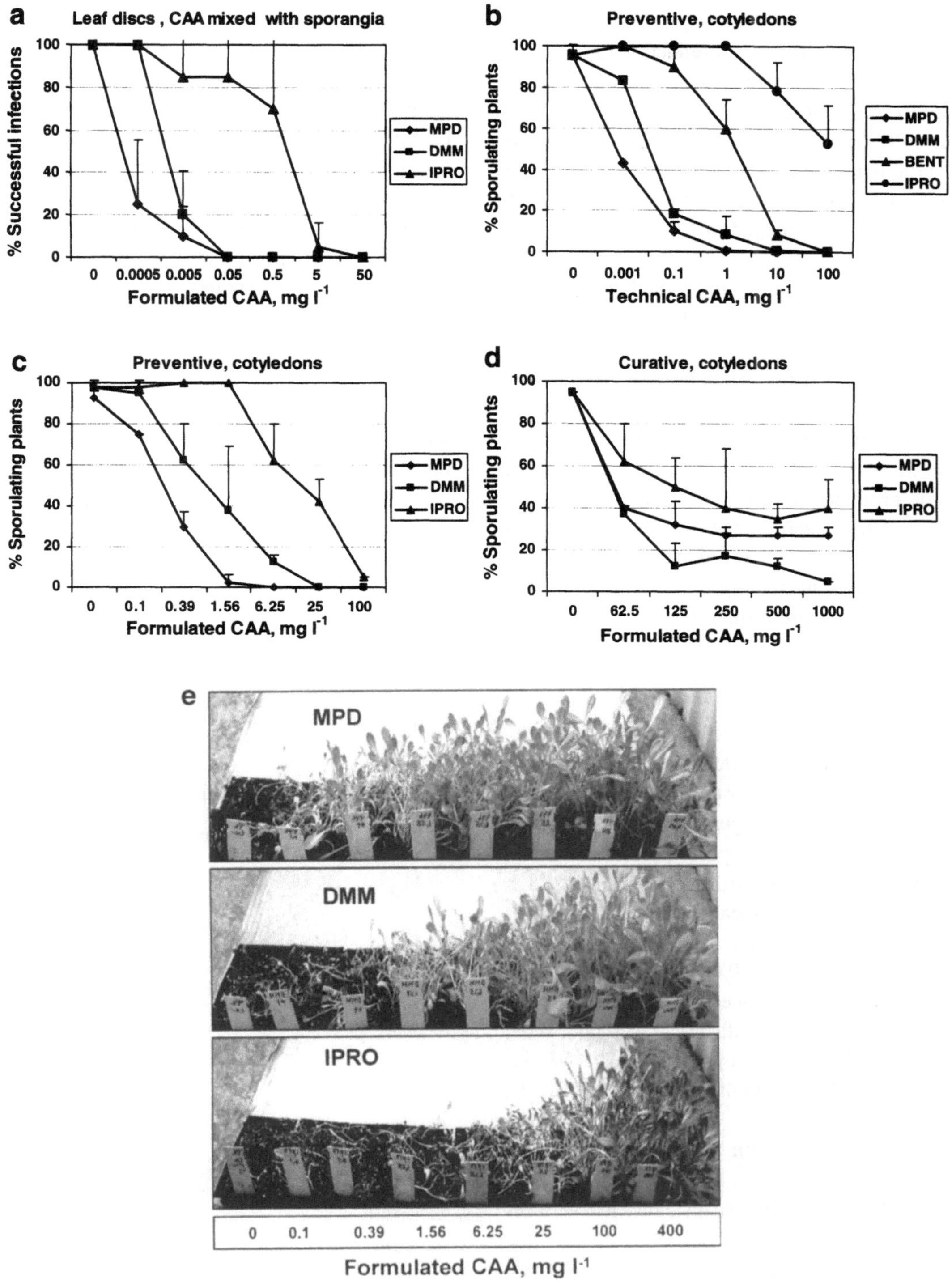

Fig. 6 Downy mildew development in lettuce plants treated with CAA (see Fig. 1 for abbreviations). **a** CAA mixed with spores of *B. lactucae* before inoculation (means and standard deviation of the mean, $n=40$). **b**, **c** CAA applied preventively 4 h before inoculation ($n=80$). **d** CAA applied curatively 1 dpi ($n=80$). Disease symptoms recorded at 7 dpi. **e** The appearance of the treated and inoculated plants shown in **c** at 21 dpi

the size of the germ-tubes (Fig. 2b, c). BENT was not available as a formulated solo product; a formulation (10% a.i.) was therefore prepared in Agsolex-8 and used for testing germination in vitro. Results confirm that BENT at 0.005 mg l^{-1} (lowest concentration tested) allowed 11% germination (88% inhibition relative to water control). The germinated spores produced a germ-tube of 11 μm compared to 170 μm in the water control.

Phalloidin staining indicated no disruption by CAA in *B. lactucae* spores (Fig. 3). Spores germinating in water (18 h, 13°C, darkness) produced a germ-tube of about 80 μm. The F-actin (red) was concentrated in the distal, growing part of the germ-tube, moving with the cytoplasm towards the tip. In CAA-treated spores, most produced no germ-tube, while some produced a germ-tube of about 3–5 μm. In non-germinating spores the red stain remained within the spore. In those that produced a small germ-tube, the red stain representing the F-actin microfilaments moved into the germ-tube (Fig. 3).

Effect of CAA on germ-tube growth

Microscopic observations (Fig. 4) made in vitro (13°C, darkness) showed that CAA (technical, 0.005–50 mg l^{-1}) added to spores at 3 h after the start of germination inhibited further extension of germ-tubes. At 3 h, when CAA was added, 50% of the spores produced a germ-tube of 40 μm (±standard error of 10 μm); at 48 h (45 h after adding CAA) mean germ-tube length in control, water-treated spores was 250±50 μm. In contrast, spores treated with ≥0.005 mg l^{-1} of technical MPD, DMM or BENT showed no extension of the germ-tubes. Spores treated with IPRO at 0.005, 0.05, 0.5, 5 and 50 mg l^{-1} produced germ-tubes of 150±30, 100±20, 80±20, 60±10, and 40±10 μm, respectively.

Effect of CAA on penetration

Spores were allowed to germinate on lettuce leaf surfaces at 13°C, and CAA added at 3 hpi when 50% of the spores had germinated with a germ-tube of 40±10 μm. Leaves were incubated for an additional 17 h at 13°C to allow penetration of the host. At 20 hpi, epifluorescence microscopy showed that penetration occurred in water-treated control leaves (Fig. 5, arrow) but not in leaves treated with CAA at 0.005–50 mg l^{-1}, except IPRO which stopped penetration at ≥5 mg l^{-1} (Fig. 5).

Temporary exposure

Spores incubated in water showed 80–90% germination (germ-tube=30–50 μm). Spores exposed to 0.5 or 1 mg l^{-1} MPD or DMM for 1 h on ice failed to germinate or infect leaves when the CAA was removed by washing with water. In contrast, spores exposed to 0.5 or 1 mg l^{-1} BENT or IPRO for 1 h germinated in vitro and produced lesions when inoculated onto detached leaves equivalent to control spores.

Effect on infection

Spores were mixed with formulated CAA and inoculated onto lettuce leaf discs laid lower surface uppermost on moist filter paper in 9 cm Petri dishes.

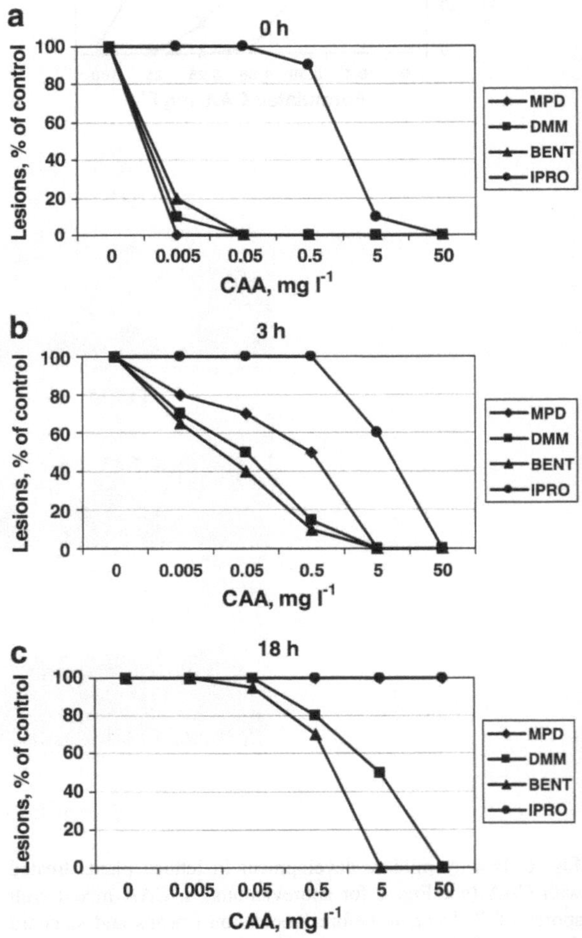

Fig. 7 Control of downy mildew lesion development in lettuce by CAA (see Fig. 1 for abbreviations) applied in **a** at 0 hpi, **b** at 3 hpi, and **c** at 18 hpi

The number of successful infections (sporulating lesions) at 10 dpi is shown in Fig. 6a. MPD was most effective while IPRO was least effective. The inhibitory concentrations of the three CAA fungicides resemble those required to inhibit spore germination in vitro (Fig. 1). Microscopic observations made at 1 dpi revealed that the failure to infect (for example in the presence of MPD) was a consequence of inhibited spore germination. In control inoculated plants, primary and secondary vesicles were seen in epidermal cells from which hyphae were emerging.

Preventive application of four technical or three formulated CAA fungicides to 7 day-old lettuce plants efficiently protected against downy mildew. Plants were sprayed with CAA at various concentrations and inoculated 4 h later. The proportion of

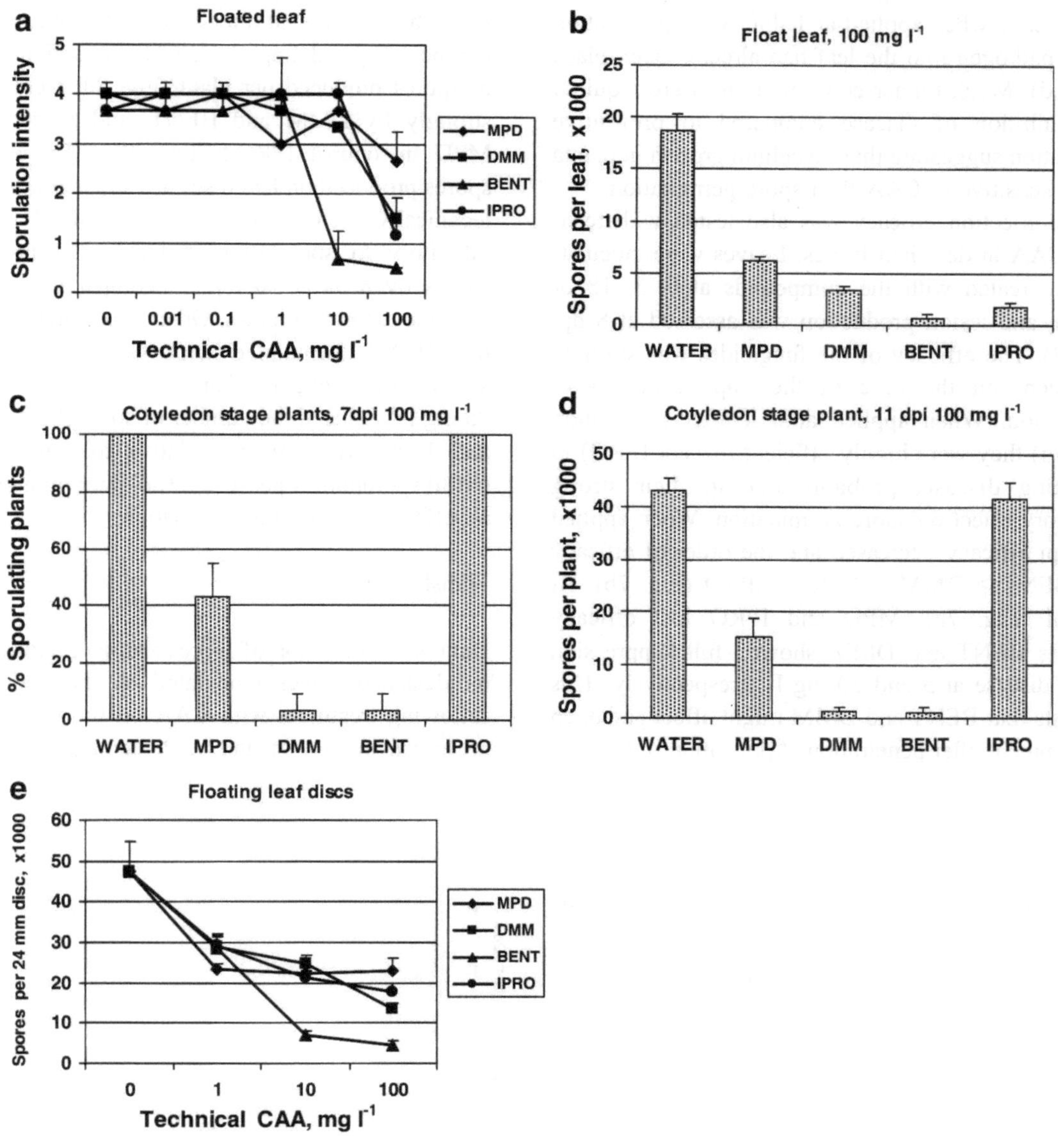

Fig. 8 Inhibition of sporulation of *B. lactucae* by CAA (see Fig. 1 for abbreviations), shown as sporulation intensity (visual assessment, means and standard deviation of the mean, $n=20$) (**a**), spores per leaf ($n=10$) (**b**), percent sporulating plants ($n=20$) (**c**), spores per plant ($n=20$) (**d**) and spores per leaf disc ($n=10$) (**e**). **a** and **b** First-formed leaves floating on CAA. **c** and **d** Whole plants at cotyledon stage, 7 and 11 dpi respectively, sprayed with CAA. **e** Leaf discs (24 mm diam) floating on CAA

plants showing sporulation of *B. lactucae* at 8 dpi is shown in Fig. 6b and c, and the appearance of the plants treated with formulated CAA at 21 dpi is shown in Fig. 6e. The fungicides differed in their efficacy in the order of MPD > DMM > BENT > IPRO.

Post-infection effects

Formulated CAA fungicides exhibited reduced control efficacy when applied at 1 dpi, when penetration of the pathogen into the leaf had already taken place (Fig. 6d). Much higher concentrations were required for inhibition of disease compared to preventive application suggesting that mycelium growth in planta is less sensitive to CAA than spore germination.

Post-infection efficacy was also tested with technical CAA in detached leaves. Leaves were inoculated and treated with the compounds at 0, 3, 18 or 45 hpi, and lesion production was assessed at 8 dpi (Fig. 7). The efficacy of the fungicides was strongly dependent on the time of their application post-inoculation. When applied at 0 h after inoculation (Fig. 7a) they were highly efficient (except IPRO) in inhibiting disease, probably due to their strong inhibitory effect on spore germination. When applied at 3 hpi efficacy decreased and the order of efficacy was BENT ≥ DMM > MPD > IPRO (Fig. 7b). At 18 hpi (Fig. 7c), MPD and IPRO lost efficacy whereas BENT and DMM showed full suppression of the disease at 5 and 50 mg l^{-1}, respectively. This suggests that BENT and DMM might affect pathogen development after penetration. Application of CAA at 45 hpi did not affect symptom production relative to untreated inoculated controls.

Effect on sporulation

BENT was most suppressive to sporulation: at 10 mg l^{-1} it caused 82% inhibition, while the other fungicides showed 0–17% inhibition at this concentration (Fig. 8a). At 100 mg l^{-1}, BENT was most suppressive and MPD least suppressive to sporulation (Fig. 8b). For whole plants at the cotyledon stage, both the number of sporulating plants (Fig. 8c) and the number of spores produced per plant (Fig. 8d) were reduced strongly by DMM and BENT, and moderately by MPD. In contrast, IPRO had no effect. The number of spores produced on leaf discs at 7 dpi and treated with technical CAA is shown in Fig. 8e. BENT was most inhibitory to sporulation, reducing the number of spores by 85% at 10 mg l^{-1} compared to 48–55% with the other fungicides. Other experiments revealed that BENT formulated with Agsolex-8 prevented sporophore emergence from stoma. BENT at 6.25 or 25 mg l^{-1} reduced the number of sporophores by 75% and 100%, respectively, relative to water controls (=130 sporophores per disc). The other fungicides had no effect on sporophore formation.

Translaminar efficacy

Data on germination of spores and sporulating lesions on detached leaves inoculated on the undersurface following treatment with CAA on the upper surface are given in Fig. 9a and b. Percentage inhibition of

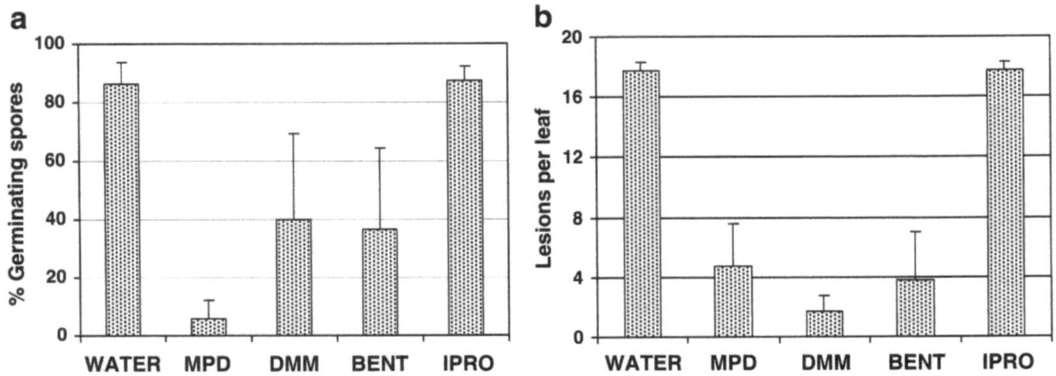

Fig. 9 Germination of *B. lactucae* spores on lower leaf surface (means and standard deviation of the mean, n=100) (**a**), and lesion development (n=10) (**b**), in detached leaves treated with CAA (see Fig. 1 for abbreviations) on upper leaf surface and inoculated on lower leaf surface

spore germination at 20 hpi was 94, 54, 58 and 0 (Fig. 9a), and percentage inhibition of lesion at 10 dpi formation was 73, 90, 79 and 0 for MPD, DMM, BENT and IPRO, respectively (Fig. 9b). The data show that while MPD was superior to the other fungicides in translocation to the opposite surface of the leaf in assessments of spore germination, DMM was superior in inhibiting lesion formation. This suggests that DMM might have post-germination effects on *B. lactucae*.

Shade-house experiments

Data from the experiments to assess the efficiency of CAA for controlling epidemics in shade-house grown plants are presented in Fig. 10. In 2006, all three concentrations were effective similarly, providing about 95% and 90% protection at 27 and 42 days after the second (last) spray, respectively (Fig. 10a and b). In 2007, a single spray of CAA provided about 90% protection at 50 days after treatment (Fig. 10c).

Discussion

Carboxylic acid amide (CAA) fungicides are shown here to control effectively (with intrinsic differences) downy mildew in lettuce in growth chamber and shade-house experiments. The toxicity of four CAA compounds towards *B. lactucae* has been compared quantitatively in vitro and in planta, by exposing the pathogen to CAA at various stages of its life-cycle: spore germination, germ-tube extension, penetration, colonisation, sporophore formation, spore production and epidemics in the field.

It appeared that spore germination was the stage in the life-cycle most sensitive to CAA. Germination in vitro and on leaf surfaces was inhibited with nanomolar concentrations of MPD (0.0005 μg ml^{-1}) and micromolar concentrations of DMM and BENT (0.005 μg ml^{-1}) or IPRO (0.5 μg ml^{-1}). Similar results were reported recently for germination in vitro of sporangia and cystospores of *P. infestans* (Cohen and Gisi 2007). Unlike cystospores or sporangia of *P. infestans* which produce no germ-tubes in CAA (Cohen and Gisi 2007), spores of *B. lactucae* produce a small germ-tube (3–5 μm) in the presence of CAA. The formation of these germ-tubes suggests that the

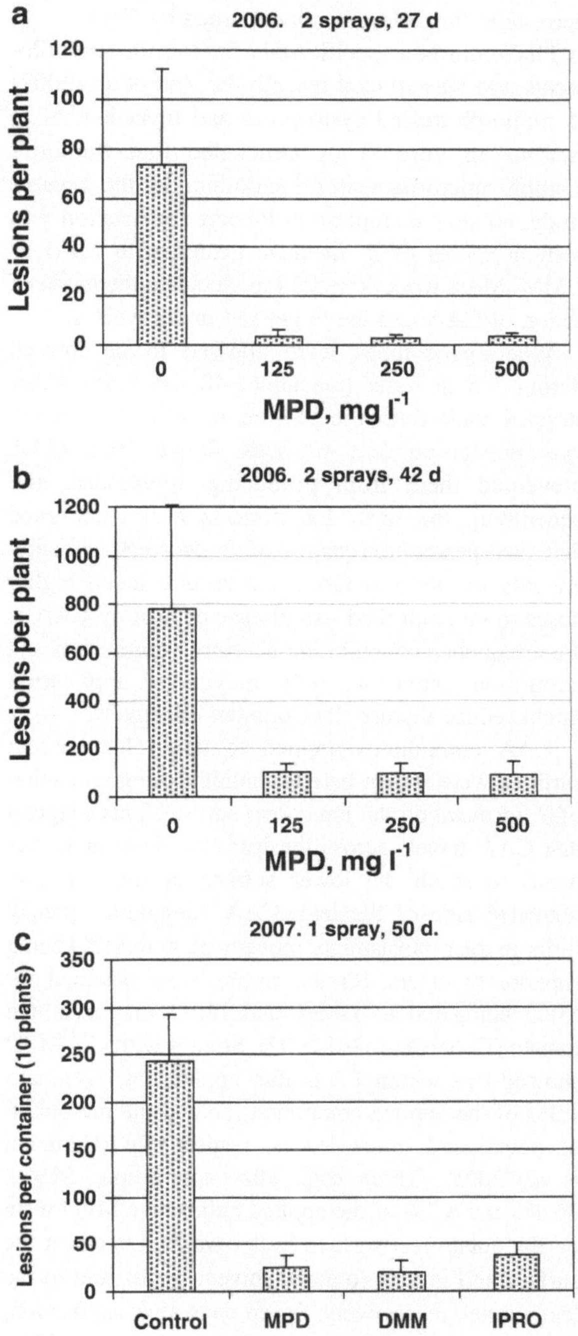

Fig. 10 Downy mildew development in shade-house grown lettuce plants treated with 125, 250 or 500 mg l^{-1} MPD (formulated) at 27 (**a**) and 42 (**b**) days after the second (last) spray. **c** Downy mildew development in shade-house grown lettuce plants treated with three CAA compounds (formulated; see Fig. 1) applied once at a dose of 500 mg l^{-1}. Lesions per ten plants in a box recorded at 50 days after spraying. n=10 boxes per dose treatment per fungicide

polarity of the spores of *B. lactucae*, required for germ-tube formation, is not disturbed by CAA.

Phalloidin is a specific stain for F-actin microfilaments and was applied recently by Zhu et al. (2007) to flumorph-treated cystospores and mycelium of *P. melonis* in vitro. They concluded that flumorph disrupts microfilament organisation. In the present study, no such disruption in F-actin organization was seen in spores of *B. lactucae* treated with MPD or DMM. More work is needed to elucidate the mode of action of CAA and the target site in oomycetes.

When germ-tubes were allowed to be formed during 3 h in water (reaching ~40 μm long), CAA stopped their further extension in vitro. Equivalent experiments on leaf surfaces showed that CAA prevented them from producing appressoria and penetrating the host. Experiments also determined that post-penetration stages of *B. lactucae* infection are less sensitive to CAA and require much higher doses to be controlled. As disease control by CAA is a consequence of inhibition of spore germination and germ-tube extension, only preventive application might reduce disease development effectively.

CAA compounds applied to upper lettuce leaf surfaces were shown here to inhibit spore germination of *B. lactucae* on the lower leaf surface. This suggests that CAA travels across the leaf (translaminar movement) to reach the lower surface of the opposite, untreated side of the leaf. CAA compounds greatly differ in their translaminar movement, with MPD being superior to others. Similar results were obtained for MPD compared to DMM and IPRO in potato and tomato (Cohen and Gisi 2007). Studies with C^{14}-MPD showed that within 1.5 h after application, 4.9% and 1.3% of the applied compound reached the mesophyll of potato and grape leaves, respectively (Hermann et al. 2005). Three days after application, 54.9%, 36.4% and 8.7% of the applied radioactive MPD were on the potato leaf surface (water wash), bound on the surface and in wax (organic solvent wash), and in the leaf extract, respectively. Seven days after application, 49% of the C^{14} MPD applied to potato leaves could be recovered from leaf surface and wax (Hermann et al. 2005). This prolonged binding of MPD to the leaf cuticle, its good rain-fastness (Hermann et al. 2005) and its pronounced translaminar mobility may ensure its prolonged availability to lower leaf surfaces. No data, however, are available for how much C^{14}-MPD is present on the opposite, lower, surface of the leaf. Miyake et al. (2003) reported that BENT has not only strong preventive activity, but also curative and penetrant activity, with excellent residual effects and rainfastness.

The biotrophic nature of *B. lactucae* does not allow testing the effects of CAA on mycelium growth or sporulation in vitro. Therefore, the interpretation of in planta experiments should consider also the incorporation of the CAA compounds into the leaf tissue. Indeed, application of CAA to detached leaves at 18 h post-inoculation, when primary and secondary vesicles have already developed in the epidermal cells, did not prevent the pathogen from completing its life-cycle by producing new spores. Nonetheless, two compounds, BENT and DMM, did partially inhibit sporulation, and BENT also suppressed sporophore development.

CAA fungicides were found to be highly effective in controlling downy mildew in lettuce in the shade-house experiments. Unfortunately, doses applied were too high to distinguish differences in efficacy among the compounds. Previous data with late blight control in potato (Cohen and Gisi 2007) showed superior activity of MPD over DMM, and of DMM over IPRO. More studies are required to evaluate the performance of CAA, including BENT, in the field.

In conclusion, CAA fungicides were found here to be extremely potent inhibitors of spore germination of *B. lactucae*, as they were in *P. infestans* (Cohen and Gisi 2007). CAA inhibited germ-tube extension when applied to germinating spores. The compounds had poor curative efficacy when applied at 1 dpi against *B. lactucae,* as was determined previously for *P. infestans* (Cohen and Gisi 2007). All the compounds, except IPRO, showed translaminar movement across lettuce leaves, with MPD superior to DMM and BENT. When applied to mature lesions, BENT was superior to the other compounds in suppressing sporulation of *B. lactucae*. CAA fungicides, therefore, should be used preventively to achieve the best control of downy mildew in lettuce.

Acknowledgement We are grateful to Dr. Alex Perelman of Bar-Ilan University for his assistance with confocal microscopy.

References

Albert, G., Curtze, J., & Drandarevski, C. (1988). *Dimethomorph (CME 151), a novel curative fungus fungicide*. Brighton Crop Protection Conference—Pests and Diseases, pp. 17–22.

Albert, G., Thomas, A., & Guehne, M. (1991). *Fungicidal activity of dimethomorph on different stages in the life cycle of Phytophthora infestans and Plasmopara viticola.* In Third International Conference on Plant Diseases, Bordeaux pp. 887–894. Paris, France: ANPP.

Anon. (2006). http://www.frac.info/work/work_CAA.htm.

Bagirova, S. F., Li, A. Z., Dolgova, A. V., Elansky, S. N., Shaw, D. S., & Dyakov, Y. T. (2001). Mutants of *Phytophthora infestans* resistant to dimethomorph fungicide. *Journal of Russian Phytopathology, 2*, 19–24.

Cohen, Y., Baider, A., & Cohen, B. H. (1995). Dimethomorph activity against oomycete fungal plant pathogens. *Phytopathology, 85*, 1500–1506.

Cohen, Y., Eyal, H., & Hanania, J. (1990). Ultrastructure, autofluorescence, callose deposition and lignification in susceptible and resistant muskmelon leaves infected with the powdery mildew fungus *Sphaerotheca fuliginea. Physiological and Molecular Plant Pathology, 36*, 191–204.

Cohen, Y., Eyal, H., Hanania, J., & Malik, Z. (1989). Ultrastructure of *Pseudoperonospora cubensis* in muskmelon genotype susceptible and resistant to downy mildew. *Physiological and Molecular Plant Pathology, 34*, 27–40.

Cohen, Y., & Gisi, U. (1996). Systemic translocation of ^{14}C-DL-3-amino-*n*-butanoic acids. *Physiological and Molecular Plant Pathology, 45*, 441–456.

Cohen, Y., & Gisi, U. (2007). Differential activity of carboxylic acid amides fungicides against various developmental stages of *Phytophthora infestans*. *Phytopathology, 97*, 1274–1283.

Dereviagina, M. K., Elanskij, S. N., & Diakov, Y. T. (1999). Resistance of *Phytophthora infestans* to the dimethomorph fungicide. *Mikologiya I Fitopatologiya, 33*, 208–213.

Gisi, U., Waldner, M., Kraus, N., Dubuis, P. H., & Sierotzki, H. (2006). Inheritance of resistance to carboxylic acid amide (CAA) fungicides in *Plasmopara viticola*. *Plant Pathology, 56*, 199–208.

Griffiths, R. G., Dancer, J., O'Neill, E., & Harwood, J. L. (2003). A mandelamide pesticide alters lipid metabolism in *Phytophthora infestans*. *New Phytologist, 158*, 345–353.

Harp, T., Cloud, G., Minton, B., & Cochran, A. (2006). Mandipropamid: A new fungicide for control of late blight and downy mildews. *Phytopathology, 96*, S185.

Harp, T., Cochran, A., Tory, D., Kuhn, P., Payan, L., Laird, D., et al. (2007). Development of Revus 2.09SC (a.i. mandipropamid) in the U.S. for control of downy mildew on leafy vegetables. *Phytopathology, 97*, S45.

Hermann, D., Bartlett, D. W., Fischer, W., & Kempf, H. J. (2005). The behaviour of mandipropamid on and in plants. *Proceedings of the BCPC International Congress of Crop Science & Technology, Glasgow, UK, 1*, 87–92.

Huggenberger, F., & Kuhn, P. J. (2006). Biological and physicochemical properties of mandipropamid, a new fungicide for the control of oomycete pathogens. *Phytopathology, 96*, S51.

Huggenberger, F., Lamberth, C., & Iwanzik, W. (2005). Mandipropamid a new fungicide against Oomycete pathogens. *Proceedings of the BCPC International Congress Crop Science & Technology, Glasgow, UK, 1*, 93–98.

Jende, G., Steiner, U., & Dehne, H. -W. (1999). Effects of iprovalicarb (SZX 0722) on the development of *Phytophthora infestans* in tomato plants. *Pflanzenchutz-Nachrichten Bayer, 52*, 49–60.

Jende, G., Steiner, U., & Dehne, H.-W. (2001). Microscopical characterization of fungicidal effects on infection structures and cell wall formation of *Phytophthora infestans*. In H. W. Dehne, U. Gisi, K. H. Kuck, P. E. Russell, & H. Lyr (Eds.), *Modern fungicides and antifungal compounds* (pp. 83–90). Bonn: AgroConcept.

Lamberth, C., Cederbaum, F., Jeanguenat, A., Kempf, H. J., Zeller, M., & Zeun, R. (2006). Synthesis and fungicidal activity of *N*-2-(3-methoxy-4-propagyloxy) phenylethyl amides. Part II: Anti-oomycetic mandelamides. *Pest Management Science, 62*, 446–451.

Matheron, M. E., & Porchas, M. (2000). Impact of azoxystrobin, dimethomorph, fluazinam, fosetyl-Al, and metalaxyl on growth, sporulation, and zoospore cyst germination of three *Phytophthora* spp. *Plant Disease, 84*, 454–458.

Miyake, Y., Sakai, J., Miura, I., Nagayama, K., & Shibata, M. (2003). Effects of a novel fungicide benthiavalicarb-isopropyl against oomycete fungal diseases. The BCPC International Congress of Crop Science and Technology, Glasgow, Scotland, UK, Nov. 2003.

Reuveni, M. (2003). Activity of the new fungicide benthiavalicarb against *Plasmopara viticola* and its efficacy in controlling downy mildew in grapevines. *European Journal of Plant Pathology, 109*, 243–251.

Sharaf, K., Lewinsohn, D., Nevo, E., & Beharav, A. (2007). Virulence patterns of *Bremia lactucae* in Israel. *Phytoparasitica, 35*, 100–108.

Stenzel, K., Pontzen, R., Seitz, T., Tiemann, R., & Witzenberger, A. (1998). SZX 722: A novel systemic oomycete fungicide. *Brighton Crop Protection Conference*, pp. 367–374.

Young, D. H., Kemmit, G. M., & Owen, J. (2005). A comparative study of XR-539 and other oomycete fungicides: Similarity to dimethomorph and amino acid amides in its mechanism of action. In H. W. Dehne, U. Gisi, K. H. Kuch, P. E. Russell, & H. Lyr (Eds.), *Modern fungicides and antifungal compounds IV* (pp. 145–152). Alton: BCPC.

Zhu, S. S., Liu, X. L., Liu, P. F., Li, Y., Li, J. Q., Wang, H. M., et al. (2007). Flumorph is a novel fungicide that disrupts microfilament organization in *Phytophthora melonis*. *Phytopathology, 97*, 643–649.

Eur J Plant Pathol (2008) 122:185–195
DOI 10.1007/s10658-008-9285-2

Beta-aminobutyric acid-induced resistance in grapevine against downy mildew: involvement of pterostilbene

Ana R. Slaughter · Mollah Md. Hamiduzzaman · Katia Gindro · Jean-Marc Neuhaus · Brigitte Mauch-Mani

Received: 29 August 2007 / Accepted: 31 January 2008

Abstract BABA, a non-protein amino acid, was used to induce resistance in grapevine against downy mildew. BABA-induced resistance was observed in the susceptible cv. Chasselas as well as in the resistant cv. Solaris. Following BABA treatment, sporulation of *Plasmopara viticola* was strongly reduced and the accumulation of stilbenes increased with time following infection. Induction of *trans*-piceide, *trans*-resveratrol and, more importantly, of *trans*-ε- and *trans*-δ-viniferin and *trans*-pterostilbene was observed in BABA-primed Chasselas. On the other hand, induction of *trans*-resveratrol, *trans* δ-viniferin and *trans*-pterostilbene was observed in BABA-primed Solaris. The accumulation of stilbenes in BABA-primed Solaris was much higher than that found in BABA-primed Chasselas. Furthermore, BABA-treatment of Solaris led to a rapid increase in transcript levels of three genes involved in the phenylpropanoid pathway: phenylalanine ammonia lyase, cinnamate-4-hydroxylase and stilbene synthase. BABA-primed Chasselas showed increased transcript levels for cinnamate-4-hydroxylase and stilbene synthase. Here we show that pre-treatment of a susceptible grapevine cultivar with BABA prior to infection with *P. viticola* primed the accumulation of specific phytoalexins that are undetectable in non-BABA-primed plants. As a result, the susceptible cultivar became more resistant to downy mildew.

Keywords Phytoalexin · *Plasmopara viticola* · Priming · Stilbenes · *Vitis vinifera*

Abbreviations

BABA	beta aminobutyric acid
BABA-IR	BABA-induced resistance

Introduction

Downy mildew caused by *Plasmopara viticola* is one of the most serious diseases in vineyards worldwide. Both susceptible and resistant cultivars can be colonised by *P. viticola* zoospores, but in resistant ones, the development of the parasite is rapidly

A. R. Slaughter · M. M. Hamiduzzaman · J.-M. Neuhaus · B. Mauch-Mani (✉)
Laboratory of Molecular and Cellular Biology, University of Neuchâtel, Rue Emile-Argand 11, Case Postale 158, CH-2009 Neuchâtel, Switzerland
e-mail: Brigitte.mauch@unine.ch

Present address:
M. M. Hamiduzzaman
Department of Plant Physiology and Botany Department, Stockholm University, Lilla Frescativagen 5, Stockholm 106 91, Sweden

K. Gindro
Swiss Federal Research Station for Plant Production of Changins-Wädenswil ACW, Case Postale 1060, CH-12 Nyon, Switzerland

inhibited. The most cultivated grape cultivars (*Vitis vinifera*) are susceptible to *P. viticola* and the control of downy mildew requires regular fungicide applications. The application of copper-containing fungicides to control downy mildew causes accumulation of this heavy metal in soil and groundwater resulting in toxic effects to the environment. Unfortunately, the replacement of such copper-based fungicides by synthetic fungicides with specific modes of action has promoted the development of resistant isolates of *P. viticola* (Matasci et al. 2008). For these reasons, alternative strategies are needed.

In grapevine, the most frequently observed and best-characterised defence reactions upon fungal infection are the deposition of lignin (Dai et al. 1995) and other phenolic compounds, increased peroxidase activity, accumulation of stilbene phytoalexins and the synthesis of pathogenesis-related (PR) proteins (Derckel et al. 1999). These defence mechanisms also seem to be present in susceptible cultivars, but in general, they are not activated or are delayed during the infection process. Because grapevine is an agriculturally and economically important crop plant, the defence mechanisms of that plant against pathogenic microorganisms, including phytoalexin production, have attracted considerable attention. Phytoalexins from the *Vitaceae* have been the subject of numerous studies, because these compounds are thought to have implications in both phytopathology and human health (Jeandet et al. 2002).

Phytoalexins are low molecular weight antimicrobial secondary metabolites (Kuc 1995). They have been shown to possess biological activity against a wide range of pathogens and can be considered as markers for plant disease resistance. Although phytoalexins display an enormous chemical diversity, phytoalexins from the *Vitaceae* seem to constitute a restricted group of molecules belonging to the stilbene family, the skeleton of which is based on the *trans*-resveratrol structure (3,5,4′-tryhydroxystilbene) (Fig. 1a) (Jeandet et al. 2002). Other compounds considered as oligomers of resveratrol and termed viniferins have also been found in grapevine as a result of infection or other stresses. Resveratrol is also glycosylated as piceide (5,4′-dihydroxystilbene-3-*O*-β-glucopyranoside) (Fig. 1b) and pterostilbene (3,5-dimethoxy-4′-hydroxystilbene) (Fig. 1c) is a dimethylated resveratrol derivative (Jeandet et al. 2002). Recently, it was shown that an isomer of ε-viniferin (Fig. 1d), δ-viniferin (Fig. 1e), is one of the major stilbenes produced from resveratrol oxidation in grapevine leaves infected by *P. viticola* (Pezet et al. 2003). Pezet et al. (2004b) tested the toxicity of these stilbenes against zoospores of *P. viticola* and found that δ-viniferin and pterostilbene were the most toxic stilbenes.

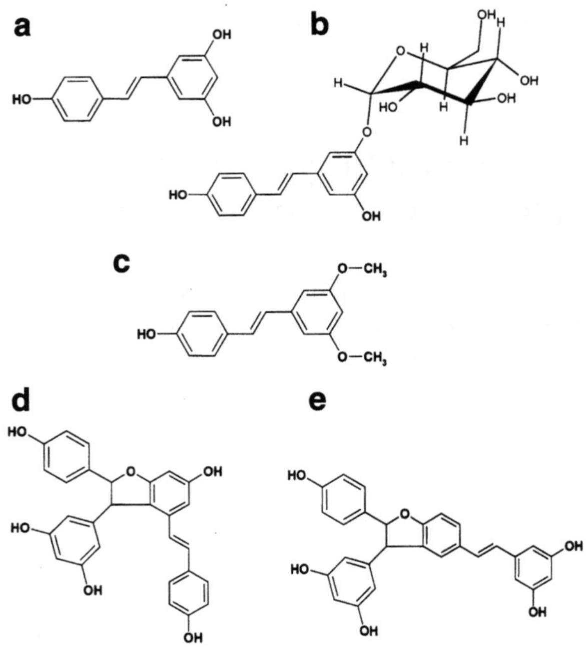

Fig. 1 Chemical structures of stilbene phytoalexins: **a** resveratrol, **b** piceide, **c** pterostilbene, **d** ε-viniferin and **e** δ-viniferin

The phenylpropanoid pathway is an important pathway in secondary plant metabolism, yielding a variety of phenolics with structural and defence-related functions. These phenolic compounds include lignins, phenolic acids, flavonoids and stilbenes. In addition, enzymes such as phenylalanine ammonia lyase (PAL; EC 4.3.1.5), cinnamate-4-hydroxylase (C4H; EC 1.14.13.11) and 4-coumarate:coenzyme A ligase (4CL, EC 6.2.1.12) are considered to be crucial to phenylpropanoid metabolism. A number of reports have shown that phenylpropanoid derivatives are capable of protecting plants against various biotic (infection by viruses, bacteria, fungi) and abiotic (low and high temperatures, UV-B light, wounding) stresses (Sgarbi et al. 2003; Solecka and Kacperska 2003). Stilbene synthase (STS) (EC 2.3.1.95) catalyses the last step of the phenylpropanoid biosynthesis pathway leading to the formation of stilbene phytoalexins. Expression of STS genes is often induced in response to biotic and abiotic stresses (Jeandet et al. 2002).

In recent years, much attention has been focused on the activation of a plant's own defence system, known as induced resistance (Sticher et al. 1997). The non-protein amino acid β-aminobutyric acid has previously been shown to induce resistance against many different oomycetes and to be effective in inducing resistance against various downy mildews (reviewed in Jakab et al 2001; Cohen 2002). It has been speculated that BABA deteriorates penetrated host cells, blocking the translocation of nutrients into the haustoria, therefore prohibiting further mycelial growth and sporangial production (Steiner and Schönbeck 1997). Experiments with ^{14}C-labelled BABA showed that it was not metabolised in tomato (Cohen and Gisi 1994) or in *Arabidopsis* (Jakab et al. 2001), ruling out the involvement of a BABA metabolite acting as an antimicrobial compound in the plant. BABA-mediated resistance is therefore most likely to be based on the activation of host resistance mechanisms. Recently, in grapevine, it has been shown that callose deposition as well as defence mechanisms depending on the phenylpropanoid and the jasmonic acid (JA) pathways all contributed to BABA-IR (Hamiduzzaman et al. 2005).

In order to obtain a clearer picture of BABA-IR in grapevine against downy mildew, we looked at the involvement of stilbene phytoalexins and the expression of genes encoding enzymes involved in the phenylpropanoid pathway. Different staining techniques were used to visualise the biochemical changes that occurred at the cellular level between BABA and water-treated resistant and susceptible cultivars. After treatment of plants with BABA and water, we performed quantitative HPLC analysis of stilbenes and investigated the level of expression of the genes involved in the biosynthesis of stilbenes (PAL, C4H and STS) by quantitative real-time PCR.

Materials and methods

Plant material

Cuttings of *V. vinifera* cvs Chasselas and Solaris, the latter resulting from a cross of Merzling x GM 6493, carried out by the Weinbauinstitut in Freiburg, Germany, are sensitive and resistant to *P. viticola*, respectively. They were obtained from the grape collection at the Swiss Federal Research Station for Plant Production in Changins, Switzerland and were grown in glasshouses. The cuttings were maintained in growth chambers (16 h light at 22°C, 8 h dark at 18°C and 60% relative humidity) until they had five fully developed leaves. They were then used for subsequent treatments and artificial inoculations.

Treatment and inoculation of plants

Cuttings from both cultivars were either soil drenched with an aqueous solution of BABA (1 mM) or with water (Hamiduzzaman et al. 2005) 2 days prior to inoculation with *P. viticola*. For the inoculum, leaves infected with *P. viticola* were harvested and sporangia were collected by vacuum aspiration, as described by Gindro et al. (2003). The abaxial leaf surfaces were inoculated by spraying an aqueous suspension of sporangia (5×10^4 sporangia ml^{-1}). The plants were then covered with transparent plastic bags and placed in growth chambers under the conditions described above. Samples were collected from both BABA- and water-treated leaves 0, 3, 7, 24, 48 and 72 h after inoculation and either used immediately for further analysis or frozen in liquid nitrogen and stored at −80°C. The experiments were carried out in duplicate.

Microscopic examination

Mycelia, sporangiophores and sporangia in inoculated leaves were stained with lactophenol-trypan blue (Keogh et al. 1980) and examined with bright field microscopy. Leaf material was placed on a glass slide in water, covered with a coverslip and examined with an epifluorescence microscope with an UV excitation filter (BP 340–380 nm, LP 425 nm). Blue fluorescence was used as an indicator of the presence of resveratrol (Dai et al. 1995). Flavonoid accumulation was visualized by using Wilson's reagent (Dai et al. 1995); leaves were immersed for 15 min, mounted in glycerol (75%) and examined under UV light with an epifluorescence microscope (BP 340–380 nm, LP 425 nm). Flavonoids fluoresced yellow and gallic acid derivatives (GAD) fluoresced blue.

Preparation of samples and HPLC analysis

At 0, 3, 7, 24, 48 and 72 h post-infection (hpi), three pieces of leaves were excised from inoculated leaves. Three replicates were made for each time point, each cultivar and each treatment. Leaf samples were weighed and placed in a microfuge tube and 50 μl of MeOH were added. The tubes were placed in a thermo-regulated shaker at 60°C for 10 min and then placed on ice for 5 min. The methanolic extracts (30 μl) were analysed for stilbenes as described by Pezet et al. (2003).

Preparation of cDNAs

Total RNA was isolated from frozen leaf tissue using a modified CTAB extraction and lithium chloride precipitation method according to Iandolino et al. (2004). The quantity of total RNA was determined with a Nano-Drop ND-1000 spectrophotometer (NanoDrop Technologies, Wilmingon, DE, USA); in addition the quality of RNA was verified by the absorbance ratios (A260/A280) of 1.8 to 2.0. For quantitative real-time PCR analysis, RNA was treated with Turbo-DNase I (Ambion) according to the manufacturer's instructions. For cDNA synthesis, 1 μg of RNA was reverse-transcribed using oligo$(dT)_{18}$ and Superscript III reverse transcriptase (Invitrogen Life Technologies) following the instructions of the manufacturer.

Real-time PCR for expression analysis of *C4H*, *PAL* and *STS*

Expression analysis of the genes *VvC4H*, *VvPAL* and *VvSTS* was done by real-time PCR, using the SYBER green method on an iCycler (Bio-Rad) real-time cycler. Each PCR reaction (20 μl) contained 0.25 mM of each primer, cDNA and 1x ABsolute QPCR SYBR Green mix (ABgene). The thermal cycling conditions were 95°C for 15 min followed by 95°C for 30 s, 56°C for 30 s, and 72°C for 35 s for 40 cycles, followed by a melt cycle from 60°C to 95°C. The primers used were as follows: VvC4H-F (5′-AGTCCAAGTCACCGAGCCTGAT-3′) and VvC4H-R (5′-TAGCAAGCCACCATGCGTTCAC-3′) for *VvC4H* (gene fragment obtained from a suppressive subtractive hybridisation SSH library constructed from *P. viticola*-infected grapevine), VvPAL-F (5′-TTGGTGCCACTTCACATAGGAG-3′) and VvPAL-R (5′-AATCTGATGCCGGAGTAGCCTT-3′) for *VvPAL*, and VvStSy-F (5′-CTCGAACCATCCGTCA GAAGAG-3′) and VvStSy-R (5′-CCTACGATTA CAGCTGCAGACC-3′) for *VvSTS*. With all cDNAs used, the above primer sets gave single PCR products, which were verified by determining the melt curves for the products at the end of each run and by analysis of the products using gel electrophoresis. The efficiency of the primers was tested in preliminary experiments with serial dilutions of cDNA samples and maintained an E value of between 0.97 and 0.98. The expression of the three genes was normalised relative to Elongation Factor 1-α (*VvEF1-α*) using the primers VvEF1-F (5′-GAACGTTGCTGTGAAG GATCTC-3′) and VvEF1-R (5′-CGCCTGTCAACCT TGGTCATGA-3′). All samples were measured in triplicate, every run included the *VvEF1-α* control for each sample, and experiments were repeated twice. The Gene-X software (Bio-Rad) was used to calculate the mean normalised expression of the genes (Vandesompele et al. 2002).

Results

Microscopic observation of the biochemical changes in infected grapevine cuttings treated with or without BABA

Successful infection with *P. viticola* in susceptible Chasselas becomes apparent about 6 days after infection as a white, downy growth, mostly on the lower side of the leaves where *P. viticola* emerged from the stomata. Sporulation of *P. viticola* was strongly reduced in BABA-primed cuttings (Fig. 2b) compared to water-treated control plants (Fig. 2a) and the infection sites on BABA-primed cuttings were surrounded by necrotic groups of cells (Fig. 2b). Extensive hyphal growth and spread were observed within 3 days inside the leaf tissue of water-treated cuttings, whereas hyphal growth and spread of *P. viticola* was reduced in BABA-primed cuttings (data not shown). The biochemical changes at the cellular level were analysed at different time points (1–5 dpi) in BABA-and water-treated plants by using different staining techniques. Infected, non-BABA-primed plants displayed only the red fluorescence of chlorophyll in both plants stained with Wilson's Reagent (Fig. 2c) or unstained plants (Fig. 2e) when visualised under UV light. However, infected, BABA-

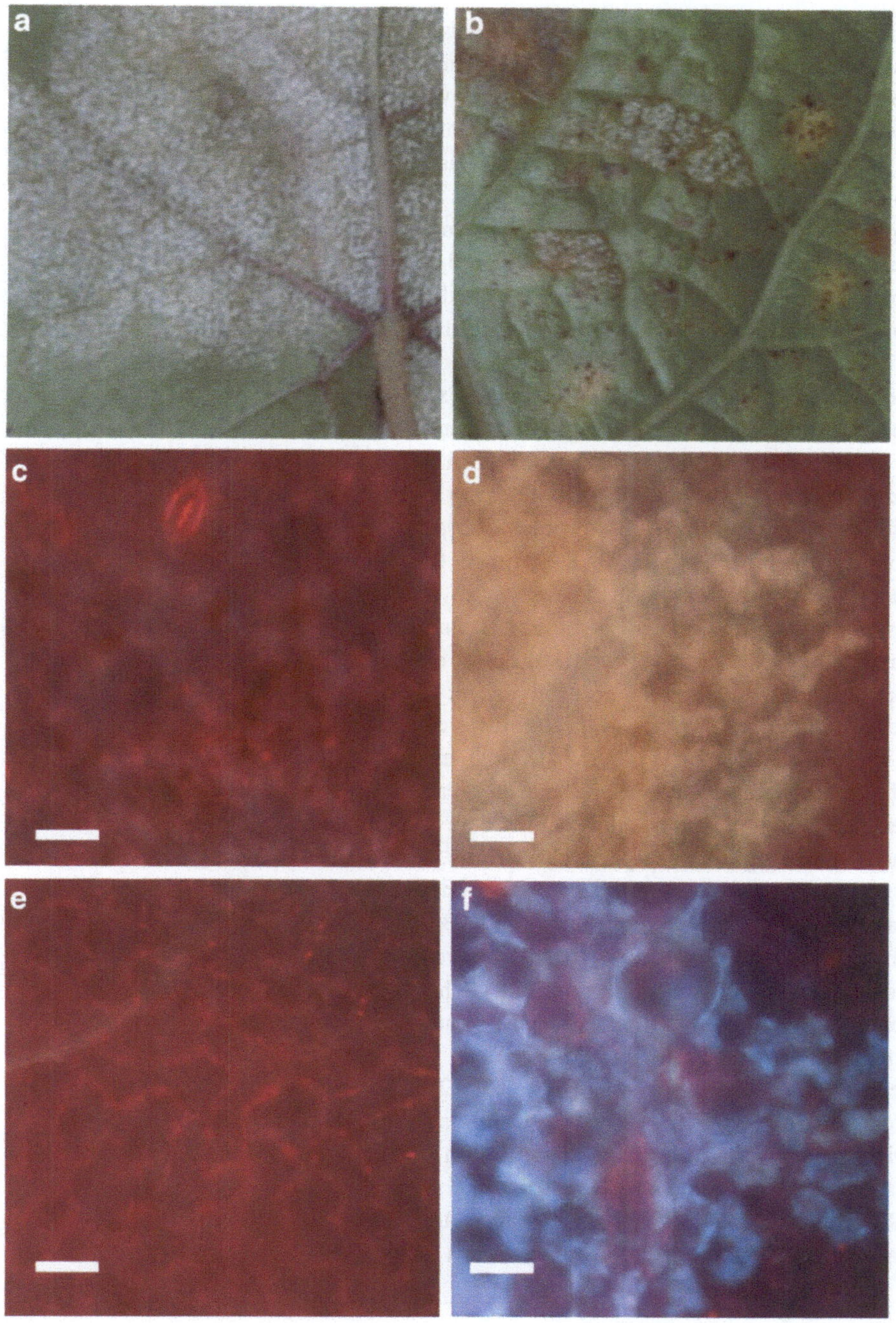

Fig. 2 Development of *P. viticola* and accumulation of stilbene compounds in grapevine (Chasselas). Plants were soil drenched with water (**a**, **c**, **e**) or 1 mM BABA (**b**, **d**, **f**) 2 days prior to challenge inoculation with sporangia of *P. viticola* (5×10^4 sporangia ml^{-1}). **a** *White* sporulation of *P. viticola* on water-treated cutting (6 dpi); **b** development of necrosis surrounding infection sites in BABA-primed cutting (6 dpi). Infected leaves were stained with Wilson's reagent and analysed by epifluorescence microscopy (**c** and **d**) or analysed by autofluorescence without prior staining (**e** and **f**). Water-treated controls show *red* chlorophyll fluorescence (**c** and **e**). *Yellow* fluorescence indicates flavonoid accumulation (**d**) and *blue* auto-fluorescence indicates presence of resveratrol in the tissues. Bars=50 μm

primed plants displayed a yellow fluorescence when stained with Wilson's Reagent (Fig. 2d) indicating the possible presence of flavonoids. The blue autofluorescence of infected, BABA-primed plants on the other hand suggests an accumulation of resveratrol in the tissues (Fig. 2f).

Accumulation of stilbene phytoalexins in *P. viticola* infected grapevine cuttings treated with and without BABA

The *trans*-form of five major stilbene phytoalexins: ε-viniferin, δ-viniferin, piceide, resveratrol and pterostilbene were quantified. *Trans*-piceide was the only phytoalexin detected in both BABA- and water-treated Chasselas and Solaris for the six investigated time points (Fig. 3a). There was a significant increase in *trans*-piceide in BABA-primed Chasselas cuttings compared to the water-treated cuttings with the highest concentration occurring at 72 h pi. In contrast, *trans*-piceide increased significantly in BABA-primed Solaris cuttings compared to water-treated cuttings with the highest concentration occurring at 24 h pi. With the BABA concentrations used for our experiments no direct induction of phytoalexins was observed and accumulation occurred only in infected or primed/infected plants. During the 72 h analysis, *trans*-resveratrol was quantitatively the most abundant stilbene produced in BABA-primed Chasselas and Solaris (Fig. 3b). In both BABA-primed Chasselas and Solaris there was a marked increase in *trans*-resveratrol starting at 24 h pi and accumulating quantitatively to about 150 and 191 μmol mg FW^{-1} for Chasselas and Solaris, respectively, at 72 h pi. Very slight amounts of trans-resveratrol were detected in water-treated Chasselas at 72 h pi (8 μmol mg FW^{-1}), whereas in water-treated Solaris, trans-resveratrol was detected at 24 h pi with a significant increase at 48 h (5 and 97 μmol mg FW^{-1}, respectively). *Trans*-ε-viniferin was only detected after 24 h pi and was not detected in water-treated Chasselas (Fig. 3c). For BABA-primed Chasselas, production started at 24 h pi, reaching a plateau at 48 h pi (28 μmol mg FW^{-1}). The highest concentration of ε-viniferin was found in water-treated Solaris at 48 h pi (97 μmol mg FW^{-1}) and then decreased to 56 μmol mg FW^{-1} at 72 h. The production of ε-viniferin increased steadily in BABA-primed Solaris starting at 24 h until 72 h where it reached the same amount as in water-treated Solaris at 48 h pi. *Trans*-δ-viniferin accumulated transiently starting at 24 h pi and peaking at 72 h only for

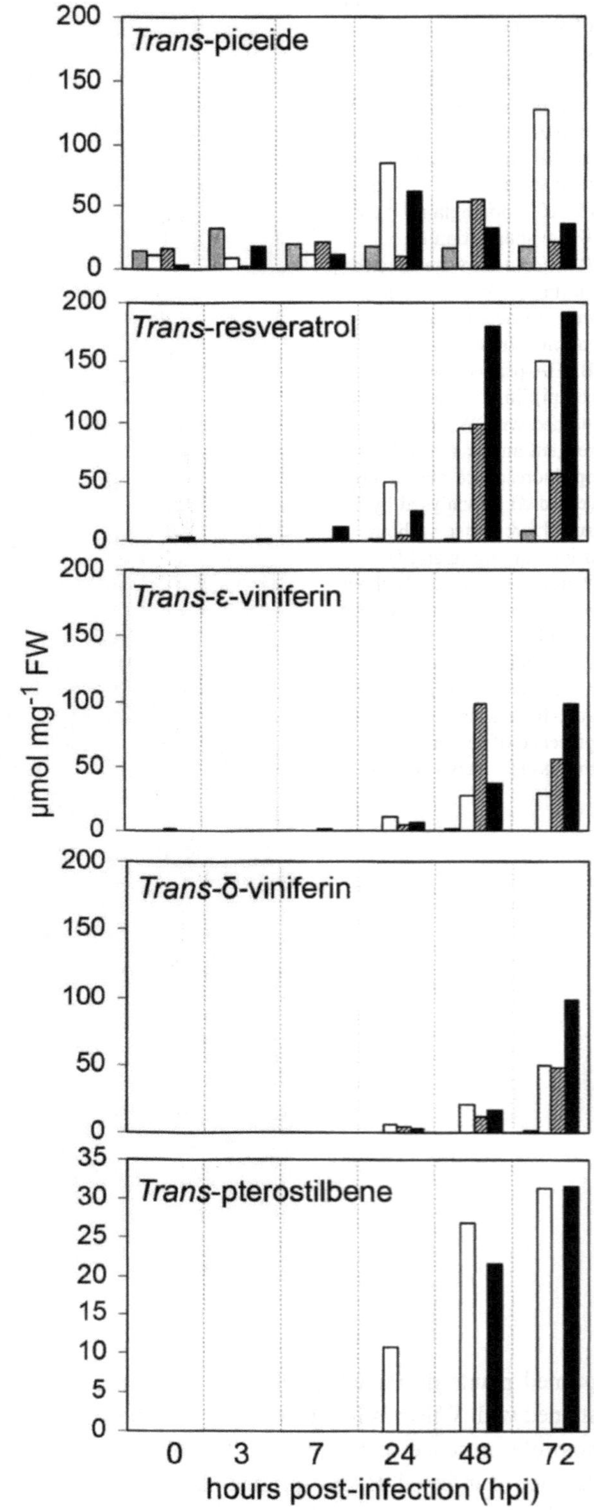

Fig. 3 Quantitative analysis of the accumulation of several ▶ stilbenes in leaf samples of a susceptible and resistant cvs Chasselas and Solaris, respectively, over time after inoculation with *P. viticola*. Chasselas cuttings were treated with water (*grey bars*) or BABA (*white bars*) and Solaris cuttings were also treated with water (*striped bars*) or BABA (*black bars*) 2 days prior to inoculation. The analysed stilbenes were *trans*-piceide, *trans*-resveratrol, *trans*-ε-viniferin, *trans*-δ-viniferin and pterostilbene. Values presented are means±standard error of the mean. The experiment was repeated twice with similar results

water-treated Solaris and BABA-primed Chasselas and Solaris (Fig. 3d). The highest concentration was found in BABA-primed Solaris at 100 μmol mg FW^{-1}, whereas that for BABA-primed Chasselas and water-treated Solaris the concentration of δ-viniferin was the same at 49 μmol mg FW^{-1} at 72 h pi. Pterostilbene was detected at relatively lower levels compared to the quantity of the other phytoalexins (Fig. 3e). It was only detected in BABA-primed cuttings, starting at 24 h pi for BABA-primed Chasselas and 12 h later for BABA-primed Solaris. Pterostilbene accumulated transiently for both BABA-primed cuttings peaking at 72 h pi with the same concentrations of 31 μmol mg FW^{-1} for both BABA-primed samples.

Response of grapevine genes involved in the phenylpropanoid biosynthesis pathway to *P. viticola* infection after treatment with and without BABA

The expression pattern of three genes involved in the phenylpropanoid biosynthesis pathway was analysed using quantitative real-time PCR: the first gene encodes phenylalanine ammonia lyase (*PAL*), the first enzyme of the pathway, the second gene encodes cinnamate-4-hydroxylase (*C4H*), the enzyme which catalyses the conversion of cinnamate into 4-hydroxy-cinnamate, a key reaction of the phenylpropanoid pathway, and the third gene encodes stilbene synthase (*STS*), the enzyme responsible for the synthesis of stilbenes such as piceides and resveratrol.

In infected, BABA-primed Solaris plants, the expression of *PAL* increased rapidly and transiently. There was no significant transcript accumulation of *PAL* in infected, BABA-primed Chasselas and in infected water-treated Chasselas and Solaris. The induction by BABA in infected Solaris peaked at 7 h with a maximum intensity 15-fold higher, than in both water-treated and BABA-primed infected Chasselas. Cinnamate-4-hydroxylase (*C4H*) was rapidly and transiently induced peaking at 7 h for both BABA-primed infected plants and for water-treated infected Chasselas. There was no significant change in expression of *C4H* in water-treated infected Solaris. In both BABA-primed infected plants *STS* was transiently induced, reaching maximum levels at 7 h (5-fold and 21-fold induction for Solaris and Chasselas, respectively). The level of expression of *STS* decreased for the first 24 h and increased again between 24 and 48 h pi. For both water-treated cultivars there was no significant accumulation of *STS* transcript levels for *STS*.

Discussion

The resistance inducer BABA has been shown to work mainly through priming of defence responses by sensitising the plants to respond faster and more adequately to the exposure to a given stress situation (Jakab et al. 2001; Conrath et al. 2002; Prime-A-Plant Group et al. 2006). Previous work in our group has shown that BABA induced resistance against *P. viticola* in both the susceptible Chasselas and the resistant Solaris cultivars of grapevine. This observed resistance in BABA-primed Chasselas plants as well as the basal resistance of Solaris depended to a large extent on the deposition of callose, which was positively correlated with BABA- and jasmonic acid-induced resistance (Hamiduzzaman et al. 2005). Here, we looked at the involvement of phytoalexins and investigated three genes involved in the phenylpropanoid biosynthesis pathway in BABA-IR in grapevine against downy mildew.

Phytoalexins have long been accepted as being important in the defence mechanisms of plants against phytopathogenic microorganisms. Previous work has demonstrated that two biochemical processes are indicative of downy mildew resistance in grapevines. One is the synthesis of callose in stomata (Gindro et al. 2003) and the second is the synthesis of resveratrol and its subsequent oxidation to ε- and δ-viniferins (Langcake 1981; Pezet et al. 2003, 2004a). The obvious priming for accumulation of different flavonoids in BABA-primed, infected grapevine plants compared to untreated water controls (Fig. 2d and f) observed at the microscopic level led us to further investigate the involvement of phytoalexins in BABA-IR at the biochemical and molecular level.

Not all stilbenes are equally toxic to *P. viticola* zoospores. Resveratrol is not a toxic compound as a consequence of its hydrophilic character (Dercks and Creasy 1989). According to Pezet et al. (2004a), piceide has never shown any toxic activity against *P. viticola* zoospores, even at concentrations greater than 1,000 μM. In addition, resveratrol was found to be glycosylated to form piceide in susceptible cultivars (Gindro et al. 2003). In our experiment,

BABA-treatment led to an induction of *trans*-piceide accumulation in the susceptible Chasselas cultivar to levels higher than found in the resistant Solaris cultivar. BABA treatment did not seem to have any effect on piceide accumulation in Solaris (Fig. 3a). On the other hand BABA-treatment significantly induced the accumulation of *trans*-resveratrol in both susceptible and resistant cultivars. After 48 h pi the amount of *trans*-resveratrol in the BABA-primed susceptible Chasselas was equal to that found in the non-treated resistant Solaris and after 72 h pi the application of BABA led to an induction of *trans*-resveratrol in both cultivars to a comparable amount (Fig. 3b). In susceptible cultivars, resveratrol is synthesised in large amounts, but it is rapidly glycosylated into the non-toxic compound piceide, which could explain the high concentration of *trans*-piceide in BABA-primed Chasselas due to the high concentration of *trans*-resveratrol. This was not the case in non-treated Chasselas plants.

The oxidation products of resveratrol, the viniferins, are active against *P. viticola*; δ-viniferins is five times more toxic than ε-viniferin (Pezet et al. 2004b). Both ε- and δ-viniferin have been found in low concentrations in susceptible cultivars (Pezet et al. 2004a). We also found both viniferins in the susceptible Chasselas but only in BABA-primed plants at 48 h pi (Fig. 3c and d). Treatment of the resistant cv. Solaris with BABA did not seem to have a significant effect on the accumulation of ε- and δ-viniferins. It is important to note that an increase in resveratrol synthesis occurred at 48 h pi for BABA-primed Chasselas providing an important pool for the synthesis of viniferins that also occurred at 48 h pi.

Pterostilbene is as toxic for *P. viticola* as δ-viniferin, but it is usually absent or its concentration is too low to contribute to resistance mechanisms (Pezet et al. 2004a). We found no detectable amounts of pterostilbene in both water-treated Chasselas and Solaris; however, BABA-treatment induced a significant increase in pterostilbene in both infected cultivars even though the amounts were lower compared to the other stilbenes analysed. In BABA-primed Chasselas the level of pterostilbene at 72 h pi was comparable to that of BABA-primed Solaris (Fig. 3e).

Pezet et al. (2004a) concluded that there were two different types of reactions that could be observed between susceptible and resistant cultivars. In susceptible cultivars, resveratrol is synthesised in large amounts after infection, but it is rapidly glycosylated into piceide, which is a non-toxic compound, whereas in resistant cultivars resveratrol is also synthesised in large amounts but it is rapidly oxidised into the more toxic viniferins. From our experiments BABA-treatment of the susceptible Chasselas cultivar made it respond more like resistant Solaris and more importantly BABA-treatment was able to prime the production of pterostilbene in both cultivars which, along with δ-viniferin, have been shown to be the most toxic stilbenes affecting zoospore mobility and disease development of *P. viticola* (Pezet et al. 2004b). BABA-treatment was also able to prime δ-viniferin accumulation in Chasselas to levels comparable to non-treated Solaris at 72 h pi but not as high as in BABA-primed Solaris (Fig. 3d).

The grapevine stilbenes are flavonoid-type phytoalexins and are formed via the phenylpropanoid pathway, and synthesis of these stilbenes only occurs if *PAL* genes and other genes encoding downstream enzymes are induced (Jeandet et al. 2002). In our experiments, only BABA-treatment of Solaris was able to prime the expression of *PAL* in the first 7 h pi. There was no induction in *PAL* expression in BABA-primed Chasselas or in water-treated Chasselas and Solaris (Fig. 4a). On the other hand BABA-treatment of Solaris also lead to the priming of the expression of *C4H* in the first 7 h pi, whereas there was no significant induction in *C4H* expression in BABA-primed compared to water-treated Chasselas for the first 7 h pi. However, 24 h pi a significant decrease in *C4H* expression was observed in the water-treated Chasselas whereas in BABA-primed Chasselas the expression level of *C4H* was comparable to levels found in BABA-primed Solaris (Fig. 4b). In tobacco, Blount et al. (2000) showed that PAL activity and levels of phenylpropanoid compounds were reduced in leaves and stems of plants in which C4H activity had been genetically down-regulated. However, C4H activity was not reduced in plants in which PAL activity had been down-regulated by gene silencing (Blount et al. 2000). These authors hypothesised that flux into the phenylpropanoid pathway is controlled, at least in part, via feedback regulation of PAL sensed through the production of cinnamic acid. Furthermore, when reduced below a threshold of 20% to 25% of wild-type activity, PAL becomes a rate-limiting step for lignin biosynthesis in tobacco (Bate et al. 1994;

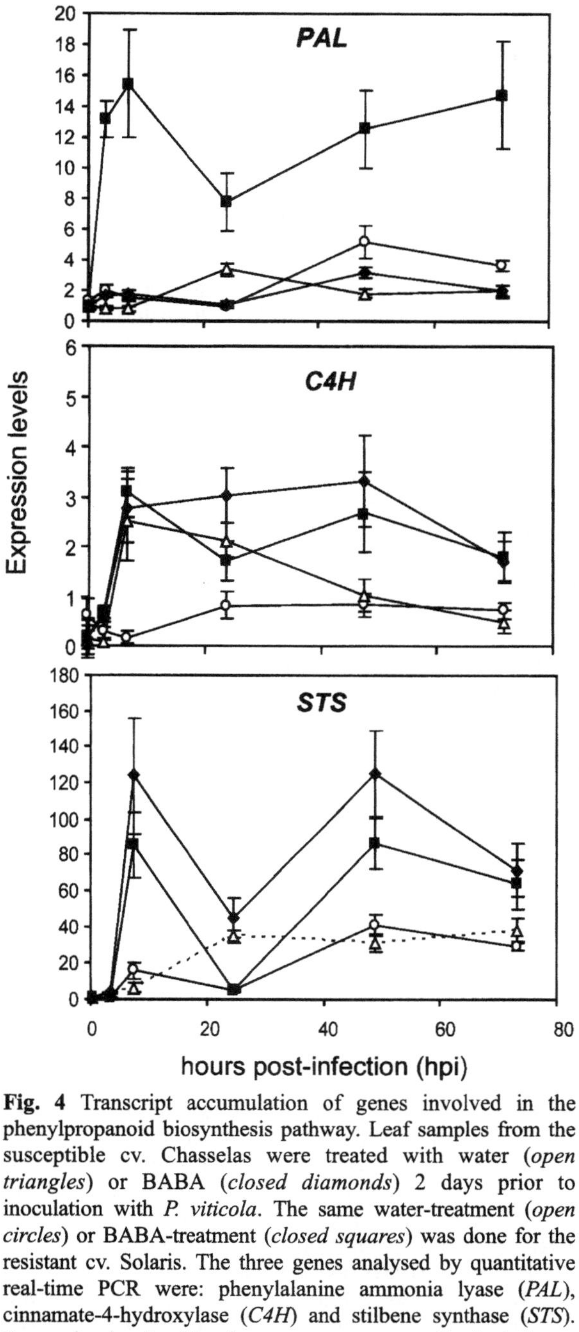

Fig. 4 Transcript accumulation of genes involved in the phenylpropanoid biosynthesis pathway. Leaf samples from the susceptible cv. Chasselas were treated with water (*open triangles*) or BABA (*closed diamonds*) 2 days prior to inoculation with *P. viticola*. The same water-treatment (*open circles*) or BABA-treatment (*closed squares*) was done for the resistant cv. Solaris. The three genes analysed by quantitative real-time PCR were: phenylalanine ammonia lyase (*PAL*), cinnamate-4-hydroxylase (*C4H*) and stilbene synthase (*STS*). Expression levels of the three genes were normalised relative to elongation factor 1-α. Values presented are means±standard error of the mean. The experiment was repeated twice with similar results

Howles et al. 1996). Reduced C4H activity is also correlated with reduced levels of lignin and phenolics (Sewalt et al. 1997). Even though BABA-treatment did not prime the expression of *PAL* in Chasselas, it primed the expression of *C4H*, which was maintained at comparable levels to those in BABA-primed Solaris thereby allowing an increase in lignin biosynthesis for BABA-primed Solaris and Chasselas but not for the water-treated Solaris and Chasselas.

Stilbene synthase is a further key branch-point enzyme in the phenylpropanoid pathway leading to the production of resveratrol. The expression of *STS* is often induced in response to biotic and abiotic stresses. We found that at the transcriptional level, BABA-treatment of both resistant and susceptible cultivars induced an accumulation of *STS* expression in two peaks: one at 7 h pi and the other after 24 h pi (Fig. 4c). There was no priming of *STS* expression in either water-treated cultivars. Similar results were previously obtained with grapevine cell suspensions elicited with cell walls of *Botrytis cinerea* (Liswidowati et al. 1991) as well as with cell walls of *Phytophthora cambivora* (Wiese et al. 1994). The induction of *STS* expression found in the BABA-primed Chasselas and Solaris cultivars corresponded to an accumulation of resveratrol whereas a low *STS* gene expression in the water-treated cultivars corresponded to low levels of resveratrol (Figs. 3b and 4c).

Grapevine plants have various possibilities to defend themselves against an attack by *P. viticola*. As shown above, the phenylpropanoid pathway leading to the accumulation of phytoalexins as well as lignification has a major influence on the expression of resistance. However, the plants have also the possibility to mount other early defenses such as callose deposition at the stomatal entry points (Gindro et al. 2003; Hamiduzzaman et al. 2005). Since the addition of the PAL inhibitor AIP does not significantly compromise BABA-IR in Solaris (Hamiduzzaman et al. 2005), it seems likely that the contribution of callose deposition can compensate for the lack of phenylpropanoid pathway products. The exact contribution of each of these defense mechanisms to final resistance is, however, not yet clear.

Our results point to a prominent role for phytoalexins as a component of BABA-IR. Of special interest is the fact that BABA treatment is able to prime the plants to accumulate a specific phytoalexin, pterostilbene, that is otherwise not present in the plants. Interestingly, pterostilbene is extremely effective in interfering with zoospore mobility and general hyphal development and its specific priming might help to develop better methods of protection of grapes against downy mildew and possibly other diseases.

Acknowledgements This project was funded by the National Centre of Competence in Research (NCCR) Plant Survival, a research programme of the Swiss National Science Foundation. We thank Mr. J. Taillens (Agroscope-RAC, Changins, Nyon) for the grapevine cutting production and Dr. S. Godard (Agroscope-RAC, Changins, Nyon) for the valuable help with the measurements of phytoalexins.

References

Bate, N. J., Orr, J., Ni, W., Meroni, A., Nadler-Hassar, T., Doerner, P. W., et al. (1994). Quantitative relationship between phenylalanine ammonia-lyase levels and phenylpropanoid accumulation in transgenic tobacco identifies a rate determining step in natural product synthesis. *Proceedings of the National Academy of Sciences of the United States of America, 91*, 7608–7612.

Blount, J. W., Korth, K. L., Masoud, S. A., Rasmussen, S., Lamb, C., & Dixon, R. A. (2000). Altering expression of cinnamic acid 4-hydroxylase in transgenic plants provides evidence for a feedback loop at the entry point into the phenylpropanoid pathway. *Plant Physiology, 122*, 107–116.

Cohen, Y. (2002). Beta-aminobutyric acid-induced resistance against plant pathogens. *Plant Disease, 86*, 448–457.

Cohen, Y., & Gisi, U. (1994). Systemic translocation of 14C-DL-3-aminobutyric acid in tomato plants in relation to induced resistance against *Phytophthora infestans*. *Physiological and Molecular Plant Pathology, 45*, 441–456.

Conrath, U., Pieterse, C. M. J., & Mauch-Mani, B. (2002). Priming in plant–pathogen interactions. *Trends in Plant Science, 7*, 210–216.

Dai, G. H., Andary, C., Mondolot-Cosson, L., & Boubals, D. (1995). Histochemical studies on the interaction between three species of grapevine, *Vitis vinifera*, *V. rupestris* and *V. rotundifolia* and the downy fungus, *Plasmopara viticola*. *Physiological and Molecular Plant Pathology, 46*, 177–188.

Dercks, W., & Creasy, L. L. (1989). The significance of stilbene phytoalexins in the *Plasmopara viticola*–grapevine interaction. *Physiological and Molecular Plant Pathology, 34*, 189–202.

Derckel, J. P., Baillieul, F., Manteau, S., Audran, J. C., Haye, B., Lambert, L., & Legendre, L. (1999). Differential induction of grapevine defences by two strains of *Botrytis cinerea*. *Phytopathology, 89*, 197–203.

Gindro, K., Pezet, R., & Viret, O. (2003). Histological study of the response of two *Vitis vinifera* cultivars (resistant and susceptible) to *Plasmopara viticola* infections. *Plant Physiology and Biochemistry, 41*, 846–853.

Hamiduzzaman, M. M., Jakab, G., Barnavon, L., Neuhaus, J.-M., & Mauch-Mani, B. (2005). b-amino butyric acid-induced resistance against downy mildew in grapevine acts through the potentiation of callose formation and JA signalling. *Molecular Plant–Microbe Interactions, 18*, 819–829.

Howles, P. A., Paiva, N. L., Sewalt, V. J. H., Elkind, N. L., Bate, Y., Lamb, C. J., et al. (1996). Over-expression of L-phenylalanine ammonia-lyase in transgenic tobacco plants reveals control points for flux into phenylpropanoid biosynthesis. *Plant Physiology, 112*, 1617–1624.

Iandolino, A. B., da Silva, F. G., Lim, H., Choi, H., Williams, L. E., & Cook, D. R. (2004). High-quality RNA, cDNA, and derived EST libraries from grapevine (*Vitis vinifera L.*). *Plant Molecular Biology Reporter, 22*, 269–278.

Jakab, G., Cottier, V., Toquin, V., Rigoli, G., Zimmerli, L., Metraux, J. P., et al. (2001). Beta-aminobutyric acid-induced resistance in plants. *European Journal of plant pathology, 107*, 29–37.

Jeandet, P., Douillet-Breuil, A. C., Bressis, R., Debord, S., Spaghi, M., & Adrian, M. (2002). Phytoalexins from the Vitaceae: Biosynthesis, phytoalexin gene expression in transgenic plants, antifungal activity, and metabolism. *Journal of Agricultural and food chemistry, 50*, 2731–2741.

Keogh, R. C., Deverall, B. J., & McLeod, S. (1980). Comparison of histological and physiological responses to *Phakopsora pachyrhizi* in resistant and susceptible soybean. *Transactions of the British Mycological Society, 74*, 329–333.

Kuc, J. (1995). Phytoalexin, stress metabolism, and disease resistance in plants. *Annual Review of Phytopathology, 33*, 275–297.

Langcake, P. (1981). Disease resistance of *Vitis* spp. and the production of the stress metabolites resveratrol, e-viniferin, a-viniferin and pterostilbene. *Physiological Plant Pathology, 18*, 213–226.

Liswidowati, F., Melchior, F., Hohmann, F., Schwer, B., & Kindl, H. (1991). Induction of stilbene synthase by *Botrytis cinerea* in cultured grapevine cells. *Planta, 183*, 307–314.

Matasci, C. L., Gobbin, D., Schärer, H.-J., Tamm, L., & Gessler, C. (2008). Selection for fungicide resistance throughout a growing season in populations of *Plasmopara viticola*. *European Journal of Plant Pathology, 120*, 79–83.

Pezet, R., Gindro, K., Viret, O., & Richter, H. (2004b). Effects of resveratrol, viniferins and pterostilbene on *Plasmopara viticola* zoospore mobility and disease development. *Vitis, 43*, 145–148.

Pezet, R., Gindro, K., Viret, O., & Spring, J.-L. (2004a). Glycosylation and oxidative dimerization of resveratrol are respectively associated to sensitivity and resistance of grapevine cultivars to downy mildew. *Physiological and Molecular Plant Pathology, 65*, 297–303.

Pezet, R., Perret, C., Jean-Denis, J. B., Tabacchi, R., Gindro, K., & Viret, O. (2003). d-viniferin, a resveratrol dehydrodimer: One of the major stilbenes synthesized by grapevine leaves. *Journal of Agricultural and Food Chemistry, 51*, 5488–5492.

Prime-A-Plant Group; Conrath, U., Beckers, G. J., Flors, V., Garcia-Agustin, P., Jakab, G., et al. (2006). Priming: Getting ready for battle. *Molecular Plant–Microbe Interactions, 19*, 1062–1071.

Sewalt, V. J. H., Ni, W., Blount, J. W., Jung, H. G., Howles, P. A., Masoud, S. A., et al. (1997). Reduced lignin content and altered lignin composition in transgenic tobacco down-regulated in expression of phenylalanine ammonia-lyase or cinnamate 4-hydroxylase. *Plant Physiology, 115*, 41–50.

Sgarbi, E., Fornassiero, R. B., Lins, A. P., & Bonatti, P. M. (2003). Phenol metabolism is differentially affected by ozone in two cell lines from grape (*Vitis vinifera* L.) leaf. *Plant Science*, *165*, 951–957.

Solecka, D., & Kacperska, A. (2003). Phenylpropanoid deficiency affects the course of plant acclimation to cold. *Physiologia Plantarum*, *119*, 253–262.

Sticher, L., Mauch-Mani, B., & Metraux, J. P. (1997). Systemic acquired resistance. *Annual Review of Phytopathology*, *35*, 235–270.

Steiner, U., & Schönbeck, F. (1997). Induced resistance. In H. Hartleb, R. Heitefuss, & H. H. Hopp (Eds.) *Resistance of crop plants against fungi* (pp. 272–297). Lubeck, Ulm: Gustav Fischer, Jena, Stuttgart.

Vandesompele, J., De Preter, K., Pattyn P., Poppe, B., Van Roy, N., De Paepe, A., et al. (2002). Accurate normalization of real-time quantitative RT-PCR data by geometric averaging of multiple internal control genes. *Genome Biology*, *3*, research0034.1–research0034.11, Retrieved from http://genomebiology.com/2002/3/7/research/0034.1.

Wiese, W., Vornam, B., Krause, E., & Kindl, H. (1994). Structural organization and differential expression of three stilbene synthase genes located on a 13 kb grapevine DNA fragment. *Plant Molecular Biology*, *26*, 667–677.

Eur J Plant Pathol (2008) 122:197–206
DOI 10.1007/s10658-008-9334-x

Effects of garlic (*Allium sativum*) juice containing allicin on *Phytophthora infestans* and downy mildew of cucumber caused by *Pseudoperonospora cubensis*

Daniela Portz · Eckhard Koch · Alan J. Slusarenko

Received: 29 January 2008 / Accepted: 8 May 2008

Abstract The volatile antimicrobial substance allicin (diallylthiosulphinate) is produced in garlic when the tissues are damaged and the substrate allicin (*S*-allyl-L-cysteine sulphoxide) mixes with the enzyme alliin-lyase (E.C.4.4.1.4). Allicin undergoes thiol-disulphide exchange reactions with free thiol groups in proteins and it is thought that this is the basis of its antimicrobial action. At 50 μg ml^{-1}, allicin in garlic juice inhibited the germination of sporangia and cysts and subsequent germ tube growth by *Phytophthora infestans* both in vitro and in vivo on the leaf surface. Disease severity in *P. infestans*-infected tomato seedlings was also reduced by spraying leaves with garlic juice containing allicin over the range tested (55–110 μg ml^{-1}) with an effectiveness ranging from approximately 45–100%. Similarly, in growth room experiments at concentrations from 50–1,000 μg ml^{-1}, allicin in garlic juice reduced the severity of cucumber downy mildew caused by *Pseudoperonospora cubensis* by approximately 50–100%. These results suggest a potential for developing preparations from garlic for use in specialised aspects of organic farming, e.g. for reducing pathogen inoculum potential and perhaps for use under glass in horticulture.

Keywords Natural fungicides · Tomato leaf blight · Plant antibiotic · Antimicrobial · Phytoanticipin

D. Portz · A. J. Slusarenko (✉)
Department of Plant Physiology,
RWTH Aachen University,
52056 Aachen, Germany
e-mail: Alan.slusarenko@bio3.rwth-aachen.de

E. Koch
Federal Research Centre for Cultivated Crops (JKI),
Institute for Biological Control,
Heinrichstr. 243,
64287 Darmstadt, Germany

Introduction

Downy mildews and diseases caused by oomycetes in general are among the most destructive and economically important agricultural problems world-wide. According to Gisi (2002) almost 17% of the world fungicides market in 1996 was for agents used in downy mildew control. Effective control by planting resistant varieties is in many cases not possible and disease management problems have been compounded by the emergence of fungicide-resistant/tolerant variants of several oomycete pathogens (Gisi 2002; Urban and Lebeda 2006, 2007; Urban et al. 2007). Furthermore, the increasing public demand for organically-grown produce, and the intended phasing out by the EU of the use of copper-containing formulations, has precipitated an urgent need for alternative control methods. In this regard resistance-inducing treatments and substances conditioning systemic acquired resistance (SAR) are considered an alternative (Mauch-Mani 2002; Körösi et al. 2007) and there is increased interest in developing treatment strategies based on natural plant defence products (Konstantinidou-Doltsinis and Schmitt 1998; Konstantinidou-Doltsinis et al. 2006; Slusarenko et al. 2008).

We have reported previously that the natural antimicrobial substance allicin, which is a volatile phytoanticipin produced in garlic (*Allium sativum*) upon wounding, is active against a broad range of phytopathogenic organisms in vitro and in planta (Curtis et al. 2004) and indeed there are several reports of garlic preparations containing allicin being used to treat plant disease (e.g. Ark and Thompson 1959; Russell and Mussa 1977). Allicin (diallylthiosulphinate) is produced in garlic when the substrate alliin (*S*-allyl-L-cysteine sulphoxide) mixes with the enzyme alliinase (alliin-lyase, E.C.4.4.1.4; see diagram below). The antimicrobial

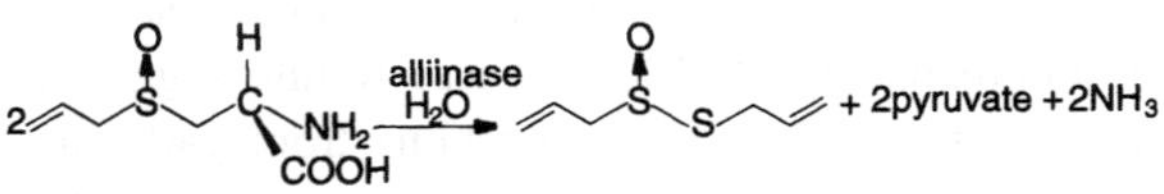

alliin allicin

activity of garlic juice had long been known and Cavallito and Bailey (1944) showed that this activity was due to allicin, which they reported to be as active against test bacteria as penicillin. Allicin crosses the cell membrane easily and undergoes thiol-disulphide exchange reactions with free thiol groups in proteins (see diagram below). It is thought that these properties are the basis of its antimicrobial action (Miron et al. 2000; Rabinikov et al. 1998). Allicin thus has several

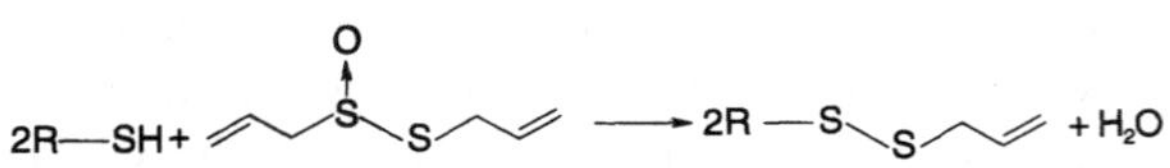

thiol allicin mixed disulphide

targets in the cell and this makes it difficult for organisms to develop resistance to it.

The use of natural products in plant protection, either directly or as starting points for targeted enhancement of desirable qualities by industry, has been reviewed recently (Slusarenko et al. 2008) and the current paper presents results using garlic juice containing allicin to combat diseases caused by the important plant pathogenic oomycetes *Phytophthora infestans* and *Pseudoperonospoa cubensis*. The effect of allicin in garlic juice was tested on the germination rate and subsequent germ tube growth of sporangia and cysts of *P. infestans* in vitro and in vivo on the surface of tomato leaves. The effectivity of allicin in garlic juice was tested in reducing leaf infection of tomato seedlings by *P. infestans*, and cucumber seedlings by *Pseudoperonospora cubensis* was also tested under growth room conditions.

Materials and methods

Cucumber/*Pseudoperonospora cubensis*

Plant cultivation

Plants were cultivated in 8 × 8 cm plastic pots filled with a 1:2 parts mixture of sand: commercial potting substrate (Fruhstorfer® Erde Typ T; Industrie-Erdenwerk Archut, Lauterbach). Twelve seeds of *Cucumis sativus* cv. Chinesische Schlange were sown in each pot. The pots were watered carefully and kept in a growth room at 20°C (cycle of 16/8 h light/dark). After 1 week the seedlings were transplanted to fresh pots (one plant per pot).

Inoculation

P. cubensis was maintained on plants grown as outlined above. Fresh inoculum was prepared from plants 10 days after previously being inoculated with *P. cubensis*. Plants were incubated overnight in a moist chamber to encourage sporulation and sporangia were harvested by washing the lower leaf surface with water. The resulting suspension was adjusted to 5×10^3 sporangia ml^{-1} using a haemocytometer.

Plants were harvested approximately 3 weeks after transplanting when the second true leaf was expanded. The upper, non-expanded leaves were excised and the first and second leaves sprayed with the treatment solution on both sides using a chromatography sprayer. Control plants were sprayed with water or with 0.2% Cuprozin Flüssig ™ (460.6 g l^{-1} copper hydroxide) (Spiess-Urania, Hamburg). After 24 h the first and second leaves were inoculated using a chromatography sprayer on both sides with a suspension of *P. cubensis* sporangia (5×10^3 ml^{-1}). The pots were then incubated overnight at 15°C in a moist chamber and the following day returned to the growth room. Disease was rated 2 weeks after inoculation by estimating the percentage of the affected leaf area. The effectivity of the treatment was calculated according to Abbott (1925):

$$\% \text{ Effectivity} = \frac{\text{affected leaf area (control)} - \text{affected leaf area (treatment)}}{\text{affected leaf area (control)}} \times 100$$

Tomato/*Phytophthora infestans*

Plant cultivation

Tomato seeds (cv. Hoffmanns Rentita®, Schmitz & Laux GmbH, Hilden, Germany) were sown in seedling trays for germination in moist potting compost covered with fine moistened sand and incubated at 22°C in a light/dark cycle of 16/8 h. After germination, 1 week-old seedlings were transferred to individual 7×7 cm pots and grown on for a further 2 weeks.

Inoculation

The *P. infestans* isolate used in this work was kindly donated by Bayer CropScience AG, Monheim. The virulence of the isolate was ensured by regular passaging through potato tuber discs. *Phytophthora infestans* was cultivated under sterile conditions on tomato juice agar (TJA) at 18°C in the dark (TJA=3 g $CaCO_3$, 12 g PDB (Difco™), 20 g agar (AppliChem GmbH), 200 ml tomato juice (Fa. Krings Fruchtsaft GmbH, Mönchengladbach) made up to a volume of 1 l and autoclaved at 121°C for 15 min). *Phytophthora infestans* inoculum was prepared by washing the surface of 8 day-old Petri plate cultures with cold (10°C), sterile deionized water and sieving through a plastic kitchen sieve. Sporangia were adjusted to a concentration of $4–5\times10^4$ ml^{-1} with a haemocytometer. Zoospores were released from sporangia after approximately 2 h at 10–12°C. After spray-inoculation, plants were placed in a seedling tray and covered with a transparent plastic lid in the growth chamber at 20°C with a light/dark cycle of 16/8 h.

Treatment with garlic juice

Unless otherwise stated, 3 week-old tomato plants were sprayed with diluted garlic juice and the leaves allowed to dry (approx. 2 h) before being spray-inoculated. As a soil drench a single application of 5 ml of the appropriate dilution of garlic juice was applied per 7×7 cm pot containing a single plant. Five to seven intact tomato seedlings were inoculated per experiment and each experiment was repeated at least three times. A representative set of results for each experiment is shown.

Preparation of garlic juice

Garlic bulbs were purchased from the supermarket and stored at 4°C in the dark until required. Axillary buds from the composite garlic bulb were peeled, weighed and a domestic juicer (Turmix Fabr. Nr. 1068, Turmix AG, 8645 Jona, Switzerland) was used to extract the juice. The juice was poured into a sterile 50 ml Falcon tube and centrifuged at 5,000 rpm (3,000×*g*) for 10 min in order to separate the majority of the pulp from the liquid (Megafuge 1.0R, Heraeus Instruments, Osterode, Germany). Floating debris was removed from the top of the liquid with a spatula and discarded. Filtering under pressure separated the remaining pulp from the pure extract (Diaphragm Vacuum Pump, Vacuubrand GmbH + Co., Wertheim, Germany). The filtrate was transferred into a second sterile 50 ml Falcon tube and sealed. The average yield was approximately 1 ml of extract from 3 g fresh weight of garlic tissue and typically contained approximately 5 mg ml^{-1} allicin (determined by HPLC). The garlic extract was used either immediately after appropriate dilution or stored undiluted at 10°C for a maximum of 2 weeks. Dilutions were carried out with de-ionized water. Appropriate amounts of stock solution to give the required end dilution in Petri plates were incorporated into agar medium kept just molten at 45°C. Plates were poured immediately after adding and mixing the stock.

Determination of allicin by HPLC

The method used was based on that of Krest and Keusgen (2002). Garlic juice was diluted 1:10 with HPLC-grade water and 1.5 ml of a 0.05 mg ml^{-1} solution (in methanol) of butyl-4-hydroxybenzoate (internal standard). To protect the column, this mixture was first filtered through a polyethersulfon-membrane (0.2 μm pore size, Steriflip, Millipore) before 20 μl were injected into the HPLC (Kontron system with diode array detector, Kontron Instruments GmbH, Neufahrn). Using the HPLC software Geminyx (version 1.91) a mixed gradient elution (solvent A, 30% (*v/v*) HPLC grade methanol adjusted to pH 2.0 with 85% (*v/v*) orthophosphoric acid; solvent B, 100% HPLC grade methanol) was performed. Spectra were recorded between 200–600 nm during elution with detection at 254 nm for the chromatogram.

Effect of garlic juice on *P. infestans* sporangium and cyst germination in vitro

Droplets (20 µl) of inoculum suspension, prepared as described above and containing sporangia and zoospores, were pipetted onto the surface of 1% agar containing 50 µg ml^{-1} allicin. Control plates contained no allicin. Plates were sealed with Micropore™-tape and incubated in a plastic container with moistened tissue paper at 18°C in the dark for 4 h. Germination rate and germ tube length were measured using a microscope (Leica DM R) at 50- to 200-fold magnification. At least 50 sporangia or encysted zoospores were scored for germination per plate and photographed using a JVC digital camera (KY-F75U) and Discus software (Version 32, Hilgers Co., Königswinter, Germany). Germ tube lengths of at least 15 germinated sporangia or cysts were measured per plate.

Effect of garlic juice on *P. infestans* sporangium and cyst germination in vivo on tomato leaves

After spraying 3-week-old tomato plants with diluted garlic juice containing 50 µg ml^{-1} allicin and allowing them to dry, leaves were excised and placed in plastic boxes (12×12 cm) on moistened tissue paper. Droplets (20 µl) of sporangial or cyst suspensions were then pipetted onto the leaves and the lids placed on the boxes for incubation for 4 h in the dark at 20°C. The leaf lamina under the droplets was then excised and stained with acid fuchsin (modified after Carmichael 1955). Excised leaf segments were fixed and decolourised for 48 h at 60°C in aqueous chloral hydrate (2.5 g ml^{-1}). Leaf segments were then stained for 1–2 h in 0.01% acid fuchsin-lactophenol solution and de-stained in 50% (*v/v*) glycerol before viewing using a confocal laser-scanning microscope (Leica TCS SP, using Leica software TCS NT) at 630- to 1000-fold magnification (excitation 543 nm; emission filter 575–640 nm, 63× PL APO w, and 100× PL FLUOTAR oil objective lenses).

Statistical treatments

Raw data were first tested for normal distribution and variance homogeneity using Sigmastat® 3.1 (SYSTAT software 2004) to a limit of $P \leq 0.05$. If the data showed normal distribution and variance homogeneity they were subjected to parametric statistic tests to show significant differences (*t*-test or one-way ANOVA) to a probability of $P \leq 0.05$. Non-normal data were analysed with either the Mann–Whitney Rank Sum Test for two groups or the Kruskal–Wallis ANOVA on Ranks for more than two groups. If these treatments pointed to a significant difference between groups, a *post hoc* test (Dunn's or Tukey's) was used to determine which groups differed significantly at the $P \leq 0.05$ level.

Results

Pseudoperonospora cubensis/*Cucumis sativus* pathosystem

Cucumber plants were sprayed with either dilutions of garlic juice, water (untreated controls) or Cuprozin™, and spray-inoculated the next day with a suspension of sporangia of *P. cubensis* (Fig. 1A). Two weeks after inoculation infected leaf areas were estimated (Fig. 1B, Table 1). Dilutions of garlic juice over a wide range of allicin concentrations (50–1,000 µg ml^{-1}) led to a reduction in disease severity which compared favourably with the degree of disease control achieved with a copper-containing commercial fungicide (Cuprozin™).

Phytophthora infestans/*Lycopersicon esculentum* pathosystem

Effects of garlic juice on P. infestans *germination and growth in vitro*

The effect of garlic juice on *P. infestans* in vitro was assessed by investigating the effects on sporangial and cyst germination and on germ tube growth. Garlic juice (50 µg ml^{-1} allicin) caused a clear reduction in

Fig. 1 Leaf of cucumber showing *A*, the spray inoculation procedure and *B*, symptoms 14 days after inoculation with *P. cubensis* (5×10^3 sporangia ml^{-1})

the germination of encysted zoospores and of sporangia under conditions where they germinate directly with a germ tube (i.e. behave like conidia) (Fig. 2). Hyphal growth from germinated sporangia or cysts was also reduced by the presence of garlic juice in the medium (50 μg ml^{-1} allicin) (Fig. 3).

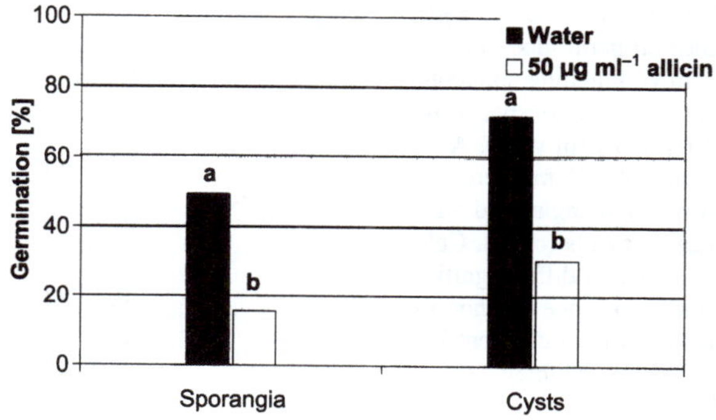

Fig. 2 Influence of garlic juice in agar (50 μg ml^{-1} allicin) on the germination of sporangia and encysted zoospores of *P. infestans* (in vitro). Means of nine replicate Petri plates of sporangia and cyst preparations. Columns which differ significantly from one another are marked with a different letter (*t*-test, $P\leq0.05$)

Effects of garlic juice on P. infestans *germination and growth* in vivo

The behaviour of sporangia and cysts on the tomato leaf surface after treatment with garlic juice is shown in Fig. 4. It can be seen that the inhibitory in vitro effects of garlic juice are mirrored in the in vivo behaviour of sporangia and cysts on the tomato leaf surface.

Effects of garlic juice on disease severity in tomato leaves inoculated with P. infestans

To assess whether the inhibitory effects of garlic juice on *P. infestans* observed in vitro and in vivo on the leaf surface translated into an effect on disease development, a systematic investigation on tomato

Table 1 Effect on disease severity of spraying garlic juice containing allicin at the concentrations shown onto leaves of 40-day-old cucumber plants 24 h prior to spray inoculation with 5×10^3 conidia ml^{-1} of *P. cubensis*

Treatment	Infected leaf area (%) ±SD			Average effectivity[a] (%)
	Experiment 1	Experiment 2	Experiment 3	
Water control	73.3±17.1	33.8±8.9	81.8±9.9	
Allicin 1000 μg ml^{-1}	N.T.[b]	0.2	0.4	>99
Allicin 500 μg ml^{-1}	N.T.	1.0	1.0	96–98
Allicin 200 μg ml^{-1}	3.7	2.8	2.0	84–94
Allicin 100 μg ml^{-1}	19.0	N.T. ±	N.T.	55
Allicin 50 μg ml^{-1}	8.2	N.T. ±	N.T.	52
Cuprozin™ (0.2%)[c]	N.T.	20.0±9.8	20.0±11.9	41–76

Plants (four per experiment, eight leaves in total) were scored 2 weeks after inoculation.

[a] According to Abbott (1925), see "Materials and methods" section

[b] *N.T.* Not tested

[c] Equivalent to 0.92 g $Cu(OH)_2$ l^{-1}

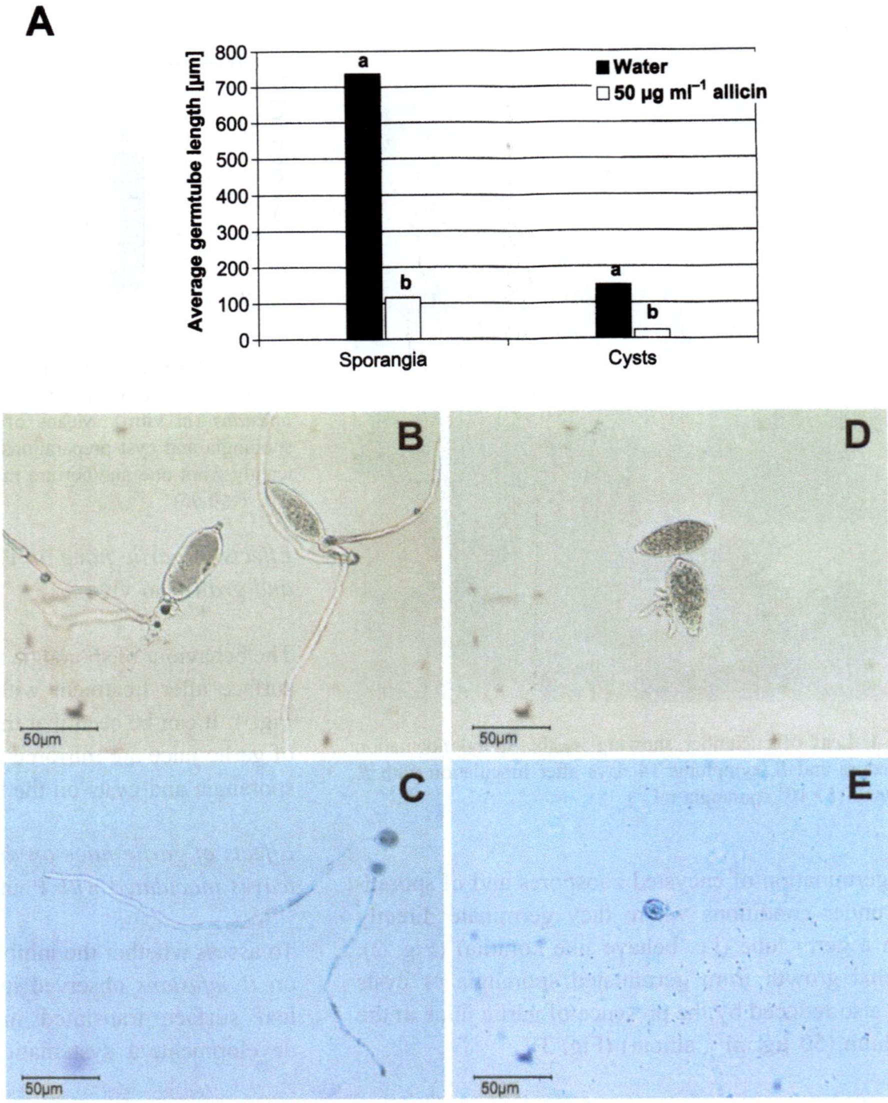

Fig. 3 Influence of garlic juice on germ tube growth from germinating sporangia and encysted zoospores of *P. infestans* (in vitro). **A** Means of ~75 measurements (sporangia) and ~135 measurements (cysts). Columns which differ significantly from one another are marked with a different letter (Mann–Whitney Test, $P \leq 0.05$). **B** Untreated sporangia. **C** Untreated cysts. **D** Sporangia on agar incorporating garlic juice to give a final concentration of 50 µg ml^{-1} allicin. **E** Cysts on agar incorporating garlic juice to give a final concentration of 50 µg ml^{-1} allicin. *Bar*= 50 µm

leaf infections was carried out. Firstly, potential phytotoxic effects of garlic juice on leaves were monitored. As shown in Table 2, spraying tomato leaves of 3-week-old plants with dilutions of garlic juice containing 200–800 µg ml^{-1} allicin led to leaf damage in category 2 (<2.5% of the leaf area showing chlorosis or necrosis), the least severe, and only at the highest concentration tested.

The effect on disease development of spraying tomato leaves with a single application of garlic juice containing a range of allicin concentrations 2 h before inoculation with *P. infestans* is shown in Fig. 5 (for a photograph showing the appearance of control and allicin-treated leaves see Fig. 6 in Slusarenko et al. 2008). In the experiment shown in Fig. 5, control tomato plants had lesions covering 77% of the leaf area 4 days after inoculation (dai). As can be seen, spraying with garlic juice very effectively reduced disease development, with a 1:50 dilution (110 µg ml^{-1} allicin) suppressing lesion development completely (Fig. 5).

The effectivity of a single pre-inoculation spray with garlic juice containing a low concentration of allicin (60 µg ml^{-1}), which did not completely suppress disease development, decreased with time but was still apparent 10 dai (data not shown). Thus, in plants treated with 60 µg ml^{-1} allicin the affected leaf area increased from 16% at 4 days to 37% at 10

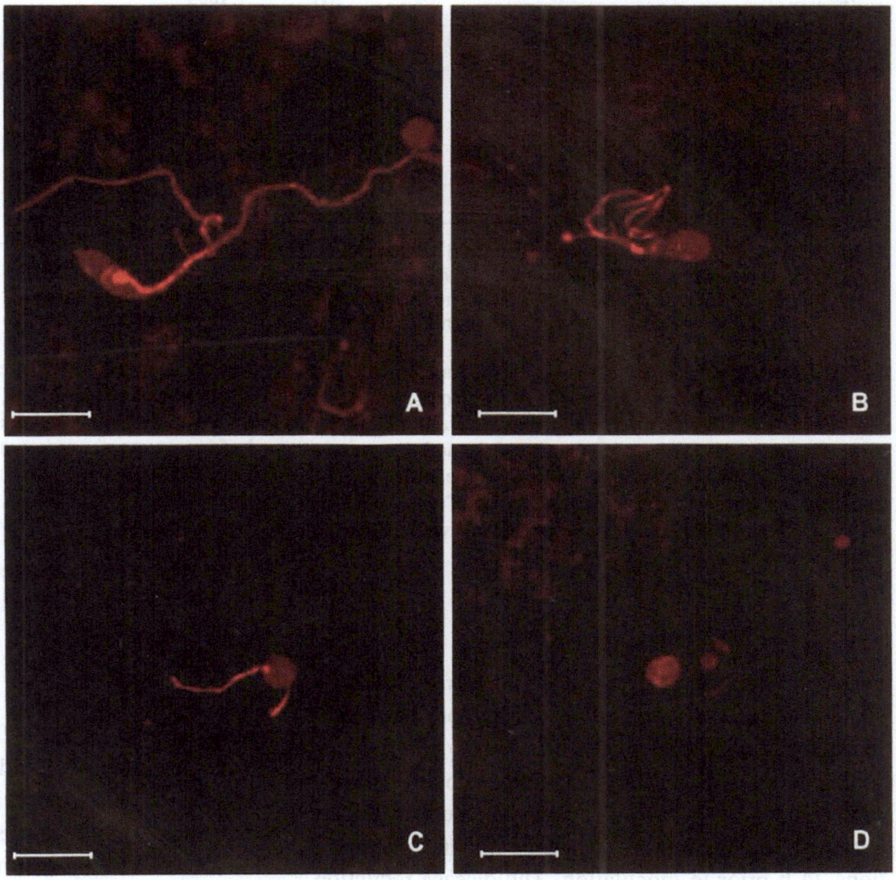

Fig. 4 Influence of garlic juice on germ tube growth from germinating sporangia and encysted zoospores of *P. infestans* on the tomato leaf surface (in vivo) shown after acid fuchsin staining under a confocal laser scanning microscope excitation, 543 nm, emission, 575–640 nm; *Scale bars* = 50 µm (A & B), 25 µm (C & D). **A** Untreated sporangia showing germination and healthy germ tube growth. **B** Sporangia on a leaf sprayed with garlic juice (50 µg ml^{-1} allicin) approximately 2 h prior to inoculation, germinated at a lower rate and has formed abnormal germ tubes with reduced growth. **C** Untreated cyst showing normal germ tube growth. **D** Ungerminated cysts on a leaf sprayed with garlic juice (50 µg ml^{-1} allicin)

dai. In the untreated controls, however, the infected leaf area was 60% after 4 days and increased to 63% by 10 dai.

The effect of various garlic juice application times in relation to the time of inoculation with *P. infestans* was investigated and it was found that the nearer to the inoculation time that allicin was sprayed, the more effective a given dosage was in suppressing disease development (Fig. 6). In contrast, spraying leaves with garlic juice 24 h after inoculation had little effect. Direct spraying onto leaves was also compared with a single application as a soil drench. As can be seen in Fig. 7, allicin was more effective when sprayed on the leaves than when applied to the soil.

Discussion

Curtis et al. (2004) previously reported that dilutions of garlic juice containing allicin were effective in reducing the production of conidiophores and

Table 2 Phytotoxicity scores for individually potted 3-week-old tomato seedlings sprayed to run-off with dilutions of garlic juice containing various concentrations of allicin

Concentration of allicin in garlic juice (µg ml^{-1})	Phytotoxicity category[a]
0	1
200	1
400	1
800	2

[a] Phytotoxicity categories (Gorog née Privitzer et al. 1988): 1= no damage, 2=<2.5% leaf area damaged (showing chlorosis or necrosis), 3=<5% leaf area damaged. The scale progresses to 9=100% leaf damage. Scores<2 are considered acceptable in screens of potential candidates for plant protection substances.

Plants were allowed to dry, and then pots were placed under plastic hoods for 4 days before the hoods were removed. Plants were incubated in a growth chamber (22°C, cycles of 18 h light 6 h dark) and scored 6 days after spraying.

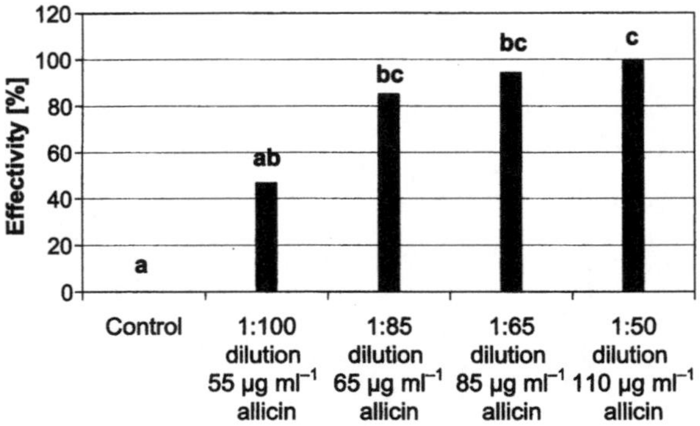

Fig. 5 Dose-dependency of disease control by allicin in garlic juice in the *P. infestans*/tomato leaf pathosystem. Three-week-old plants (cv. Hoffmans Rentita) were sprayed with garlic juice, the leaves allowed to dry (approx. 2 h) and then spray-inoculated with $4–5\times10^4$ sporangia ml^{-1}. The effectivity of treatment (Abbot 1925) is shown at 4 dai. Columns which differ significantly from one another are marked with a different letter (Dunn's Test, $P\leq0.05$)

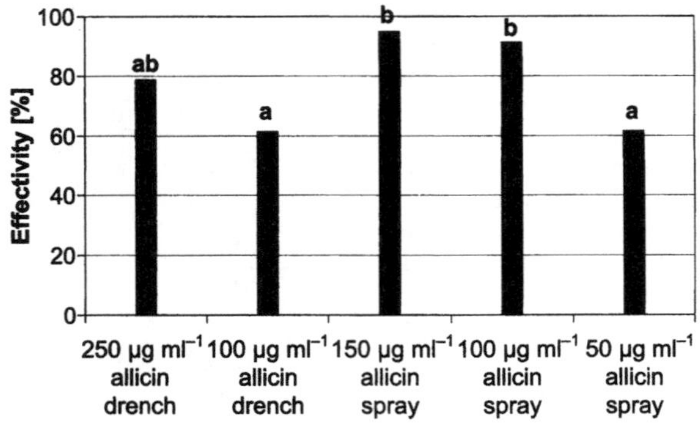

Fig. 7 Comparison of the effectivity of garlic juice containing allicin as a soil drench or a foliar spray in the *P. infestans*/tomato pathosystem at 4 dai. Columns which differ significantly from one another are marked with a different letter (Tukey's Test, $P\leq0.05$)

oospores in downy mildew of *Arabidopsis* caused by *Hyaloperonospora parasitica*. In the present study these observations are extended to show that macroscopic disease symptoms of cucumber downy mildew can be markedly reduced by spraying the leaves with garlic juice containing a range of allicin concentrations 24 h prior to inoculation. The disease reduction compared very favourably with a commercial copper-fungicide treatment and suggests that development of garlic products for at least small-scale application such as in glasshouse situations might be feasible and desirable as an alternative to standard treatments (Fig. 1, Table 1). Resistance of *P. cubensis* against conventional fungicide treatments is increasing (Urban and Lebeda 2006, 2007; Urban et al. 2007) and because allicin appears to have a multi-site mode of action (Portz et al. 2005; Slusarenko et al. 2008) it will presumably be difficult for pathogens to mutate to resistance against it, thus conferring a strong advantage on allicin-based disease treatments.

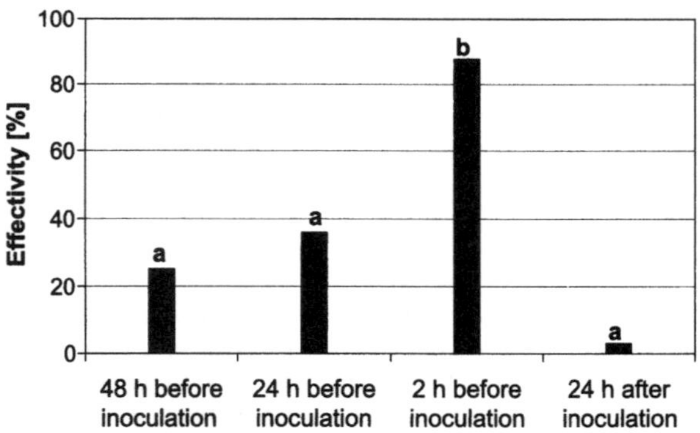

Fig. 6 Influence of time between treatment with garlic juice (70 µg ml^{-1} allicin) and time of inoculation on effectivity in the *P. infestans*/tomato pathosystem at 4 dai. Columns which differ significantly from one another are marked with a different letter (Dunn's Test, $P\leq0.05$)

The inhibitory effect of allicin on the vegetative mycelial growth of *P. infestans* and the reduction of potato tuber colonization by allicin in the gas phase have been reported previously (Curtis et al. 2004). Now, the inhibitory effects of garlic juice containing allicin on the germination of sporangia and encysted zoospores and subsequent reduction in germ tube growth, both in vitro and on the tomato leaf surface (Figs. 2, 3, 4) are reported. These effects presumably contribute to the reduction in infection seen in inoculated tomato seedlings (Fig. 5). The tomato leaf/*P. infestans* pathosystem was used in preference to potato/*P. infestans* because it is easier to work with in the laboratory. Nevertheless, since it appears that the effect of allicin is directly against the pathogen, rather than via an induced resistance mechanism (Curtis et al. 2004), it seems likely that a similar degree of control might be expected in the potato/*P. infestans* pathosystem, particularly in view of the effects of allicin in reducing tuber colonisation at least under controlled conditions (Curtis et al. 2004). The effectivity of garlic juice in reducing disease in tomato leaves was very high and approached 100%

at an allicin concentration of 110 μg ml^{-1} (Fig. 5). In fungicide screening, substances are usually only considered for further development if they do not cause leaf damage above category 2 (<2.5% leaf area affected) on a scale of 1–9 (Gorog née Privitzer et al. 1988) (see Table 2). Garlic juice was assessed at various dilutions for phytotoxicity, and disease control was achieved at allicin concentrations well below those where phytotoxicity was observed (Table 2, Fig. 5). Thus, allicin in garlic juice would not be excluded in a conventional screening programme based on this criterion.

The effectivity of the allicin treatment in reducing disease on tomato seedlings is more pronounced in the early stages after treatment. If allicin is working mainly via a reduction of successful infections by killing a certain proportion of the spores and subsequently by suppressing germ tube growth from surviving propagules, then a time-lag in disease development would be expected until inoculum levels had reached those present before the sanitation treatment. However, the dynamics of disease development in fungicide-treated plants are difficult to model and disease development often deviates from the ideal mathematical description (Jeger 1987). In control plants not treated with allicin, the disease level 4 dai was already high and this increased only marginally in subsequent days. In the allicin-treated plants the affected leaf area increased from 16% at 4 days to 37% by 10 dai. Thus, even a single treatment with allicin at a dose (60 μg ml^{-1}) below that necessary to completely eradicate disease (~110 μg ml^{-1}, see Fig. 5) is already effective at reducing the rate of disease progress over a substantial time period.

The data presented in Fig. 6 show the effectivity of a single allicin treatment in relation to the time of inoculation and support a low-persistence, contact-fungicide type of effect for allicin. Thus, the effectivity of the treatment increases with decreasing time before inoculation (from 48 to 24 h), is maximal when inoculation takes place approximately 2 h after treatment with garlic juice, and is least effective at later times after inoculation (e.g. 24 h) when the pathogen has already penetrated the leaf and is perhaps less easily accessed by allicin. In this regard the kinetics of allicin behaviour on the leaf surface, and its uptake by the leaf, are aspects which need further investigation.

In downy mildew of *Arabidopsis* it was shown that treatment of the plant with garlic juice did not lead to the accumulation of SAR markers (Curtis et al. 2004) and the authors suggested that garlic juice was exerting its antimicrobial effect directly on the pathogen rather than via inducing SAR in the plant. The data presented in Fig. 6 for tomato support this conclusion and extend it to a further pathosystem.

Interestingly, in the tomato/*P. infestans* pathosystem, applying garlic juice as a soil drench was also effective at reducing disease levels, although a better degree of control was achieved with lower concentrations of allicin as a direct spray on the leaves (Fig. 7). As stated earlier, allicin appears to act directly against the pathogen and it is unclear whether the disease reduction after applying garlic juice as a soil drench is due to the action of allicin against the pathogen via the gas phase, or whether allicin is also taken up via the roots and transported systemically within the plant. Allicin is readily membrane-permeable (Miron et al. 2000; Rabinikov et al. 1998; Slusarenko et al. 2008) and could therefore enter the symplast in the roots, but whether it is transported within the plant is unknown at present. In this regard, it is perhaps important to mention that it is difficult to quantify allicin in the gas phase because the temperature of the injection port in the GC is too high and leads to modifications producing other polysulphides (Block 1992).

The potential for allicin in garlic juice to be used as an effective control agent against diseases caused by oomycetes is clear, although there is scope for optimisation of treatment regimes, and field testing is certainly necessary. Very clearly, transfer from the laboratory to the field/glasshouse is a stumbling block which many otherwise promising compounds fail to negotiate successfully (Slusarenko et al. 2008). This may also prove true for allicin in garlic juice. Also, it will be necessary to carry out organoleptic assessment of harvested plant parts to ensure the absence of undesireable flavour notes in any development of garlic products for plant protection. Neither garlic juice nor allicin are named presently as plant protection substances specifically permitted for organic farming in the EU (Directive 2092/91). However, it is not likely that these substances, which are a common foodstuff or a component thereof, have properties that would not allow them to be added to the list in the future. Furthermore, chemical modification of allicin, which has an activity comparable to several conventional antibiotics (Cavallito and Bailey 1944; Curtis et al. 2004; Slusarenko et al. 2008), to enhance its desirable properties and reduce its undesirable ones, might even lead to a new multi-target plant protection compound useable in conventional agriculture and horticulture.

Acknowledgements RWTH Aachen University provided a student assistantship (D.P.) and financial support. Technical assistance by Ulrike Noll (Aachen) and Monika Eitzen-Ritter (Darmstadt) is gratefully acknowledged. Ales Lebeda and Nikolaus Schlaich are thanked for critical reading of the manuscript.

References

Abbott, W. S. (1925). A method for computing the effectiveness of an insecticide. *Journal of Econonomic Entomology, 18*, 265–267.

Ark, P. A., & Thompson, J. P. (1959). Control of certain diseases of plants with antibiotics from garlic (*Allium sativum* L.). *Plant Disease Reporter, 43*, 276–282.

Block, E. (1992). The organosulfur chemistry of the genus *Allium* – implications for the organic chemistry of sulfur. *Angewandte Chemie International Edition in English, 31*, 1135–1178.

Carmichael, J. W. (1955). Lacto-fuchsin: A new medium for mounting fungi. *Mycologia, 47*, 611.

Cavallito, C. J., & Bailey, H. J. (1944). Allicin, the antibacterial principle of *Allium sativum* I. Isolation, physical properties and antibacterial action. *Journal of the American Chemical Society, 66*, 1950–1951.

Curtis, H., Noll, U., Störmann, J., & Slusarenko, A. J. (2004). Broad-spectrum activity of the volatile phytoanticipin allicin in extracts of garlic (*Allium sativum* L.) against plant pathogenic bacteria, fungi and Oomycetes. *Physiological and Molecular Plant Pathology, 65*, 79–89.

Gisi, U. (2002). Chemical control of downy mildews. In P. T. N. Spencer-Phillips, U. Gisi, & A. Lebeda (Eds.), *Advances in downy mildew research* (pp. 119–159). Dordrecht: Kluwer Academic Publishers.

Gorog née Privitzer, K., Kocsis née Bagyi, M., Orosz, G., Torok, T., Laszlo, K., Toth, I., et al. (1988). Herbicide compositions containing a nitrile derivative as antidote. *US Patent* 4787933.

Jeger, M. J. (1987). Modelling the dynamics of pathogen populations. In M. S. Wolfe, & C. E. Caten (Eds.), *Populations of plant pathogens: Their dynamics and genetics* (pp. 91–107). Oxford: Blackwell Scientific Publishers.

Konstantinidou-Doltsinis, S., Markellou, E., Fanouraki, M. N., Kasselaki, A.-M., Koumaki, C. M., Schmitt, A., et al. (2006). Efficacy of MilsanaÒ, a formulated plant extract from *Reynoutria sachalinensis*, against powdery mildew of tomato (*Leveillula taurica*) (Lév.) Arn. *Biocontrol, 51*, 375–392.

Konstantinidou-Doltsinis, S., & Schmitt, A. (1998). Impact of treatment with plant extracts from *Reynoutria sachalinensis* (F. Schmidt) Nakai on disease severity of powdery mildew and yield in cucumber under Greek conditions. *Crop Protection, 17*, 649–656.

Körösi, K., Lazar, N., & Viranyi, F. (2007). Resistance response to downy mildew (*Plasmopara halstedii*) in sunflower (*Helianthus annuus*) activated by chemical inducers. In A. Lebeda, & P. T. N. Spencer-Phillips (Eds.), *Advances in downy mildew research* (vol. 3 (pp. 237–241). Czech Republic: Olomouc, Palacky University (Publishers).

Krest, I., & Keusgen, M. (2002). Biosensoric flow-through method for the determination of cysteine sulfoxides. *Analytica Chimica Acta, 469*, 155–164.

Mauch-Mani, B. (2002). Host resistance to downy mildew diseases. In P. T. N. Spencer-Phillips, U. Gisi, & A. Lebeda (Eds.), *Advances in downy mildew research* (pp. 59–83). Dordrecht: Kluwer Academic Publishers.

Miron, T., Rabinikov, A., Mirelman, D., Wilchek, M., & Weiner, L. (2000). The mode of action of allicin: Its ready permeability through phospholipid membranes may contribute to its biological activity. *Biochimica et Biophysica Acta, 1463*, 20–30.

Portz, D., Noll, U., & Slusarenko, A. J. (2005) Allicin from garlic (*Allium sativum* L.): A new look at an old story. Proceedings of the 14th International Reinhardsbrunn Symposium, "*Modern Fungicides and Antifungal Compounds IV*". Eds. H.-W. Dehne, U. Gisi, K. H. Kuck, P. E. Russell & H. Lyr. British Crop Protection Council, Alton, U.K. pp. 227–234.

Rabinikov, A., Miron, T., Konstantinovski, L., Wilchek, M., Mirelman, D., & Weiner, L. (1998). The mode of action of allicin: Trapping of radicals and interaction with thiol-containing proteins. *Biochimica et Biophysica Acta, 1379*, 233–244.

Russell, P. E., & Mussa, A. E. A. (1977). The use of garlic (*Allium sativum*) extracts to control foot rot of *Phaseolus vulgaris* caused by *Fusarium solani* f. sp. *phaseoli*. *Annals of Applied Biology, 86*, 369–372.

Slusarenko, A. J., Patel, A., & Portz, D. (2008). Control of plant diseases by natural products: Allicin from garlic as a case study. *European Journal of Plant Pathology, 121*, 313–322.

Urban, J., & Lebeda, A. (2006). Fungicide resistance in cucurbit downy mildew—methodological, biological and population aspects. *Annals of Applied Biology, 149*, 63–75.

Urban, J., & Lebeda, A. (2007). Variation for fungicide resistance in Czech populations of *Pseudoperonospora cubensis*. *Journal of Phytopathology, 155*, 143–151.

Urban, J., Lebeda, A., & Pejchar, M. (2007). Differential sensitivity to fungicides in Czech populations of cucumber downy mildew (*Psedoperonospora cubensis*). In A. Lebeda, & P. T. N. Spencer-Phillips (Eds.),*Advances in downy mildew research* (vol. 3 (pp. 251–255). Czech Republic: Olomouc, Palacky University (Publishers).

MIX
Papier aus verantwortungsvollen Quellen
Paper from responsible sources
FSC® C105338
FSC
www.fsc.org

If you have any concerns about our products,
you can contact us on
ProductSafety@springernature.com

In case Publisher is established outside the EU,
the EU authorized representative is:
Springer Nature Customer Service Center GmbH
Europaplatz 3, 69115 Heidelberg, Germany

Printed by Libri Plureos GmbH
in Hamburg, Germany